239
Topics in Current Chemistry

Topics in Current Chemistry

Recently Published and Forthcoming Volumes

Functional Molecular Nanostructures
Volume Editor: Schlüter, A.D.
Vol. 245, 2004

Natural Product Synthesis II
Volume Editor: Mulzer, J.H.
Vol. 244, 2004

Natural Product Synthesis I
Volume Editor: Mulzer, J.H.
Vol. 243, 2004

Immobilized Catalysts
Volume Editor: Kirschning, A.
Vol. 242, 2004

Transition Metal and Rare Earth Compounds III
Volume Editor: Yersin, H.
Vol. 241, 2004

The Chemistry of Pheromones and Other Semiochemicals II
Volume Editor: Schulz, S.
Vol. 240, 2004

The Chemistry of Pheromones and Other Semiochemicals I
Volume Editor: Schulz, S.
Vol. 239, 2004

Orotidine Monophosphate Decarboxylase
Volume Editors: Lee, J.K., Tantillo, D.J.
Vol. 238, 2004

Long-Range Charge Transfer in DNA II
Volume Editor: Schuster, G.B.
Vol. 237, 2004

Long-Range Charge Transfer in DNA I
Volume Editor: Schuster, G.B.
Vol. 236, 2004

Spin Crossover in Transition Metal Compounds III
Volume Editors: Gütlich, P., Goodwin, H.A.
Vol. 235, 2004

Spin Crossover in Transition Metal Compounds II
Volume Editors: Gütlich, P., Goodwin, H.A.
Vol. 234, 2004

Spin Crossover in Transition Metal Compounds I
Volume Editors: Gütlich, P., Goodwin, H.A.
Vol. 233, 2004

New Aspects in Phosphorus Chemistry IV
Volume Editor: Majoral, J.-P.
Vol. 232, 2004

Elemental Sulfur and Sulfur-Rich Compounds II
Volume Editor: Steudel, R.
Vol. 231, 2003

Elemental Sulfur and Sulfur-Rich Compounds I
Volume Editor: Steudel, R.
Vol. 230, 2003

New Aspects in Phosphorus Chemistry III
Volume Editor: Majoral, J.-P.
Vol. 229, 2003

Dendrimers V
Volume Editors: Schalley, C.A., Vögtle, F.
Vol. 228, 2003

Colloid Chemistry II
Volume Editor: Antonietti, M.
Vol. 227, 2003

Colloid Chemistry I
Volume Editor: Antonietti, M.
Vol. 226, 2003

Modern Mass Spectrometry
Volume Editor: Schalley, C.A.
Vol. 225, 2003

The Chemistry of Pheromones and Other Semiochemicals I

Volume Editor: Stefan Schulz

With contributions by
T. Ando · S.-I. Inomata · R. Jurenka · C. I. Keeling · K. Mori ·
E. Plettner · G. Pohnert · K. N. Slessor · M. Yamamoto

Springer

The series *Topics in Current Chemistry* presents critical reviews of the present and future trends in modern chemical research. The scope of coverage includes all areas of chemical science including the interfaces with related disciplines such as biology, medicine and materials science. The goal of eaxch thematic volume is to give the nonspecialist reader, whether at the university or in industry, a comprehensive overview of an area where new insights are emerging that are of interest to a larger scientific audience.

As a rule, contributions are specially commissioned. The editors and publishers will, however, always be pleased to receive suggestions and supplementary information. Papers are accepted for *Topics in Current Chemistry* in English.

In references *Topics in Current Chemistry* is abbreviated Top Curr Chem and is cited as a journal.

Visit the TCC content at http://www.springerlink.com/

Library of Congress Control Number: 2004112158

ISSN 0340-1022
ISBN 3-540-20828-3 **Springer Berlin Heidelberg New York**
DOI 10.1007/b96138

Springer is a part of Springer Science+Business Media
springeronline.com

Printed in Germany

Cover design: KünkelLopka, Heidelberg/design & production GmbH, Heidelberg
Typesetting: Fotosatz-Service Köhler GmbH, Würzburg

Printed on acid-free paper 02/3020/ra – 5 4 3 2 1 0

Volume Editor

Prof. Dr. Stefan Schulz
Institut für Organische Chemie
Technische Universität Braunschweig
Hagenring 30
38106 Braunschweig
Germany
stefan.schulz@tu-bs.de

Editorial Board

Topics in Current Chemistry Also Available Electronically

For all customers who have a standing order to Topics in Current Chemistry, we offer the electronic version via SpringerLink free of charge. Please contact your librarian who can receive a password for free access to the full articles by registration at:

springerlink.com

If you do not have a subscription, you can still view the tables of contents of the volumes and the abstract of each article by going to the SpringerLink Homepage, clicking on "Browse by Online Libraries", then "Chemical Sciences", and finally choose Topics in Current Chemistry.

You will find information about the

- Editorial Board
- Aims and Scope
- Instructions for Authors
- Sample Contribution

at http://www.springeronline.com using the search function.

Preface

Chemistry becomes particularly interesting when reaching out to other disciplines and this has been documented impressively over the last few decades by many cooperative efforts with biology. One of these fields which is now over forty years old is what most chemists know as pheromone research, started in 1959 by Butenandt with the identification of the first pheromone, bombykol. But pheromones are only part of the larger area of inter-individual chemical communication in general. This means of transportation of information is not used to a great extent by human beings, but vastly exploited by other living organisms. The research on these subject is part of Chemical Ecology, a discipline which tries to understand why secondary metabolites are produced by a certain organism and what their effects and functions are in an ecological perspective. This field goes beyond the normally anthropocentric view of traditional natural product research with its focus on application for human welfare.

To understand a given chemical communication system normally one needs a close and fruitful cooperation between chemists and biologists, making this field particularly interesting for many scientists. Chemists are involved in this research by identifying the compounds which provoke behavioral or physiological changes in the receiver, synthesizing them or their analogs, often in enantiomerically pure form, to prove a structure and to provide material for biological testing, working on the biosynthesis, and doing research on the large biomolecules which are needed for formation or processing of the exogenous signal molecules.

The terminology used in this research area is not well established in the chemical community; while the term pheromone is widely known, semiochemical is not. Nevertheless, chemicals used in the communication between individuals are correctly called semiochemicals; recently the equivalent term infochemicals was introduced. These compounds can be further divided into pheromones, compounds used in communication between individuals of the same species, while allelochemicals serve interspecific communication. Most pheromones are releasers, i.e. they provoke a behavioral change in the receiver. More rare are primers, which provoke physiological changes. Allelochemicals can be divided into kairomones, which are advantageous for the receiver, while the emitter benefits from allomones. Synomones are advantageous for both the emitter and the receiver.

Many different functions of pheromones have been found since Butenandt. Aggregation pheromones attract both sexes to a special location, while sex pheromones are offered by one sex only to attract or arouse the other one. Trail pheromones used by ants mark food trails and alarm pheromones change the state of alertness of conspecifics. These are only some of the functions pheromones can have, and similar different functions can be found in allelochemicals as well.

This two volume book tries to give an overview from a chemical perspective about the progress made during the last decade in semiochemical research. Synthesis, a key field of organic chemistry, is covered in many chapters, but the most innovative work is presented concisely in the first chapter by K. Mori, the focus of which is on the synthesis of pheromones, which is mostly target oriented and only rarely used to invent new methodology. Most work has been done on insect pheromones, which is reflected by the selected synthesis and the number of chapters devoted to insects in this book. This chapter is followed by a review on the lepidopteran pheromones by T. Ando et al. This order is the best investigated so far, primarily because of the great economic importance of moths and butterflies. Some species are ideally suited to serve as model organisms in studies going beyond the identification of pheromones to signal perception (see the chapter by Leal) and biosynthesis (see the chapter by Jurenka).

Pheromone identification is still difficult because the structure of unique compounds present in small amounts in mixtures of similar molecules has to be elucidated. This topic will be discussed in detail by Ando as well as by others, showing nicely the recent progress in analytical techniques. The following chapter by R. Jurenka deals with insect pheromone biosynthesis with special emphasis on lepidopteran pheromones and also covers genetic aspects. The subsequent chapter by C. Keeling et al. describes the hymenopteran semiochemicals (bees and ants), describing pheromones and allelochemicals. The hymenoptera add a certain flavor to the scene, because now the complexity of social insects with their many interactions comes into play, as well as the multi-level (multi-trophic) signals used by parasitoids.

The first volume ends with a chapter by G. Pohnert on chemical defence in the marine environment. Defense compounds, which can be regarded as allomones, are often, but not always, more complex than other semiochemicals and may have unique modes of action. The biological mechanisms are not always easy to unravel, which is shown by some of examples. The reader may be tempted to compare the chemical complexity with that of terrestrial insect defence, which can be found in the second volume chapter by D. Daloze and J.-C. Braekman. Insects thus do not only produce interesting pheromones, but also complex allelochemicals for their own protection.

The second volume starts with biochemistry and new insights into pheromone perception and transport by W. Leal. These findings show that specificity is not only achieved by uniquenss of compounds or blends, but also by the perceiving receptors and transport molecules. The following chapters on bugs and beetles by J. Millar and W. Francke and K. Dettner also cover methods used for identification besides target-oriented synthetic approaches and discussion on the application of pheromones in insect control. The book will close with two chapters on highly complex and relatively simple organisms, namely mammals and bacteria. B. Burger points out the difficulty in working with behaviorally complex animals for establishing biological activity of certain compounds or mixtures. Furthermore, complexity can also be found in exocrine secretions of mammals which poses specific problems to the analytical chemist. P. Williams et al. explore chemical communication in bacteria. Microbiologists often use different terminology than zoologists, but from my point of view quorum-sensing-factors are still pheromones or at least semiochemicals. This exciting new field shows extensive progress and facilitates the application of biotechnological methods more easily than in more complex animals.

Hopefully the reader will get an overview of the recent work in the field after reading the chapters. Nevertheless, many exciting subjects have not been included, especially when recent reviews exist, as is the case for semiochemistry of arachnids (spiders and mites) or cockroaches. Further interesting subjects are pheromones of fish, reptiles, amphibians, algae, fungi, yeast, insect-plant interactions, etc. The research described here lays the foundation for further progress in the future, which will definitely benefit from the technological advances seen during the last years in chemistry and biology. A better understanding of the role and function of secondary metabolites may hopefully be obtained.

Braunschweig, August 2004 Stefan Schulz

Contents

Contents of Volume 240
The Chemistry of Pheromones and Other Semiochemicals II

Volume Editor: Stefan Schulz
ISBN 3-540-21308-2

Topics in Current Chemistry (2004) 239: 1–50
DOI 10.1007/b95448

Pheromone Synthesis

Kenji Mori (✉)

The University of Tokyo, 1-20-6-1309 Mukogaoka, Bunkyo-ku, Tokyo 113-0023, Japan
kjk-mori@arion.ocn.ne.jp

Abstract Various aspects of pheromone synthesis were reviewed by analyzing 61 synthetic examples from 9 different structural categories of pheromones. Syntheses executed with new methodologies such as organoborane reactions, organotransition metal chemistry including ring-closing olefin metathesis, asymmetric epoxidations and dihydroxylations, and enzymatic reactions are selected to illustrate the usefulness of modern new reactions.

Keywords Asymmetric dihydroxylation · Isoprenoidal pheromones · Pheromone acetals · Pheromone epoxides · Ring-closing olefin metathesis

List of Abbreviations

BINAL	1,1′-Bi-2-naphthol
Cy	Cyclohexyl
DHP	3,4-Dihydro-2*H*-pyran
DNB	3,5-Dinitrobenzoyl
dppp	1,3-Bis(diphenylphosphino)propane
EDC	1-[3-(*N*,*N*-Dimethylamino)propyl-3-ethylcarbodiimide hydrochloride
imid or Im	Imidazole
KAPA	Potassium 3-aminopropylamide
MPLC	Medium pressure liquid chromatography
Ms	Methanesulfonyl
NMO	*N*-Methylmorpholine *N*-oxide
PPL	Pig pancreatic lipase
Red-Al	Sodium bis(2-methoxyethoxy)aluminum hydride
Sia	3-Methyl-2-butyl
TBS	*tert*-Butyldimethylsilyl
THEX	2,3-Dimethyl-2-butyl
TPAP	Tetra-*n*-propylammonium perruthenate

1 Introduction

The purpose of synthesis in pheromone science is (1) to establish the proposed structure including absolute configuration and (2) to provide samples in amounts sufficient for biological studies and practical pest control.

Pheromones are often low molecular weight organic compounds of diverse structural types. Double-bonds and stereogenic centers in pheromone molecules further enrich their diversities by rendering stereoisomerism possible. Accordingly, pheromone synthesis must deal with the common problems of organic synthesis: carbon-carbon bond formation, functional group transformation, and diastereo- and/or enantioselective preparation of the desired stereoisomer. In this chapter, modern solutions for these problems are illustrated with 61 examples of pheromone synthesis. You can see that many new reactions such as Sonogashira, Stille-Kosugi, or Miyaura-Suzuki couplings, asymmetric dihydroxylation and ring-closing olefin metathesis are cleverly employed in preparing pheromones. Among almost 900 synthetic literatures of interest in the period of 1990 to early 2003, 63 examples are selected. The choice was made rather arbitrary according to my own taste. I intend, however, to show you almost all of the important and modern aspects of organic synthesis as applied to pheromone science.

Pheromone synthesis was thoroughly reviewed in the past. In 1989, synthetic methods useful in preparing optically active pheromones were reviewed by Mori [1]. More comprehensive reviews are also available [2, 3]. The present chapter follows the style adopted in Mori's encyclopedic review [3] to classify the pheromones according to the compound types. In comparison to 1229 references listed in [3], only 101 references are listed in this chapter. I apologize to those whose syntheses have not been reviewed here. Today, however, electronic information service has developed remarkably so as to make any literature search possible. Those who want to carry out pheromone synthesis must carefully search the existing information to avoid duplication with others.

There are three reviews emphasizing the importance of synthesis in pheromone science [4], in semiochemicals research [5], and in chemical ecology [6]. Stereochemistry-pheromone activity relationships are also discussed in the above three reviews, and more thoroughly in three other reviews [7–9].

2 General Methods

Four new or improved methods useful in the synthesis of aliphatic insect pheromones were reported. ω-Bromo-1-alkanols are important building blocks in pheromone synthesis. Chong found that α,ω-diols gave good yield of ω-bromo-1-alkanols when heated with hydrobromic acid and toluene at reflux (Scheme 1) [10].

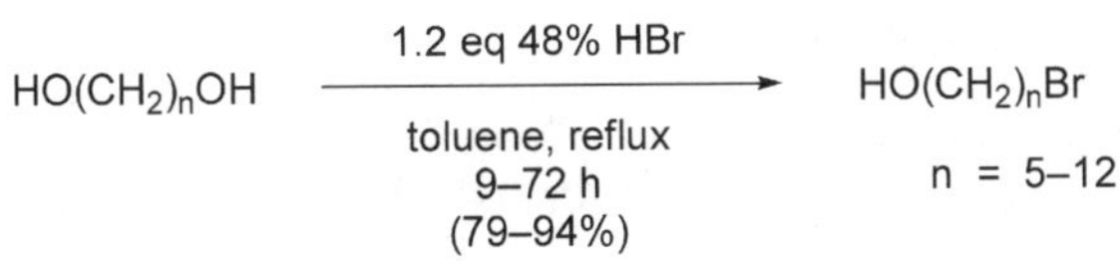

Scheme 1

Alkylation of 1-alkynes with alkyl halides was carefully examined by Chong [11]. Alkynes could be alkylated easily in the absence of HMPA by treatment with *n*-butyllithium followed by *n*-alkyl iodide in THF. In the case of bromides, a catalytic amount of tetra(*n*-butyl)ammonium iodide or sodium iodide should be added (Scheme 2).

Three new methods for the conversion of alkynes to (*Z*)-alkenes were reported, although Lindlar semi-hydrogenation still remains as the most convenient method. Copper (I) hydride reagent could reduce alkynes to (*Z*)-alkenes as shown in Scheme 3 [12]. Yoon employed nickel boride prepared on borohydride exchange resin for selective hydrogenation of alkynes to (*Z*)-alkenes (Scheme 4) [13].

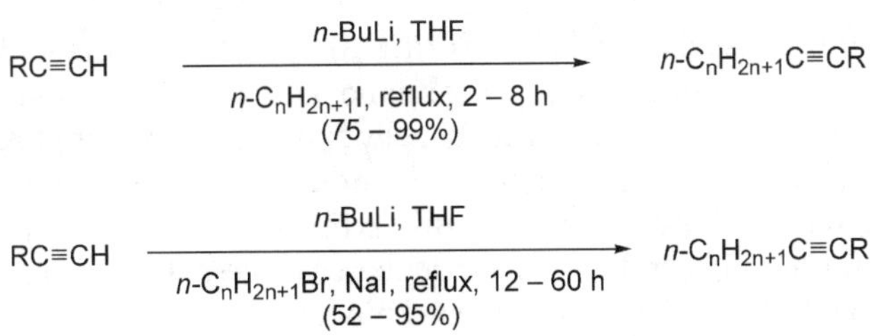

Scheme 2

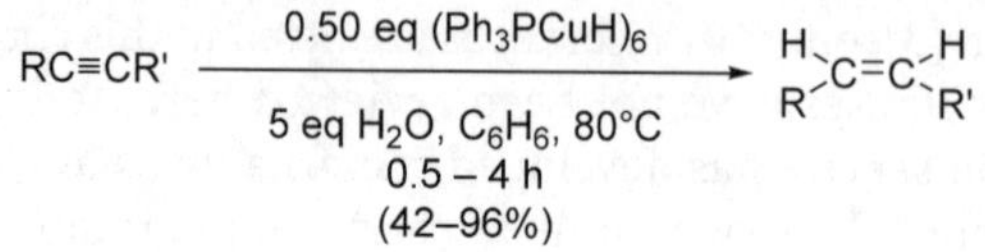

Scheme 3

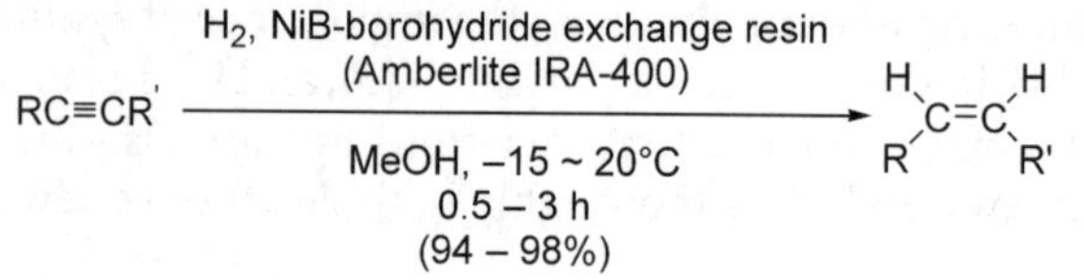

Scheme 4

F. Sato developed titanium (II)-based *cis*-reduction of alkynes as shown in Scheme 5 [14], and the method was applied to the synthesis of pheromones by Kitching (Scheme 6) [15]. This titanium (II)-based reaction gives pure (*Z*)-alkenes. Kitching summarized the contemporary methods for the synthesis of skipped polyynes and their reduction to skipped polyenes [15].

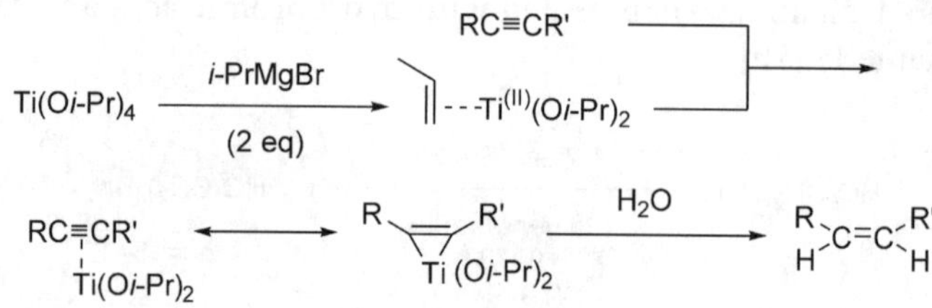

Scheme 5

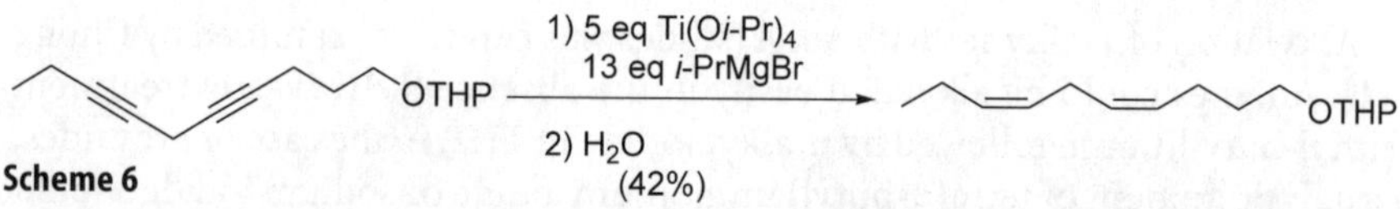

Scheme 6

Various enzymes, especially esterases and lipases, are employed in the enantioselective syntheses of pheromones [16, 17]. Examples will be given later.

3 Synthesis of Alkane Pheromones

Scheme 7 summarizes the synthesis of (7*R*,11*S*)-7,11-dimethylheptadecane (**1**), the female sex pheromone of the spring hemlock looper (*Lambdina athasaria*) by Mori [18]. Enantiopure alkanes are usually synthesized by coupling enantiopure building blocks derived from natural products or compounds prepared by asymmetric synthesis. Even among hydrocarbons, chirality is very important for pheromone activity, and in this particular case *meso*-**1** was bioactive, while neither (7*R*,11*R*)-**1** nor (7*S*,11*S*)-**1** showed bioactivity.

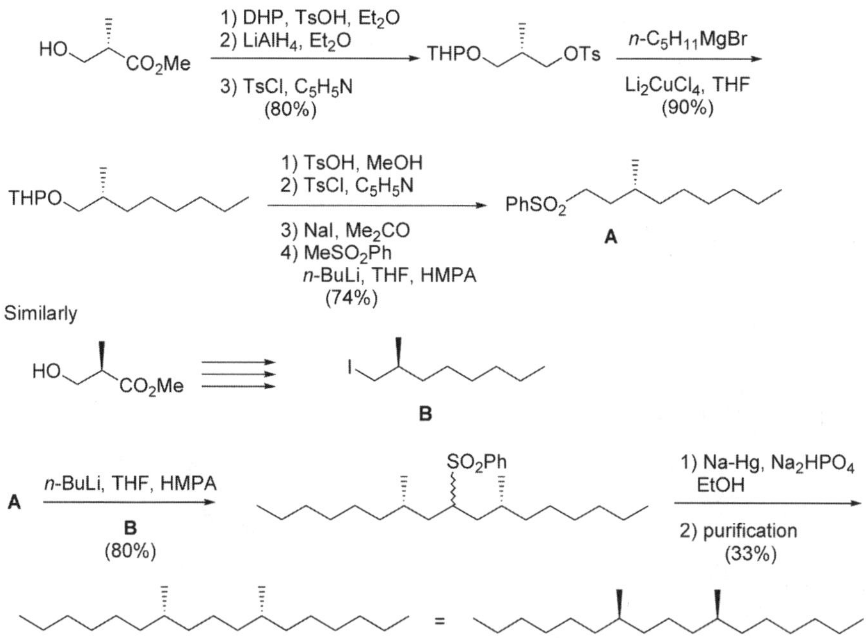

Scheme 7 (7*R*,11*S*)-**1**

4
Synthesis of Olefinic Pheromones

Recent trend in the synthesis of olefinic pheromones is the use of transition metal-catalyzed cross coupling reaction for carbon-carbon bond formation. Scheme 8 summarizes a synthesis of the termite trail marker pheromone, (3*Z*,6*Z*)-3,6-dodecadien-1-ol (**2**) by Oehlschlager [19]. The key-step is the palladium-catalyzed cross-coupling of allylic chloride **A** and alkenylalane **B**.

Scheme 9 shows Svatoš's synthesis of (8*Z*,10*E*)-tetradeca-8,10-dienal (**3**), the female pheromone of the horse-chestnut leafminer (*Caneraria ohridella*) [20]. Palladium-catalyzed coupling of **A** with 1-pentyne was the key-step.

Synthesis of (10*E*,12*Z*)- and (10*Z*,12*Z*)-10,12-hexadecadienal (**4** and **5**) by Oehlschlager is summarized in Scheme 10 [21]. These two aldehydes **4** and **5** are the female sex pheromone components of melonworm (*Diaphania hyalinata*).

Palladium-catalyzed coupling of the borane derived from **A** with 1-bromo-1-pentyne to give **B** as well as the coupling of iodoalkene **C** with alkyne **D** were the two key-steps.

Two new syntheses of bombykol (**6**), the female sex pheromone of the silkworm moth (*Bombyx mori*), were reported [22, 23]. Scheme 11 shows Negishi's synthesis of **6** based on organoborane chemistry [22], and Uenishi's synthesis of **6** based on palladium and nickel catalyses is summarized in Scheme 12 [23]. Both syntheses afforded bombykol of >98% purity.

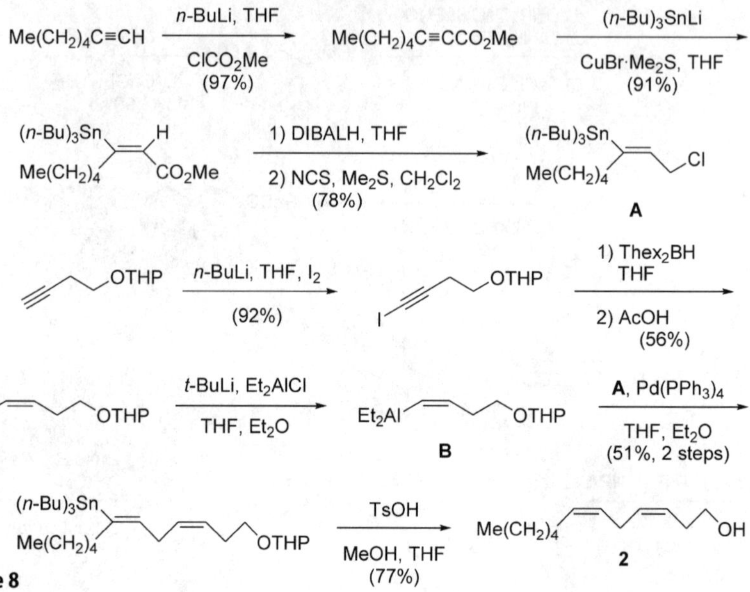

Scheme 8

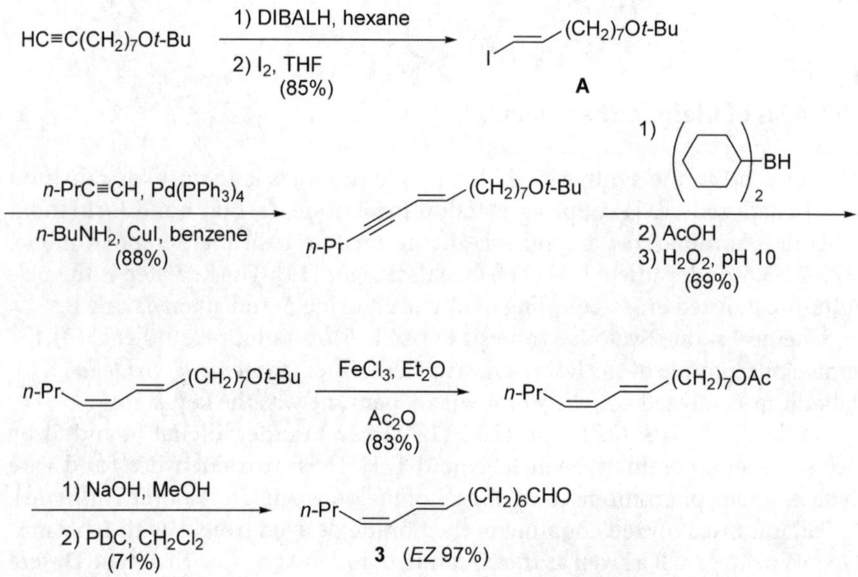

Scheme 9

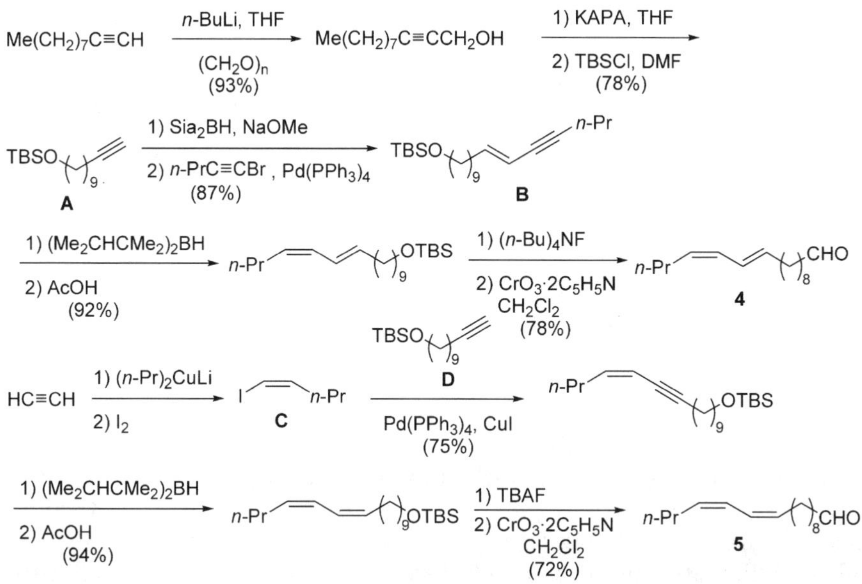

Scheme 10

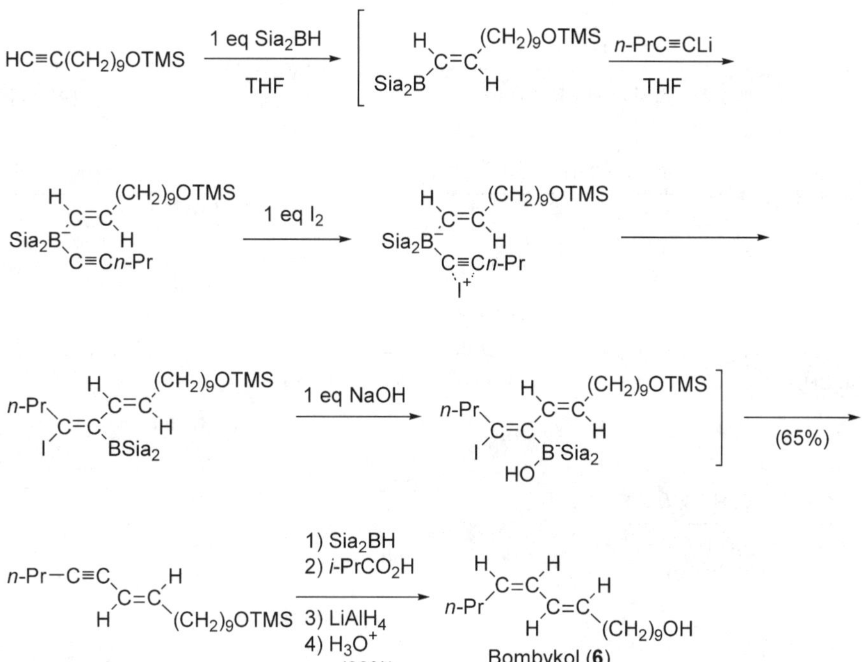

Scheme 11

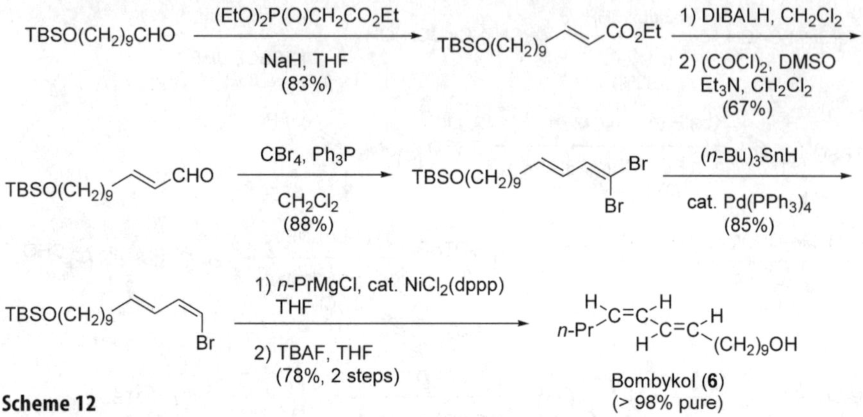

Scheme 12

A new group of allenic hydrocarbons was isolated from five Australian melonthine scrab beetles (*Antitrogus consanguineus*), and synthesized by Kitching as shown in Scheme 13 [24]. This synthesis allowed the assignment of (*R*)-absolute configuration to the natural 9,10-pentacosadiene (7, 89% ee) in *A. consanguineus* [24]. Similarly, (*R*)-configuration was assigned to natural 9,10-tricosadiene (86% ee) [24].

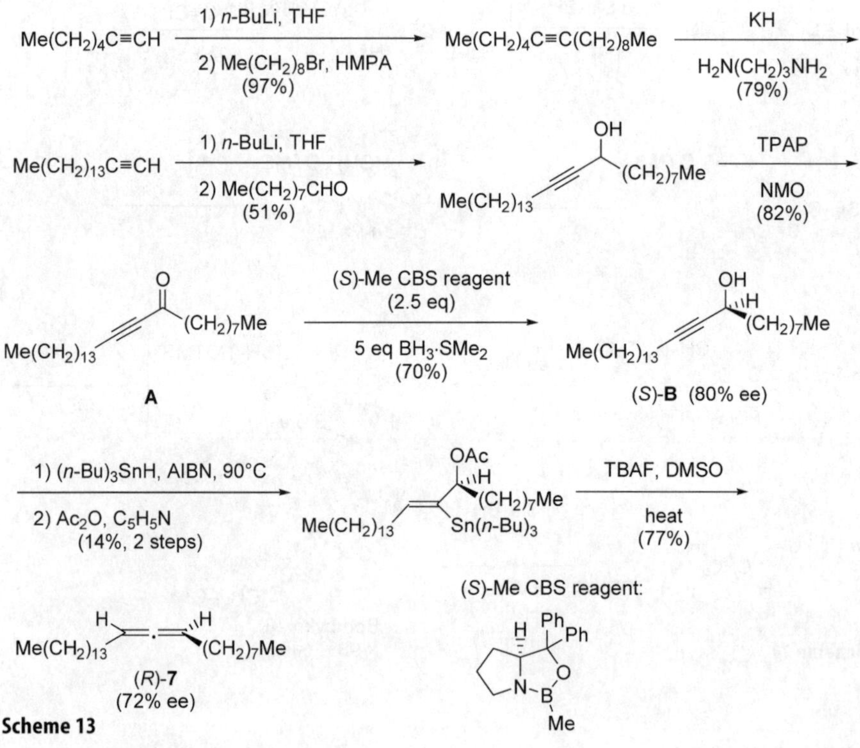

Scheme 13

Scheme 14 illustrates Linstrumelle's synthesis of (9*Z*,11*E*)-9,11,13-tetradecatrienyl acetate (**8**), the pheromone of the pyralid moth, *Stenoma cecropia* [25]. The key steps were palladium and copper-catalyzed Sonogashira couplings (**A**+**B** and **C**+**D**). Another noteworthy feature in this synthesis was the use of activated zinc dust in aqueous methanol for the reduction of the triple bonds of **E** to give two double bonds of **8**.

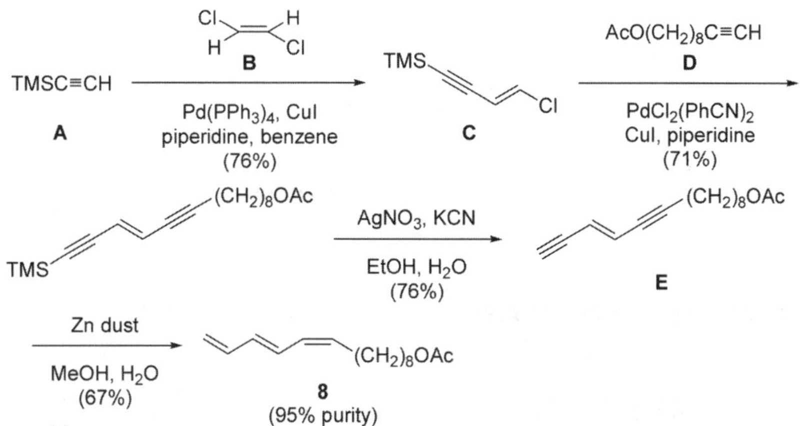

Scheme 14

Kitching employed the titanium (II)-based *cis*-reduction of alkynes in their synthesis of (3*E*,8*Z*,11*Z*)-3,8,11-tetradecatrienyl acetate (**9**), the pheromone of the moth *Scrobipalpuloides absoluta* as shown in Scheme 15 [15].

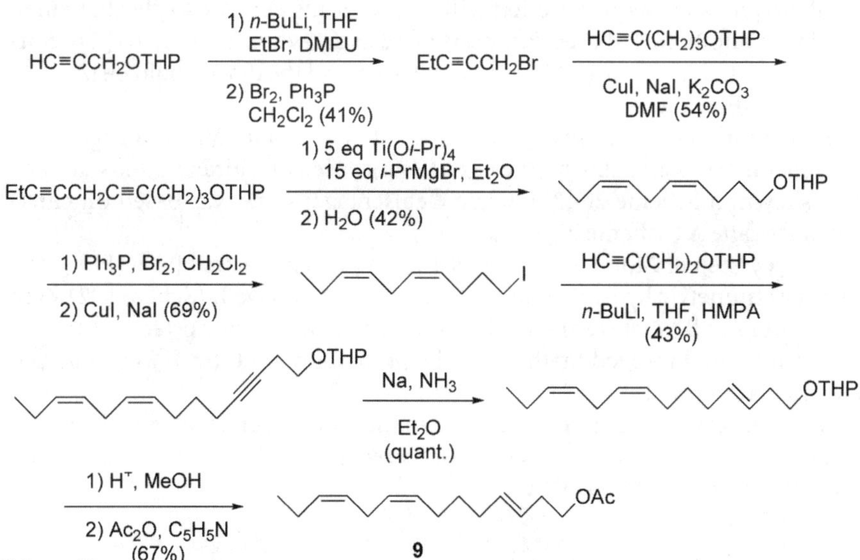

Scheme 15

In 1992, nitrate esters **10** and **11** were identified as the female sex pheromone of the cotton leaf perforator (*Bucculatrix thurberiella*) [26]. Synthesis of (*Z*)-9-tetradecenyl nitrate (**10**) and (*Z*)-8-tridecenyl nitrate (**11**) is shown in Scheme 16 [26]. A 100:2 blend of **10** and **11** is highly attractive for male *B. thurberiella*.

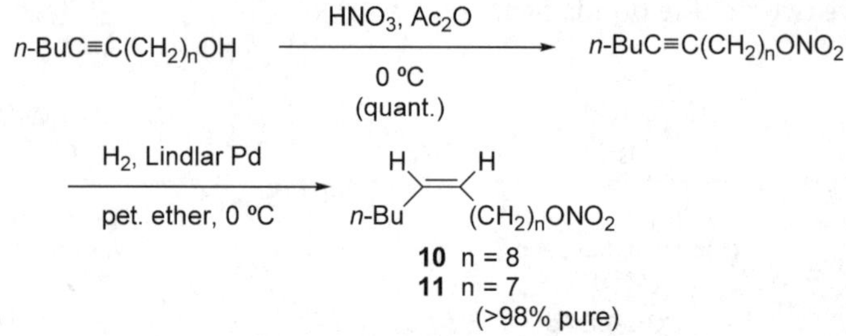

Scheme 16

5 Synthesis of Epoxy Pheromones

5.1 Synthesis of (7*R*,8*S*)-Disparlure

(7*R*,8*S*)-(+)-Disparlure (**12**) is the female sex pheromone of the gypsy moth (*Lymantria dispar*). Advent of Sharpless asymmetric dihydroxylation (AD) allowed several new syntheses of **12** possible. Sharpless synthesized **12** as shown in Scheme 17 [27]. Scheme 18 summarizes Ko's synthesis of **12** employing AD-mix-α [28]. He extended the carbon chain of **A** by Payne rearrangement followed by alkylation of an alkynide anion with the resulting epoxide to give **B**. Keinan developed another AD-based synthesis of **12** as shown in Scheme 19 [29]. Mitsunobu inversion of **A** to give **B** was the key step, and the diol **C** could be purified by recrystallization.

Two additional syntheses of **12** employed organostannane and organoborane chemistry. Marshall's synthesis of **12** was based on chiral organostannane **A** as shown in Scheme 20 [30], while Oehlschlager's one was based on chiral organoborane **A** (Scheme 21) [31].

The key-step of Mori's synthesis of **12** was pig pancreatic lipase (PPL)-catalyzed asymmetric hydrolysis of *meso*-diacetate **A** to give **B** (Scheme 22) [32]. Purification of **B** (90.8% ee) afforded pure **C**, which was converted to **12**.

Recently, an improved method for the preparation of **C** by lipase-catalyzed asymmetric acetylation of (±)-**C** was reported [33].

A new analytical method [34] for the enantiomeric purity of epoxide pheromones as well as a new method [35] for stereochemical inversion of saturated epoxides were reported by Oliver.

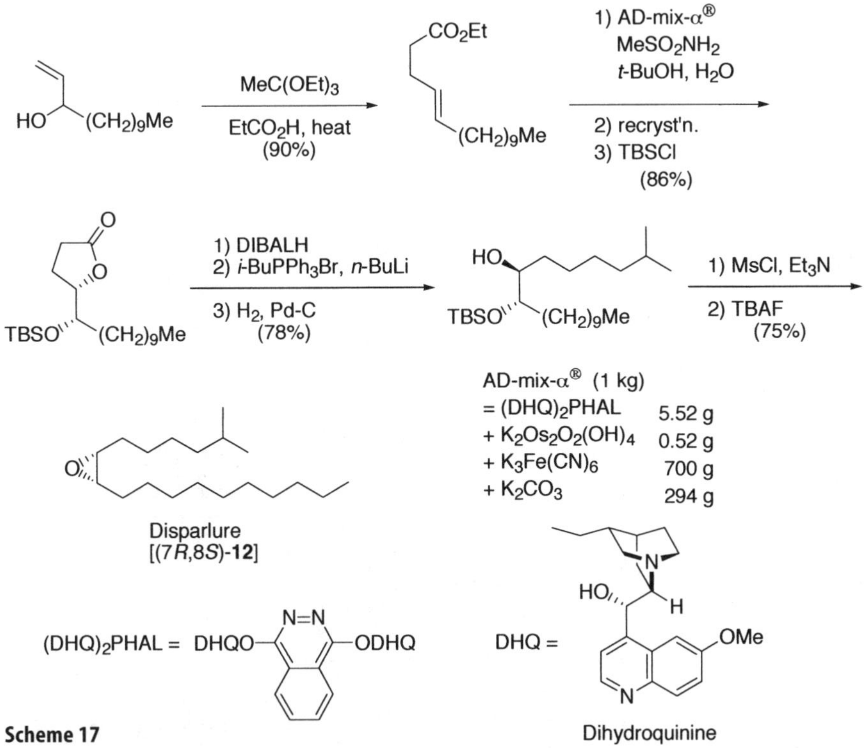

Scheme 17

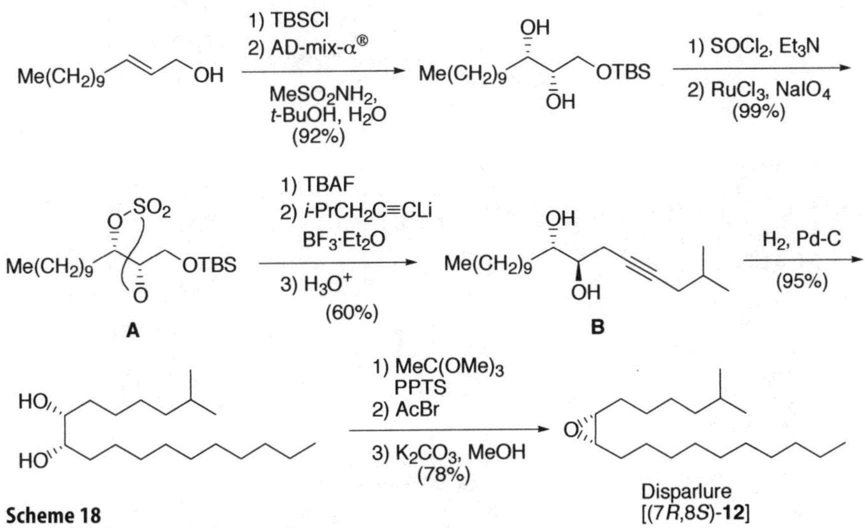

Scheme 18

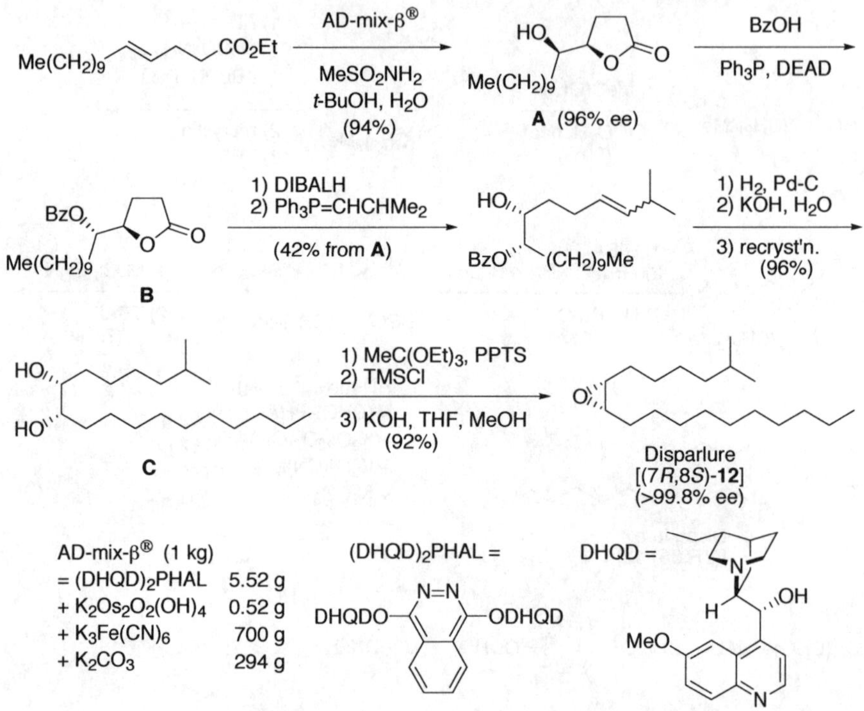

Scheme 19

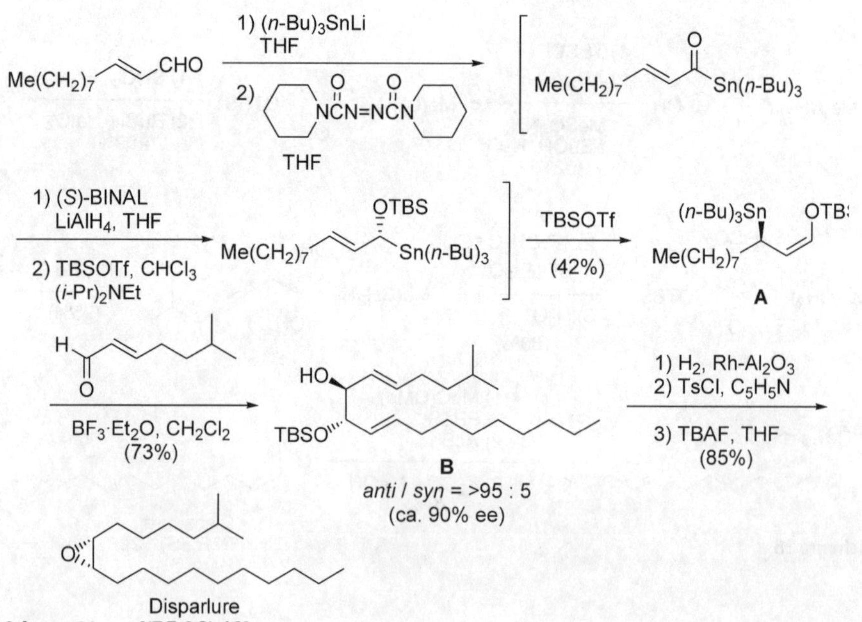

Scheme 20 [(7*R*,8*S*)-**12**]

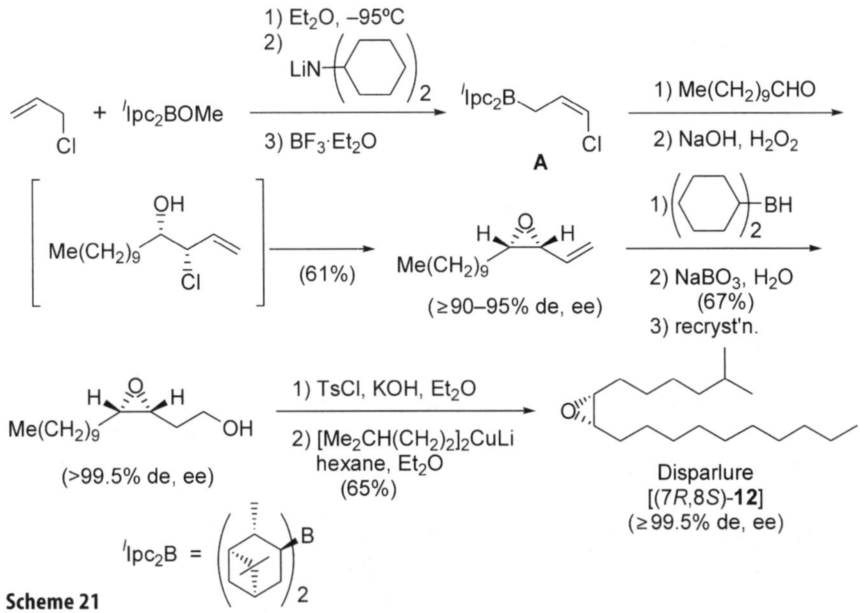

Scheme 21

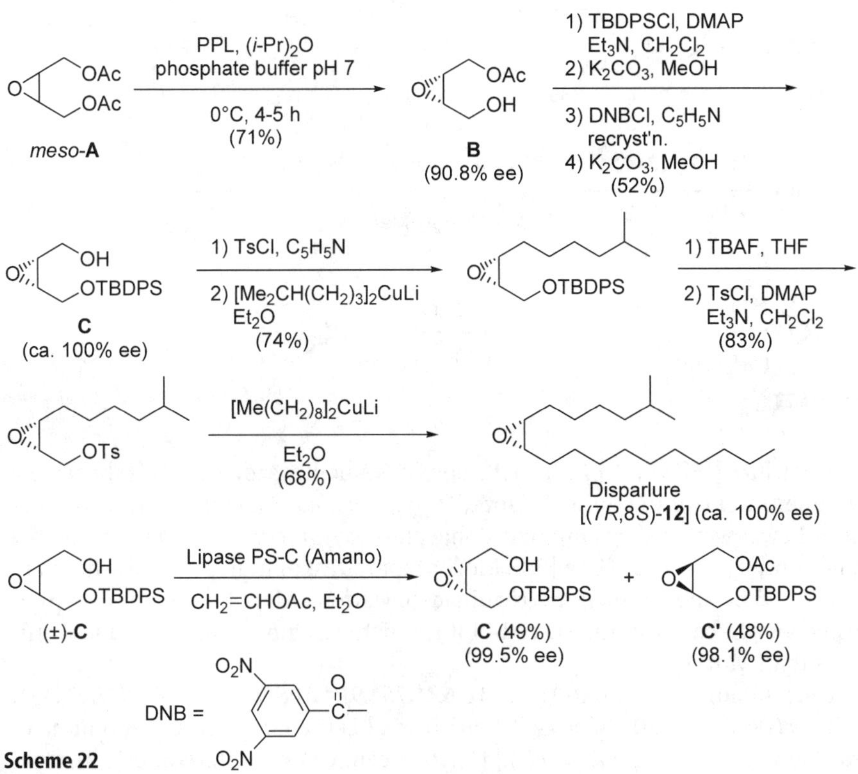

Scheme 22

5.2
Synthesis of Olefinic Epoxy Pheromones

(6*Z*,9*S*,10*R*)-9,10-Epoxyhenicos-6-ene (**13**) is the female sex pheromone of moths such as ruby tiger moth (*Phragmatobia fuliginosa*), fruit-piercing moth (*Oraesia excavata*), and painted apple moth (*Teia anartoides*). Scheme 23 summarizes Shi's synthesis of **13** based on Sharpless asymmetric dihydroxylation (AD) [36]. Mori synthesized **13** employing lipase to prepare **A** (Scheme 24) [37]. Alkylation of the acetylide anion **C** was possible neither with tosylate nor with iodide, but triflate **B** could alkylate **C** to give **D**.

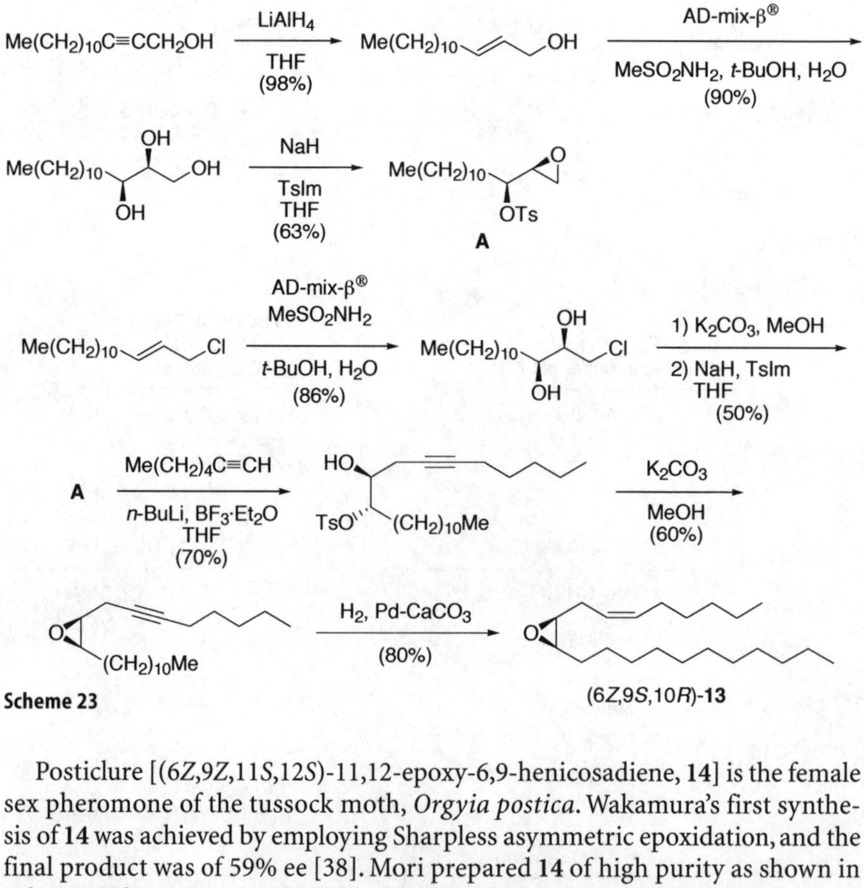

Scheme 23 (6*Z*,9*S*,10*R*)-**13**

Posticlure [(6*Z*,9*Z*,11*S*,12*S*)-11,12-epoxy-6,9-henicosadiene, **14**] is the female sex pheromone of the tussock moth, *Orgyia postica*. Wakamura's first synthesis of **14** was achieved by employing Sharpless asymmetric epoxidation, and the final product was of 59% ee [38]. Mori prepared **14** of high purity as shown in Scheme 25 basing on asymmetric dihydroxylation (AD) [39]. Kumar also published an AD-based synthesis of **14** [40], which was more lengthy and less efficient than Mori's [39].

Leucomalure [**15**, a mixture of (3*Z*,6*R**,7*S**,9*R**,10*S**)- and (3*Z*,6*R**,7*S**,9*S**,10*R**)-*cis*-6,7-*cis*-9,10-diepoxy-3-henicosene] is the female sex pheromone of the Satin moth (*Leucoma salicis*) [41]. Its racemic and diastereomeric mixture

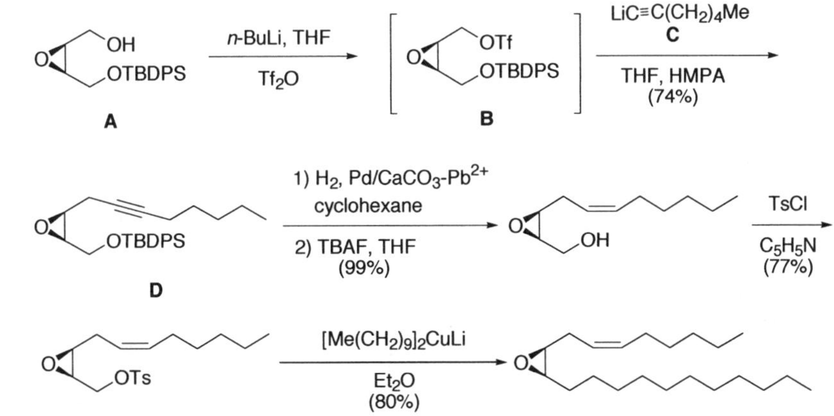

Scheme 24 (6*Z*,9*S*,10*R*)-**13**

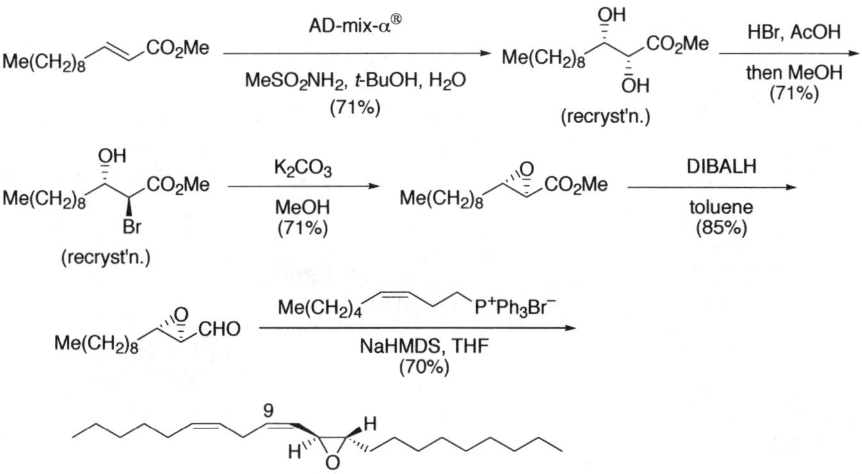

Scheme 25 Posticlure [(11*S*,12*S*)-**14**] (9*Z* / 9*E* = 93 : 7, 99.9% ee)

was prepared by epoxidation of (6*Z*,9*Z*)-6,9-henicosadien-3-yne followed by semi-hydrogenation of the triple bond [41]. Separation of the epoxidation products of (3*Z*,6*R*,7*S*,9*Z*)- or (3*Z*,6*S*,7*R*,9*Z*)-6,7-epoxy-3,9-henicosadiene by HPLC on a chiral stationary phase enabled Ando to prepare four stereoisomers of **15** [42]. After the synthesis of the two diastereoisomers of (±)-leucomalure [43], Mori synthesized all of the four possible stereoisomers of **15** as shown in Scheme 26 [44]. Bioassay of these four isomers against the Satin moth in Hungary will clarify the stereochemistry of the bioactive enantiomer of **15**.

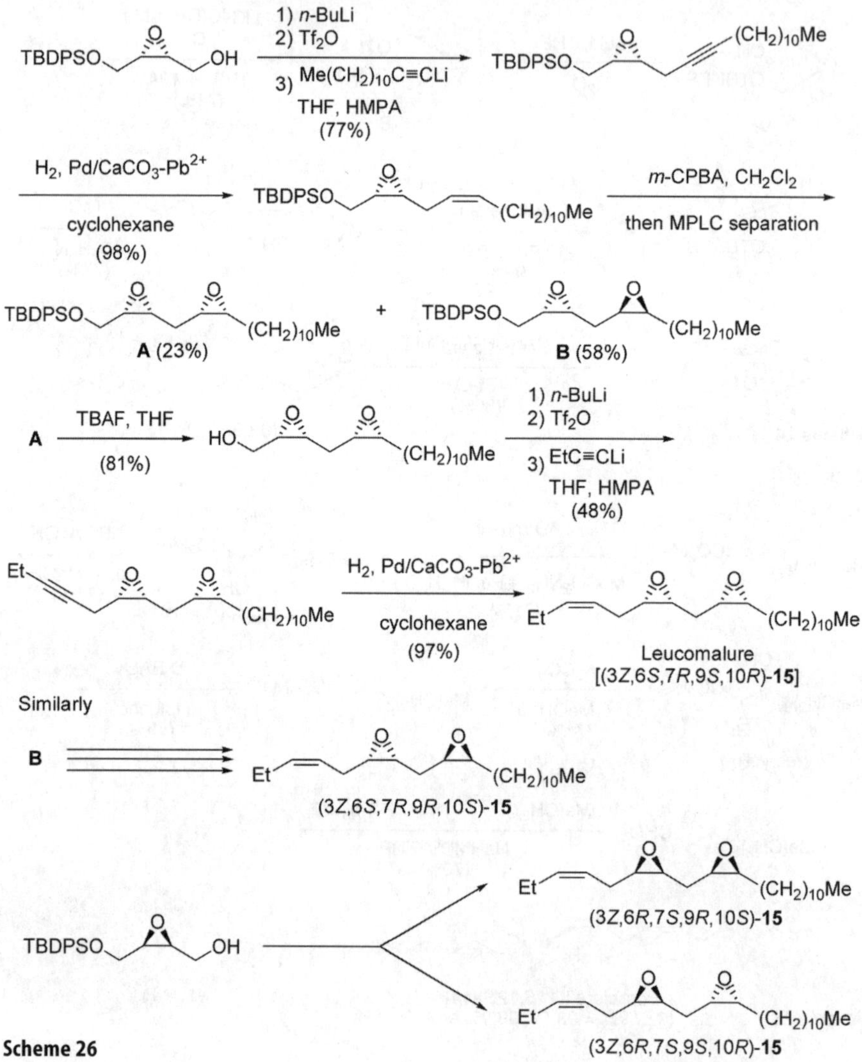

Scheme 26

6
Synthesis of Non-Isoprenoidal Pheromone Alcohols and their Esters

Scheme 27 summarizes Yamamoto's synthesis of lardolure (**16**), the aggregation pheromone of the acarid mite, *Lardoglyphus konoi* [45]. The key-step was the transformation of acetal **A** to hemiacetal **B**.

(2*S*,12*S*)-2,12-Diacetoxytridecane (**17**) is a component of the female pheromone of pea midges (*Contarinia pisi*). Kitching synthesized **17** as shown in Scheme 28 by employing Jacobsen's hydrolytic kinetic resolution of terminal epoxides with a (salen)Co(OAc) complex, (*S*,*S*)-**B** [46]. By this reaction bis-

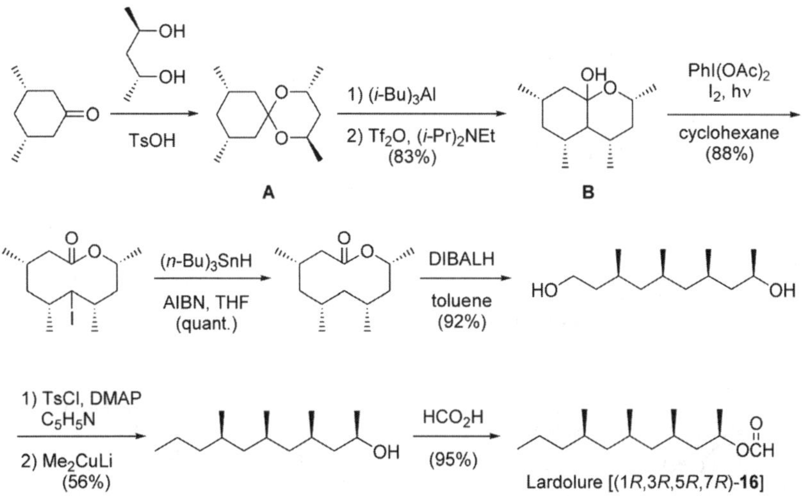

Scheme 27

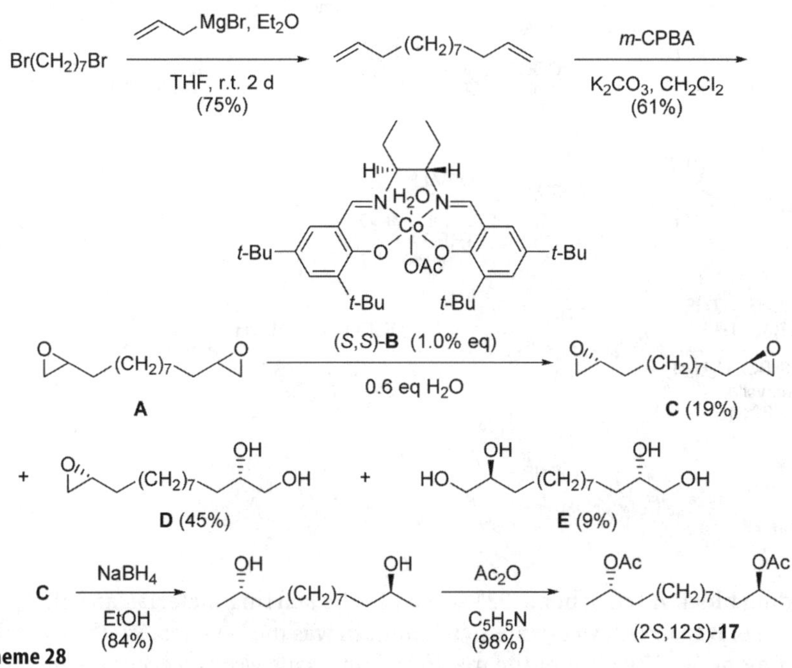

Scheme 28

epoxide **A** afforded **C**, **D**, and **E**. Bis-epoxide **C** was converted to **17**. Kitching used Jacobsen's kinetic resolution to synthesize six other pheromones.

Four different syntheses were reported for the enantiomers of 1-methyl-2-cyclohexen-1-ol, a component of the aggregation pheromone of *Dendroctonus pseudotsugae* [47–50].

7
Synthesis of Non-Isoprenoidal Pheromone Ketones, Acids and Esters

Several female sex pheromones of pine bast scales were identified and synthesized. Scheme 29 summarizes Mori's synthesis of the pheromone [(*R*)-**18**] of the Israeli pine bast scale (*Matsucoccus josephi*) [51]. Enzymatically prepared

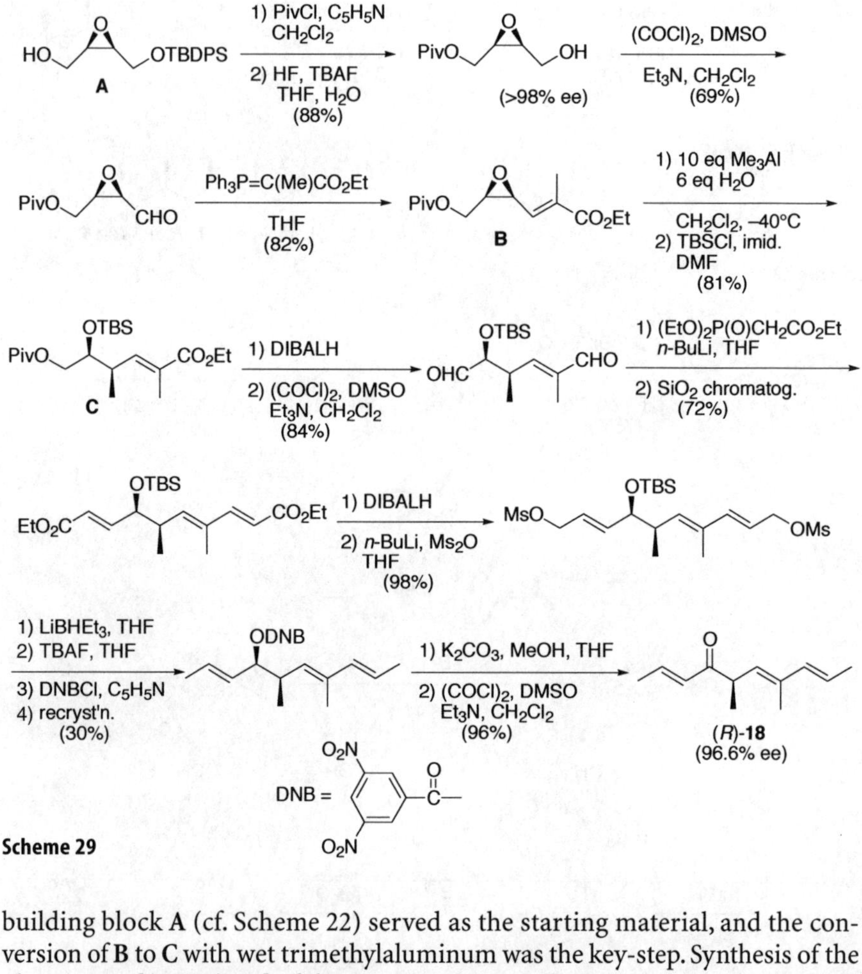

Scheme 29

building block **A** (cf. Scheme 22) served as the starting material, and the conversion of **B** to **C** with wet trimethylaluminum was the key-step. Synthesis of the pheromone [(3*S*,7*R*)-**19**] of the maritime pine scale (*Matsucoccus feytaudi*) as well as that [(6*R*,10*R*)-**20**] of the Japanese pine scale (*Matsucoccus matsumurae*) from the same starting material **A** were shown in Scheme 30 [52, 53].

Mori later developed a shorter synthesis of these pheromones by employing a Weinreb amide **A** (Scheme 31) as the common intermediate [54]. The products **18**–**20**, however, were less enantiomerically pure than those previously synthesized from the epoxy alcohol **A** of Schemes 29 and 30.

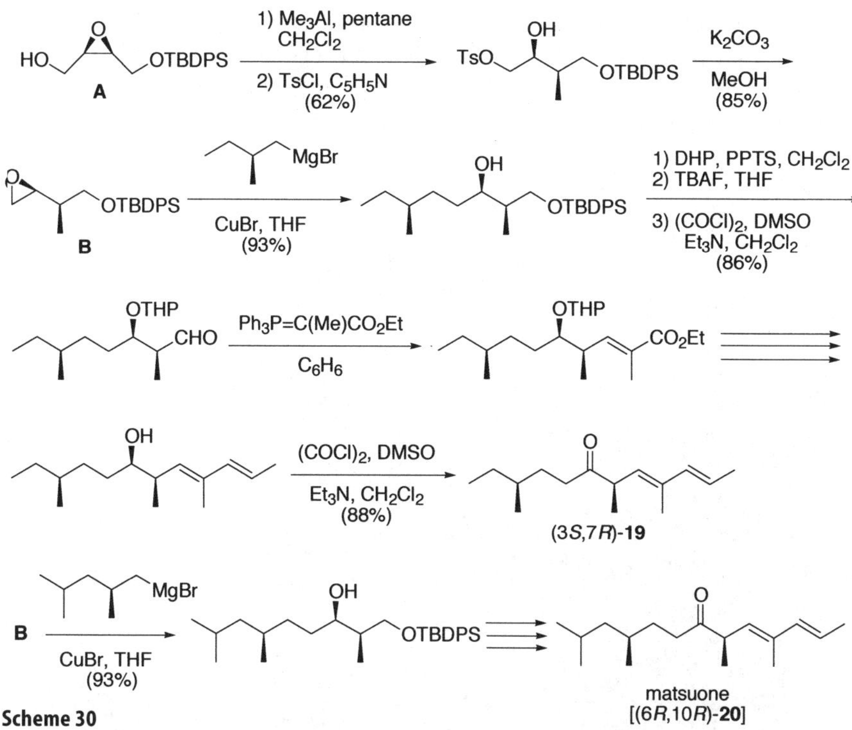

Scheme 30

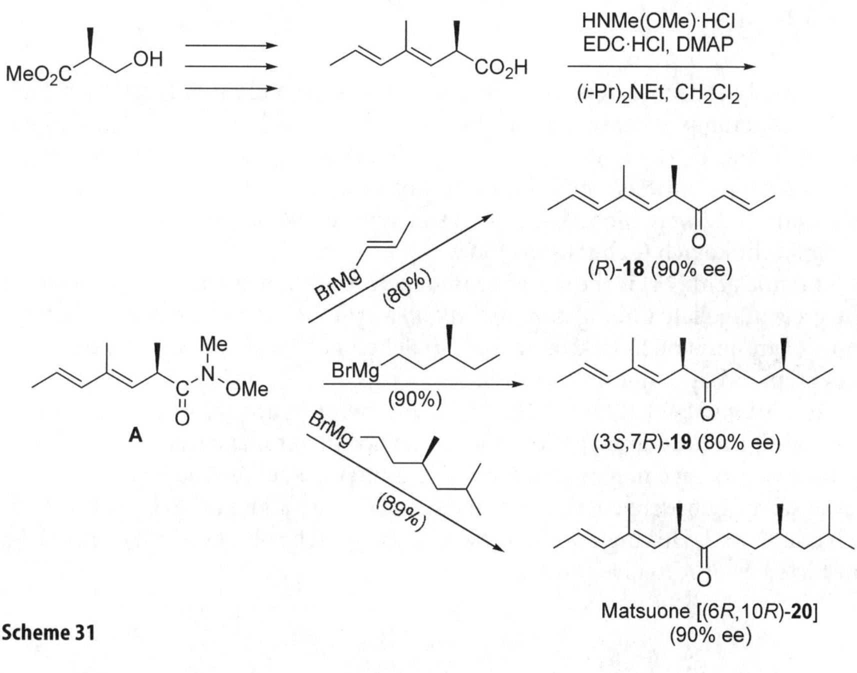

Scheme 31

Scheme 32 summarizes Matteson's synthesis of serricornin [(4*S*,6*S*,7*S*)-**21**], the female sex pheromone of the cigarette beetle (*Lasioderma serricorne*) via boronic esters [55]. Due to the highly stereoselective nature of boronic ester chemistry, this synthesis of **21** was quite efficient.

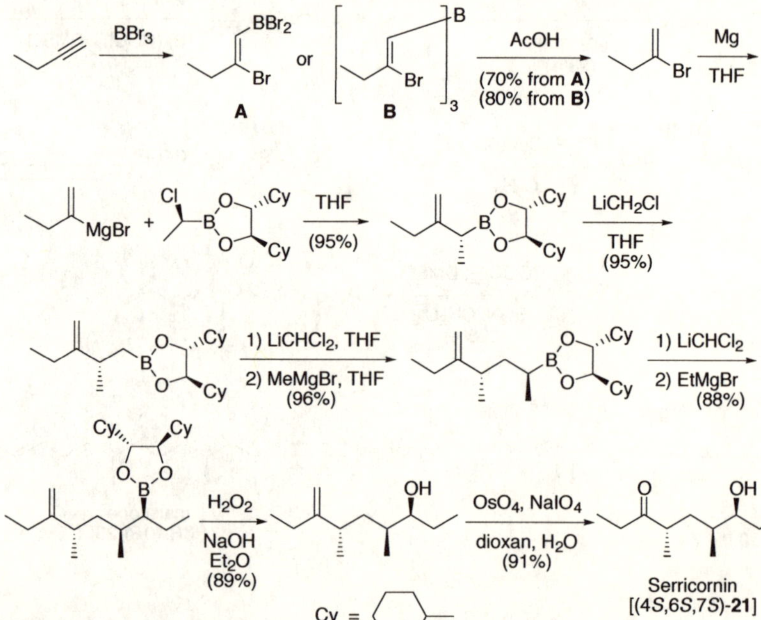

Scheme 32

A myxobacterium *Stigmatella aurantiaca* uses stigmolone (**22**) as its fruiting body inducing pheromone. Both the enantiomers of **22** were synthesized by Mori [56] and also by Enders [57], and found to be equally bioactive. Scheme 33 shows Mori's synthesis, and Scheme 34 summarizes that of Enders. Since both (*R*)- and (*S*)-**22** were bioactive, (±)-**22** was synthesized in a very efficient manner by Kulinkovich (Scheme 35) [58].

Lurlenic acid (**23**) is the sex pheromone produced by the female gametes of the green flagellate *Chlamydomonas allensworthii* to attract the male gametes. Mori's synthesis of **23** is summarized in Scheme 36 [59]. The aglycone part **C** was prepared by coupling **A** with **B**.

Sitophilate [1-ethylpropyl (2*S*,3*R*)-2-methyl-3-hydroxypentanoate, **24**] is the male-produced aggregation pheromone of the granary weevil (*Sitophilus granarius*). Its racemate is also bioactive. Almeida synthesized (±)-**24** by diastereoselective hydrogenation of the Baylis-Hillman adduct **A** to **B** as shown in Scheme 37 [60]. This high diastereoselectivity could be observed only with TBS-protected ester **A** to give pure **B**.

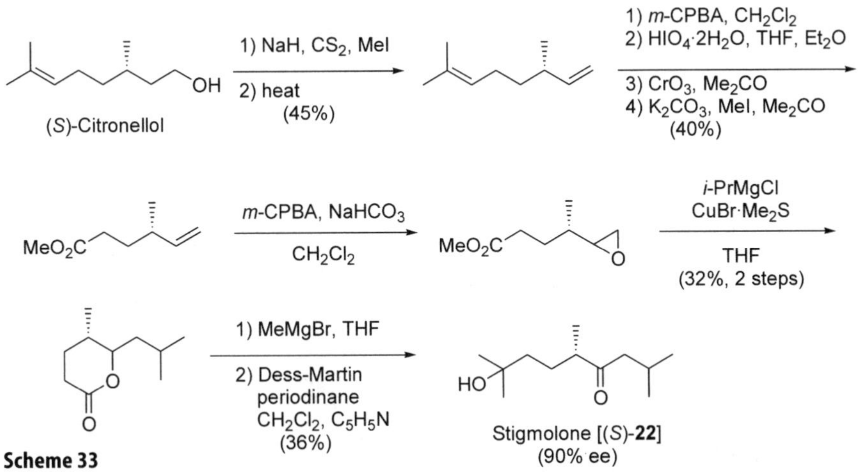

Scheme 33

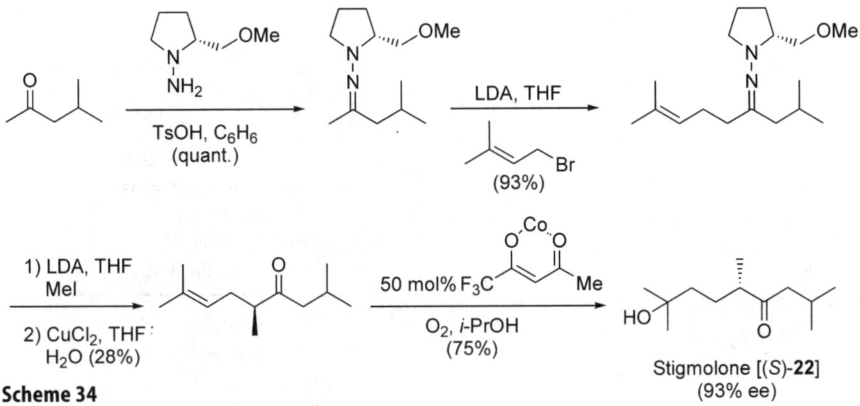

Scheme 34

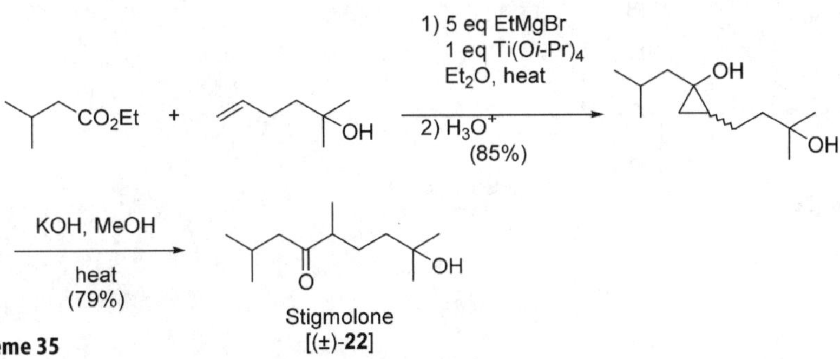

Scheme 35

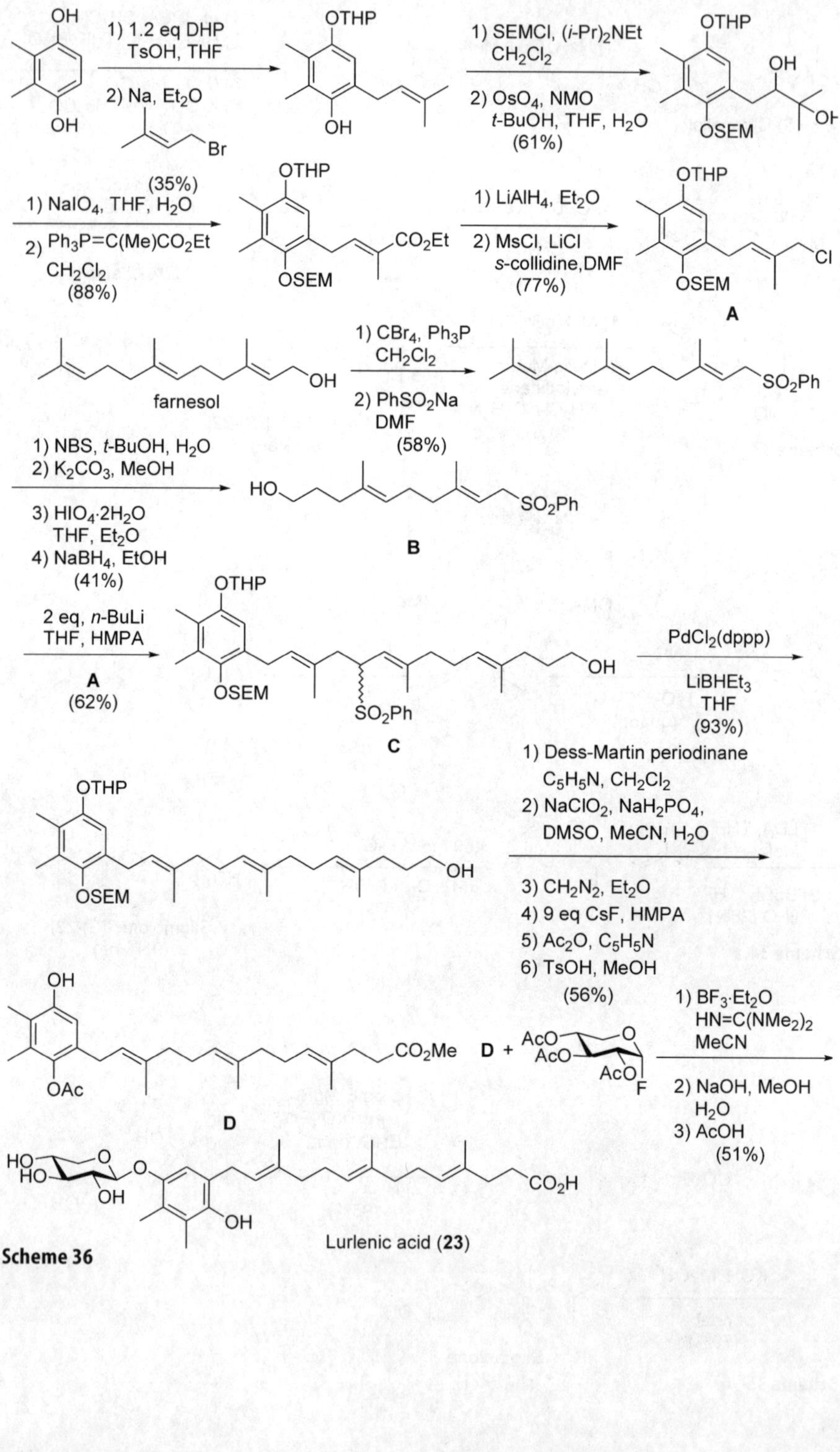
1) 1.2 eq DHP
TsOH, THF
2) Na, Et2O
Br
(35%)
1) SEMCl, (i-Pr)2NEt
CH2Cl2
2) OsO4, NMO
t-BuOH, THF, H2O
(61%)
1) NaIO4, THF, H2O
2) Ph3P=C(Me)CO2Et
CH2Cl2
(88%)
1) LiAlH4, Et2O
2) MsCl, LiCl
s-collidine,DMF
(77%)
A
farnesol
1) CBr4, Ph3P
CH2Cl2
2) PhSO2Na
DMF
(58%)
1) NBS, t-BuOH, H2O
2) K2CO3, MeOH
3) HIO4·2H2O
THF, Et2O
4) NaBH4, EtOH
(41%)
B
2 eq, n-BuLi
THF, HMPA
A
(62%)
C
PdCl2(dppp)
LiBHEt3
THF
(93%)
1) Dess-Martin periodinane
C5H5N, CH2Cl2
2) NaClO2, NaH2PO4,
DMSO, MeCN, H2O
3) CH2N2, Et2O
4) 9 eq CsF, HMPA
5) Ac2O, C5H5N
6) TsOH, MeOH
(56%)
D
1) BF3·Et2O
HN=C(NMe2)2
MeCN
2) NaOH, MeOH
H2O
3) AcOH
(51%)
Lurlenic acid (23)

Scheme 36

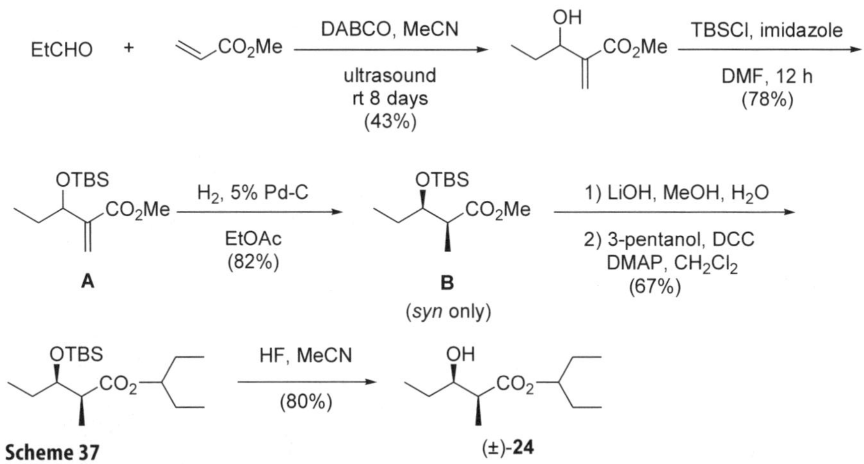

Scheme 37

8
Synthesis of Hemiacetals and Lactones as Pheromones

Hemiacetal **25** [(3*R*,4*S*,1′*E*)-3,4-bis(1′-butenyl)tetrahydro-2-furanol] is the male pheromone of the spined citrus bug (*Biprorulus bibax*). Scheme 38 summarizes Mori's synthesis of **25** [61]. Claisen rearrangement (**A**→**B**) and lipase-catalyzed asymmetric acetylation [*meso*-**C**→(5*S*,6*R*)-**D**] were the two key steps of the synthesis. Further purification of **D** was executed at the stage of its crystalline derivative **E**. In this particular case, the unnatural (3*S*,4*R*,1′*E*)-**25** was as active as the natural (3*R*,4*S*,1′*E*)-**25**. Accordingly, a more efficient synthesis of (±)-**25** was achieved by the rearrangement of **F**, avoiding the use of toxic HMPA [62].

Enantiomerically pure japonilure [(4*R*,5*Z*)-5-tetradecen-4-olide, **26**] is commercially important to attract the Japanese beetle (*Popillia japonica*), because only pure (4*R*,5*Z*)-**26** is bioactive, while (±)-**26** is totally inactive. Synthesis of (4*R*,5*Z*)-**26** via enzymatic resolution of racemic intermediates was first reported by Sugai [63], and then further studied by Fukusaki [64] to establish the process as summarized in Scheme 39. Lipase PS (Amano) from *Pseudomonas* sp. and lipase OF (Meito) from *Candida cylindracea* were shown to be the enzymes of choice.

(*R*)-5-Hexadecanolide (**27**) is the pheromone produced by the queens of the oriental hornet (*Vespa orientalis*). Yamamoto synthesized (*R*)-**27** by employing an interesting asymmetric process (**A**→**B**) followed by a multi-step conversion of **B** to (*R*)-**27** via **C** (Scheme 40) [65].

Scheme 41 summarizes Couladouros's synthesis of the oviposition attractant pheromone of the Southern house mosquito (*Culex pipiens fatigans*), (5*R*,6*S*)-6-acetoxy-5-hexadecanolide (**28**) [66]. The key-steps are: (i) *E*-selective Schlosser olefination (**A**→**B**), asymmetric dihydroxylation (**B**→**C**), and lactonization of carbonate **C** to the desired δ-lactone with inversion at C-5.

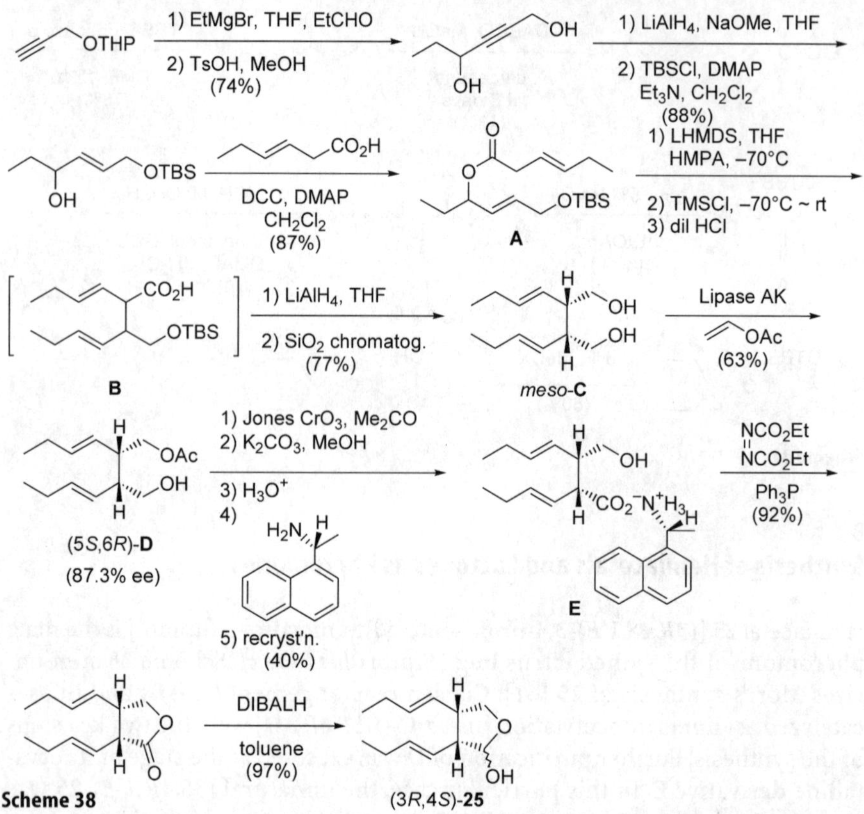

Scheme 38

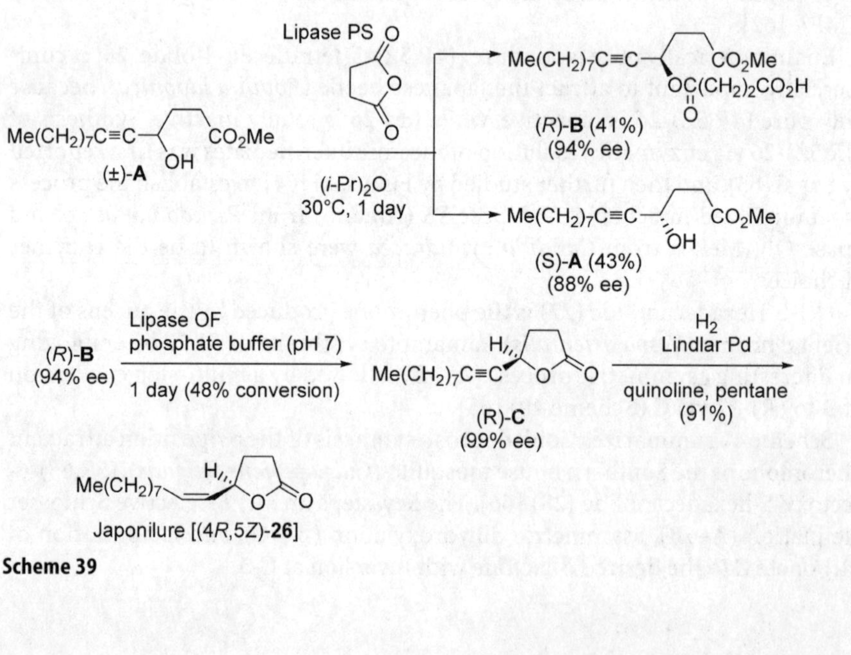

Scheme 39

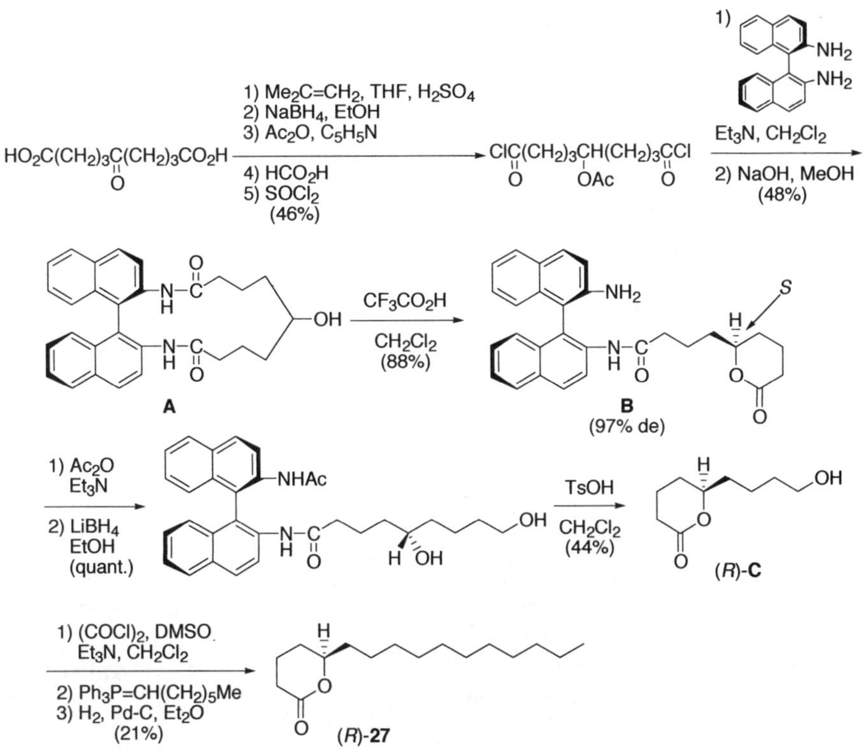

Scheme 40

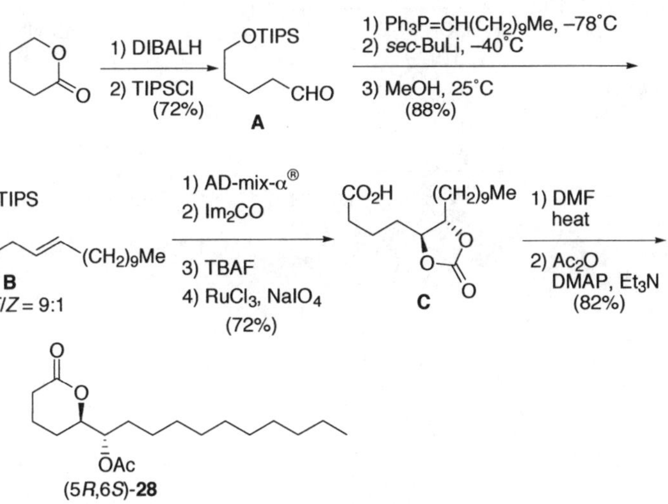

Scheme 41

Advent of ring-closing olefin metathesis reaction has influenced the synthetic strategy of pheromones very much. For example, the racemate of a macrolide pheromone such as (5*Z*,13*S*)-5-tetradecen-13-olide (**29**) (aggregation pheromone of the flat grain beetle, *Cryptolestes pusillus*) could readily be synthesized as shown in Scheme 42 [67]. Unfortunately, (±)-**29** prepared by this method was a mixture of *cis*, *trans*-isomers (*Z*/*E*=69:31). (*S*)-5-Hexadecanolide (**27**) was also synthesized by means of ring-closing olefin metathesis as shown in Scheme 43 [68].

(4*R*,9*Z*)-9-Octadecen-4-olide (**30**) is the female pheromone of the currant stem girdler (currant sawfly, *Janus integer*). Both the enantiomers of **30** were synthesized as summarized in Scheme 44 [69]. Lipase-catalyzed asymmetric acetylation of (±)-**A** was the key-step to give unreacted (*R*)-**A** and acetylated (*S*)-**B**.

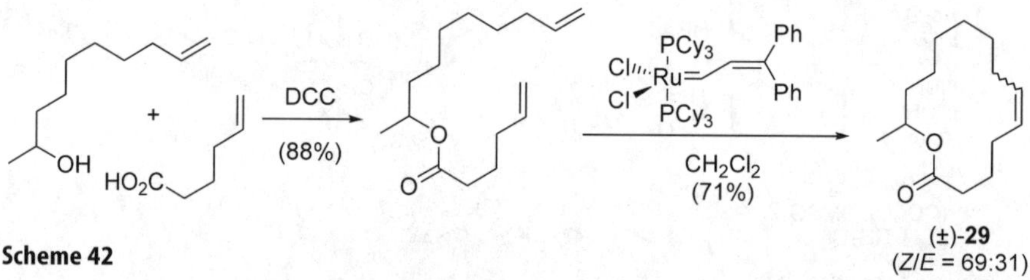

Scheme 42

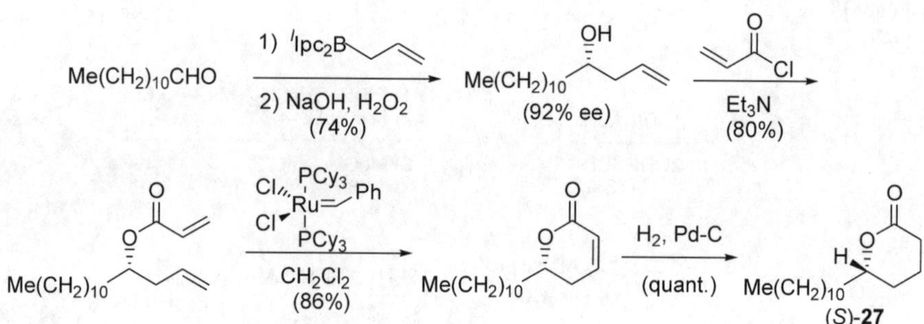

Scheme 43

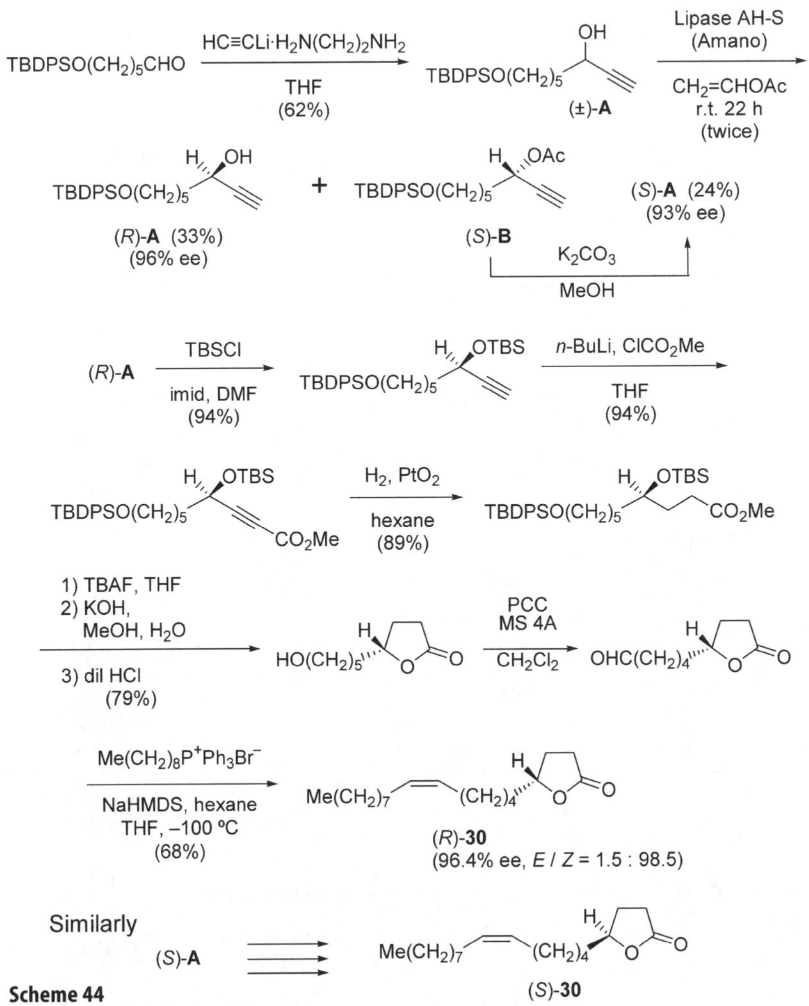

Scheme 44

9 Synthesis of Isoprenoids as Pheromones

9.1 Synthesis of Isoprenoidal Hydrocarbons as Pheromones

Scheme 45 summarizes Mori's synthesis of the male-produced sex pheromone [(1*R*,3*R*,7*S*)-3-methyl-α-himachalene (**31**)] of the sandfly (*Lutzomyia longipalpis*) in Jacobina, Brazil [70]. This sandfly is the vector of *Leishmania chagasi*, the causative protozoan parasite of visceral leishmaniasis in South America. The key-steps of the synthesis of **31** were the asymmetric methylation of **A** to give **C** via **B** and the intramolecular Diels-Alder reaction of **D** to give **E**.

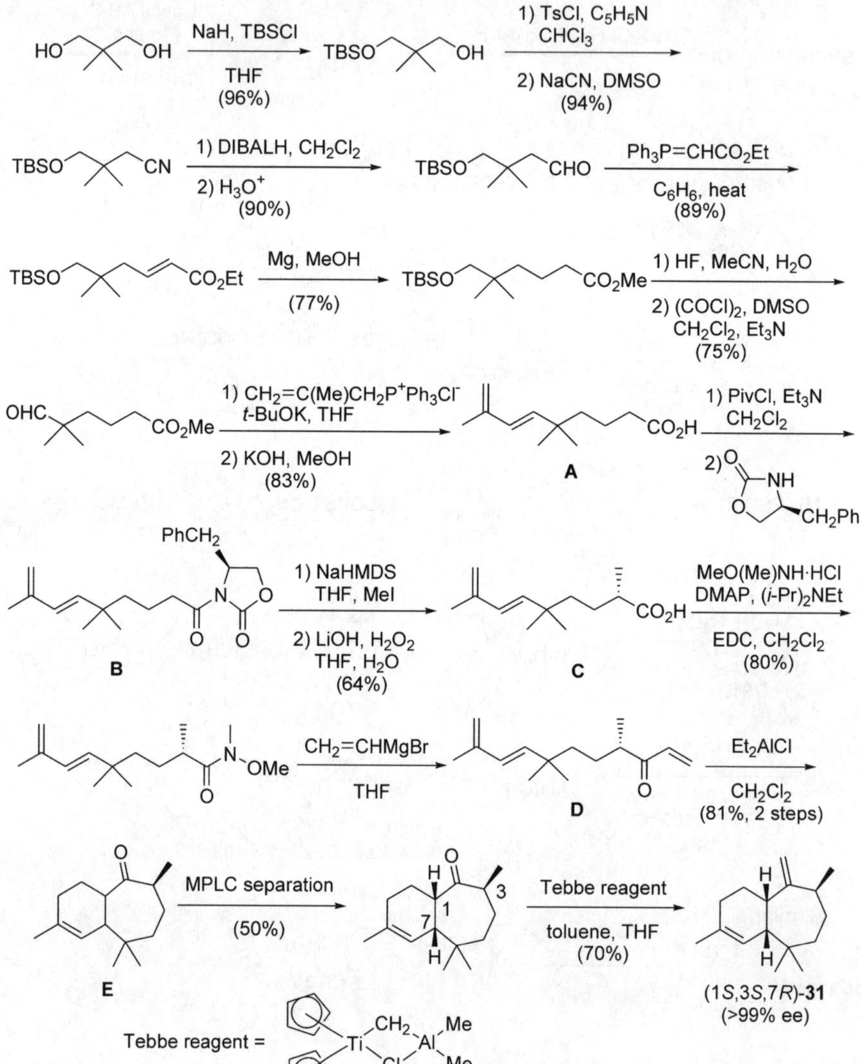

Scheme 45

The same sandfly *Lutzomyia longipalpis* in Lapinha, Brazil, produces (*S*)-9-methylgermacrene-B (**32**) as the male-produced sex pheromone, which was also synthesized by Mori as shown in Scheme 46 [71]. The key-step of this synthesis was the intramolecular cyclization of **A** to give **B**.

The aggregation pheromone of the broad-horned flour beetle (*Gnatocerus cornutus*) was reported to be (1*R*,4*R*,5*S*)-(+)-acoradiene (**33**) by Tebayashi et al. [72]. Scheme 47 shows Mori's synthesis of (1*R*,4*R*,5*S*)-**33** [73]. The key-step was the ring-closing olefin metathesis of **A** to give **B**. An X-ray analysis of **C** confirmed the structure shown. The product (1*R*,4*R*,5*S*)-**33**, however, was different

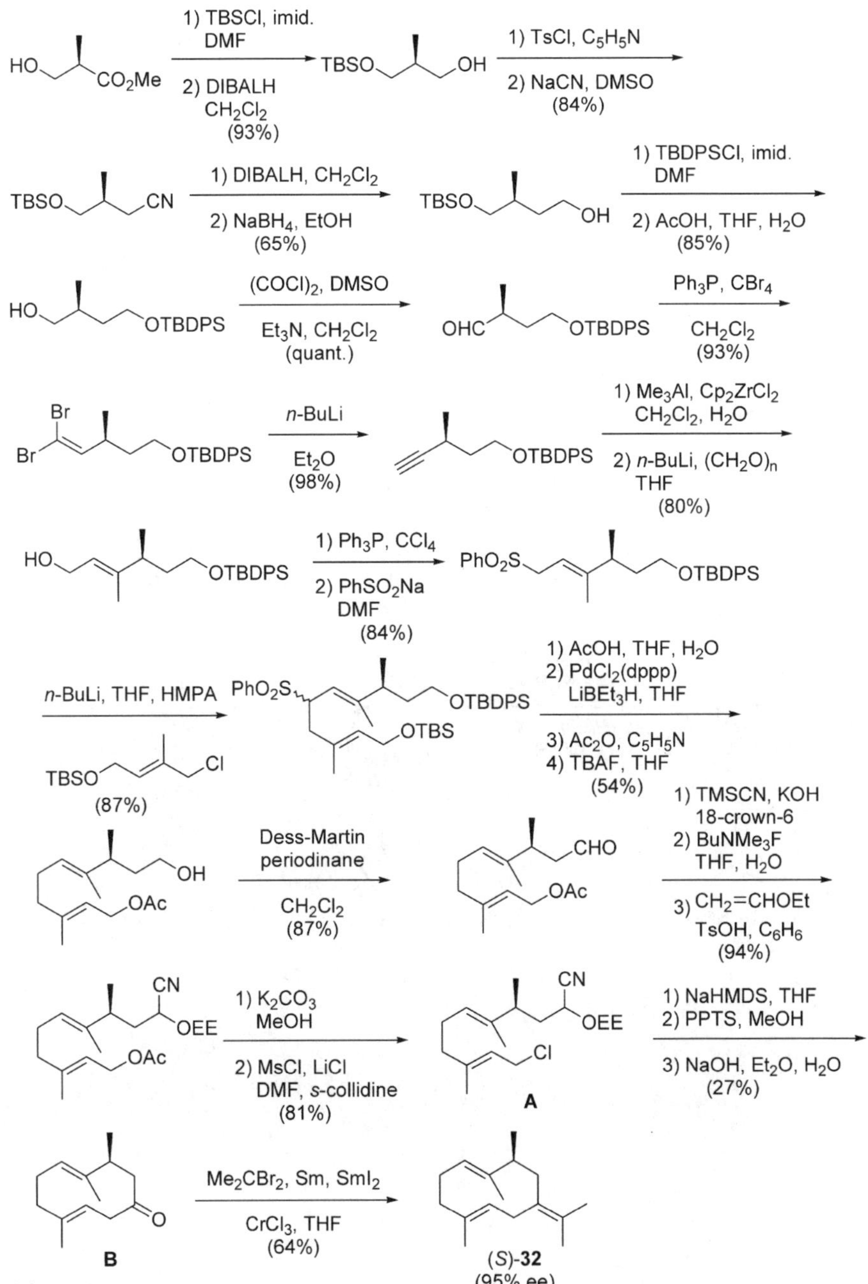
1) TBSCl, imid.
DMF
2) DIBALH
CH2Cl2
(93%)
HO
CO2Me
TBSO
OH
1) TsCl, C5H5N
2) NaCN, DMSO
(84%)
CN
1) DIBALH, CH2Cl2
2) NaBH4, EtOH
(65%)
1) TBDPSCl, imid.
DMF
2) AcOH, THF, H2O
(85%)
OTBDPS
(COCl)2, DMSO
Et3N, CH2Cl2
(quant.)
OHC
Ph3P, CBr4
CH2Cl2
(93%)
Br
n-BuLi
Et2O
(98%)
1) Me3Al, Cp2ZrCl2
CH2Cl2, H2O
2) n-BuLi, (CH2O)n
THF
(80%)
1) Ph3P, CCl4
2) PhSO2Na
DMF
(84%)
PhO2S
n-BuLi, THF, HMPA
OTBS
Cl
(87%)
1) AcOH, THF, H2O
2) PdCl2(dppp)
LiBEt3H, THF
3) Ac2O, C5H5N
4) TBAF, THF
(54%)
OAc
Dess-Martin
periodinane
CH2Cl2
(87%)
CHO
1) TMSCN, KOH
18-crown-6
2) BuNMe3F
THF, H2O
3) CH2=CHOEt
TsOH, C6H6
(94%)
OEE
1) K2CO3
MeOH
2) MsCl, LiCl
DMF, s-collidine
(81%)
A
1) NaHMDS, THF
2) PPTS, MeOH
3) NaOH, Et2O, H2O
(27%)
O
B
Me2CBr2, Sm, SmI2
CrCl3, THF
(64%)
(S)-32
(95% ee)

Scheme 46

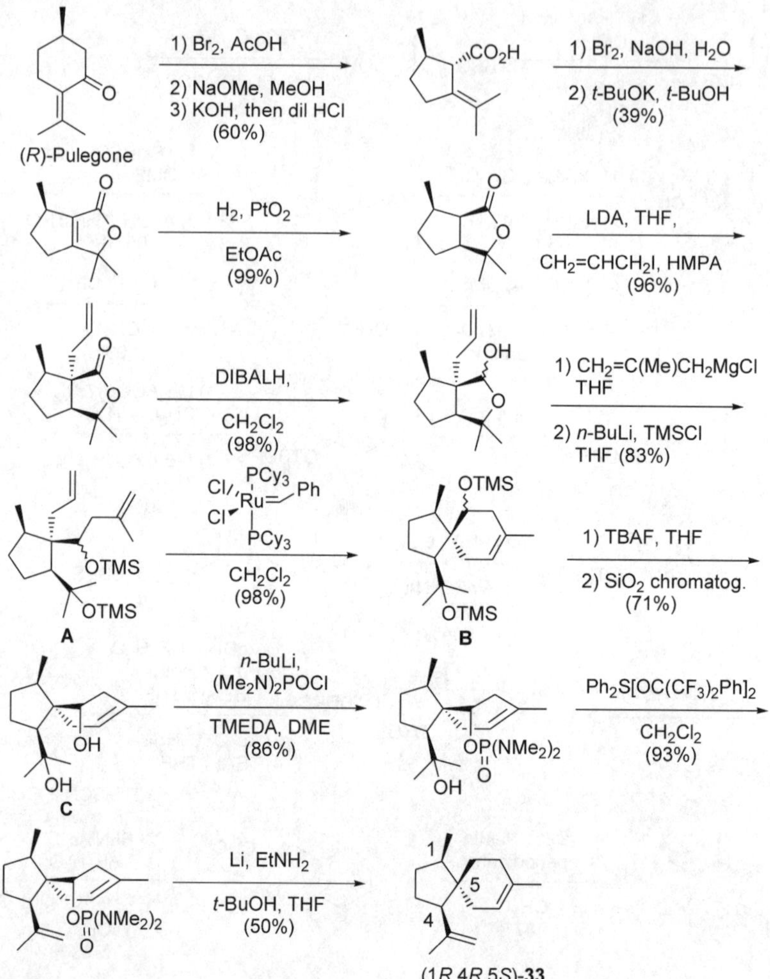
1) Br2, AcOH
2) NaOMe, MeOH
3) KOH, then dil HCl
(60%)
(R)-Pulegone
CO2H
1) Br2, NaOH, H2O
2) t-BuOK, t-BuOH
(39%)
H2, PtO2
EtOAc
(99%)
LDA, THF,
CH2=CHCH2I, HMPA
(96%)
DIBALH,
CH2Cl2
(98%)
OH
1) CH2=C(Me)CH2MgCl
THF
2) n-BuLi, TMSCl
THF (83%)
OTMS
A
PCy3
Cl
Ru
Ph
CH2Cl2
(98%)
B
1) TBAF, THF
2) SiO2 chromatog.
(71%)
C
n-BuLi,
(Me2N)2POCl
TMEDA, DME
(86%)
OP(NMe2)2
Ph2S[OC(CF3)2Ph]2
CH2Cl2
(93%)
Li, EtNH2
t-BuOH, THF
(50%)
(1R,4R,5S)-33

Scheme 47

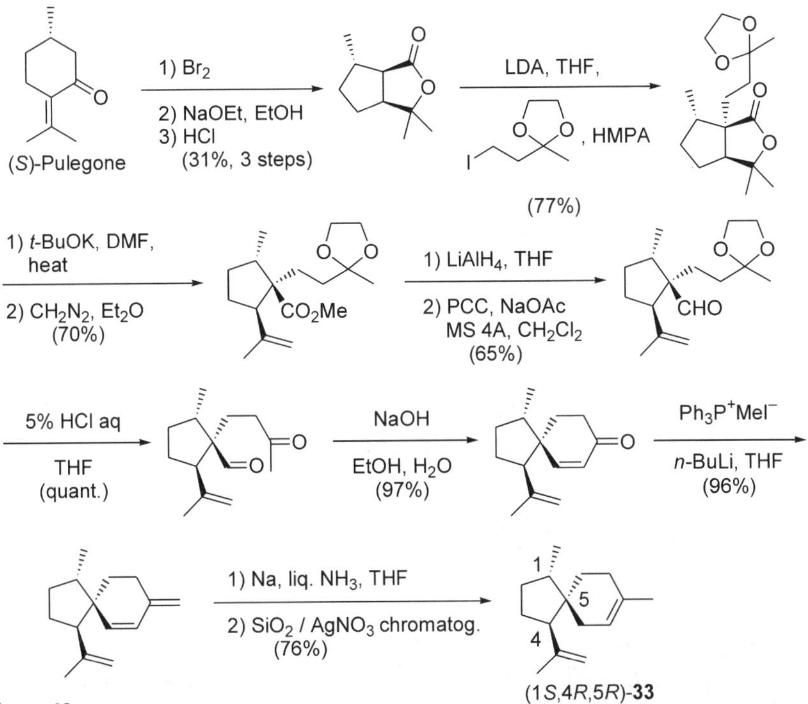

Scheme 48

from the natural pheromone [73]. Mori subsequently synthesized (1*S*,4*R*,5*R*)-(+)-α-acoradiene (**33**) as shown in Scheme 48 [74]. The natural pheromone was identical with (1*S*,4*R*,5*R*)-**33**.

9.2 Synthesis of Isoprenoidal Alcohols as Pheromones

A number of new and asymmetric syntheses of (*S*)-(–)-ipsenol (**34**) and (*S*)-(+)-ipsdienol (**35**), the pheromone of *Ips* bark beetles, were reported. Scheme 49 summarizes the synthesis of ipsenol by Riedeker and Steiner [75], which enabled them to prepare 56 g of (*S*)-**34**. They employed chiral auxiliary **B** derived from D-glucose.

Brown's synthesis of **34** and **35** utilized *B*-2′-isoprenyldi-*l*-isopinocampheylborane **B** as a reagent for asymmetric isoprenylation of aldehydes as shown in Scheme 50 [76].

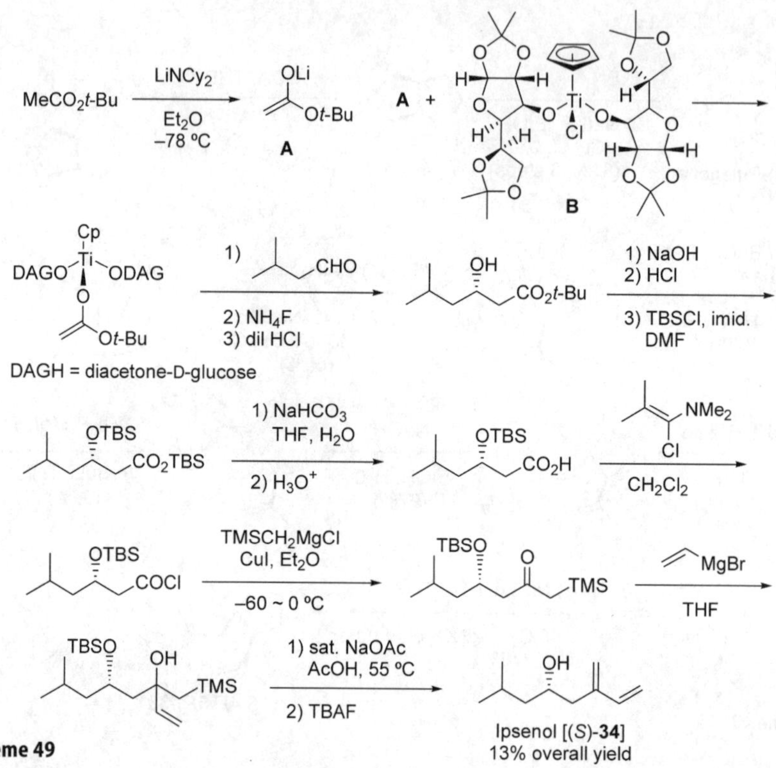

Scheme 49

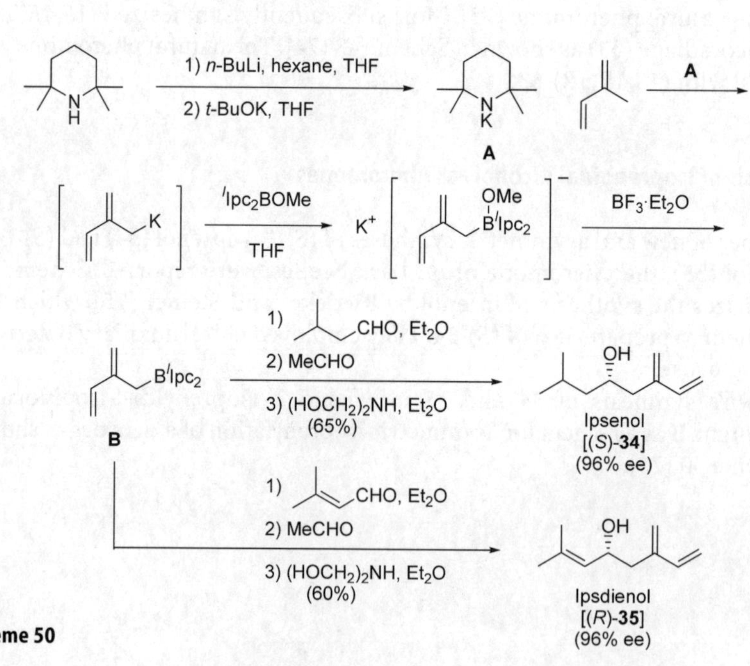

Scheme 50

Scheme 51 summarizes Mikami's synthesis of (*R*)-**35**, employing the carbonyl-ene reaction of isoprene (**A**) with glyoxylate (**B**) to give **C** as catalyzed by a modified binaphthol-titanium complex [77].

By using allylation reaction of aldehydes with chiral allylic boronate **A** in Scheme 52, Lebreton synthesized (*R*)-**35** and (*S*)-**34** via (*R*)-**B** and (*S*)-**C**, respectively [78].

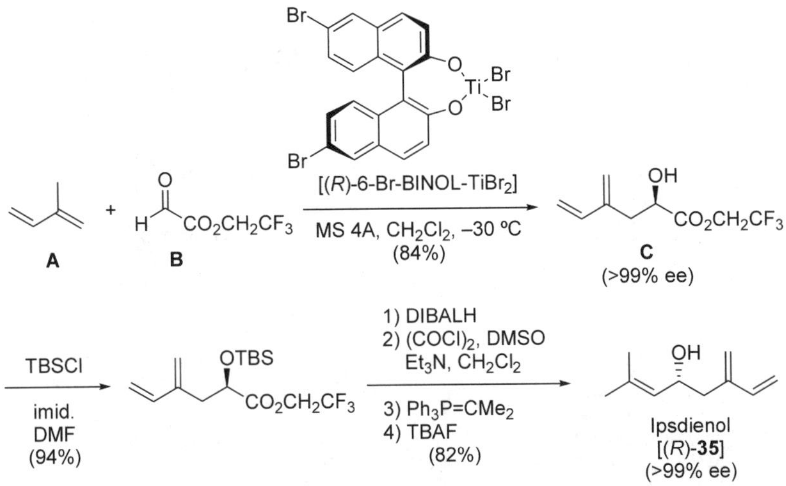

Scheme 51

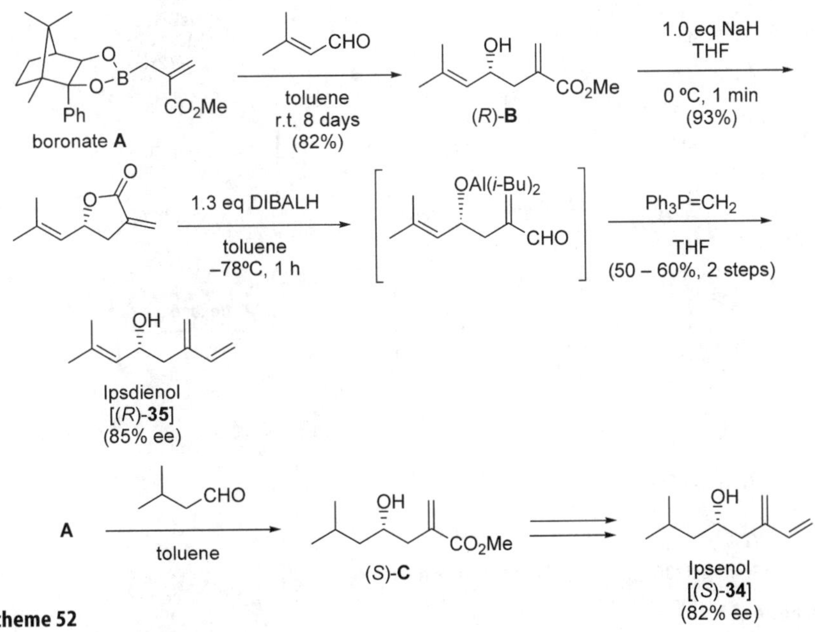

Scheme 52

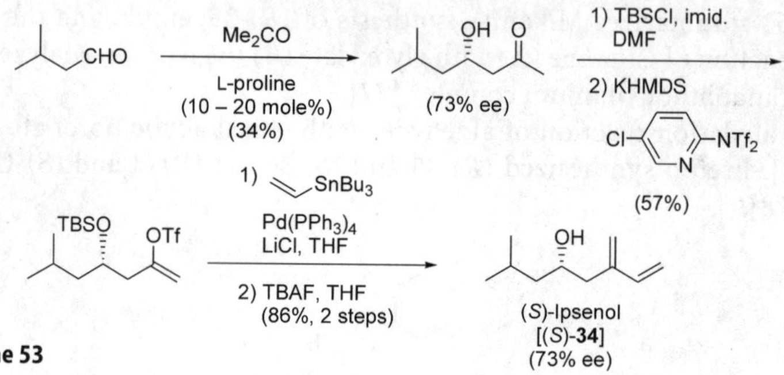

Scheme 53

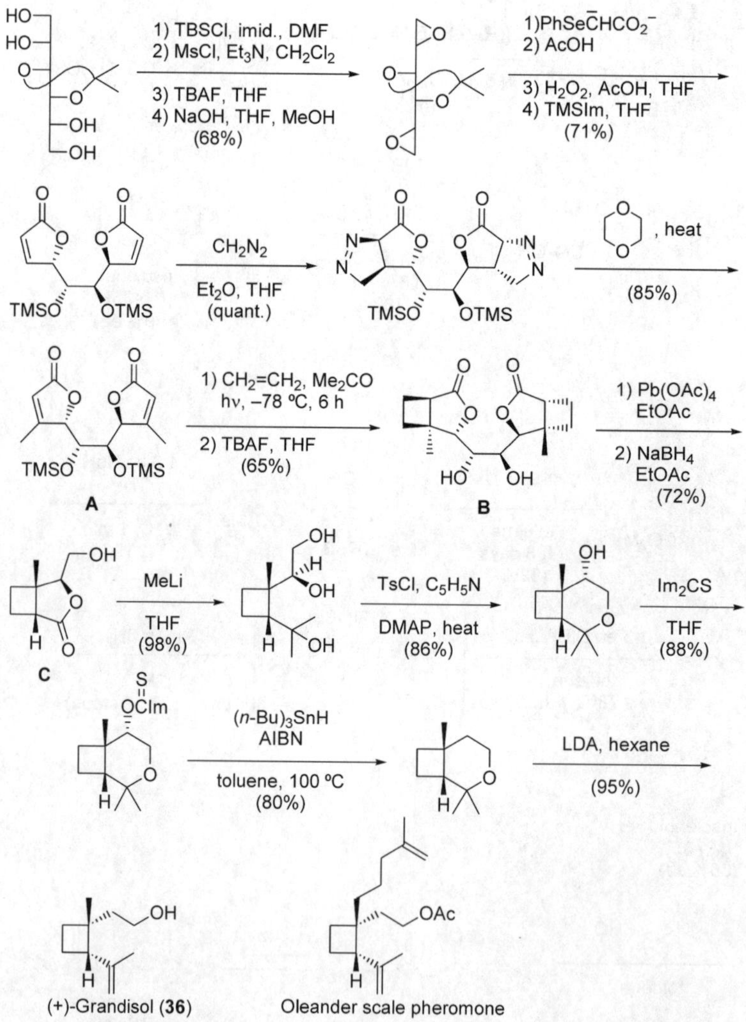

Scheme 54

As shown in Scheme 53, L-proline-catalyzed asymmetric aldol reaction between 3-methylbutanal and acetone was used by List for the synthesis of (*S*)-**34** [79].

Scheme 54 summarizes Font's synthesis of (+)-grandisol (**36**), the male pheromone of the cotton boll weevil (*Anthonomus grandis*) [80]. The key-step is the double [2+2] photocycloaddition of ethylene to bis(α,β-butenolide) **A** to give **B**, which yielded **C** after glycol cleavage. The recently identified pheromone of the oleander scale (*Aspidiotus nerii*) possesses a structure similar to that of grandisol (Scheme 54), and its synthesis was reported by Ducrot [81] and also by Guerrero [82, 83].

9.3
Synthesis of Isoprenoidal Aldehydes and Ketones as Pheromones

Mori reported an improved synthesis of (3*S*,4*R*,6*E*,10*Z*)-faranal (**37**), the trail pheromone of the Pharaoh's ant (*Monomorium pharaonis*) [84]. As summarized in Scheme 55, the key-reaction was the coupling of iododiene **A** with iodide **E**. The geometrically pure **A** was prepared by the zirconocene-mediated carboalumination reaction, and **E** was prepared from **B** by the asymmetric cleavage of its epoxy ring to give **C** (77% ee), which could be purified via its crystalline 3,5-dinitrobenzoate **D**.

Scheme 56 summarizes Mori's synthesis of (*S*)-vesperal (**38**), the female sex pheromone of the longhorn beetle (*Vesperus xatarti*) [85]. (*R*)-Limonene yielded (*S*)-**38** by utilizing organoselenium chemistry.

(*S*)-3,7-Dimethyl-2-oxo-6-octene-1,3-diol (**39**) was recently identified as the aggregation pheromone of the Colorado potato beetle (*Leptinotarsa decemlineata*), and synthesized by Oliver et al., starting from (*S*)-linalool [86]. An improved synthesis of (*S*)-**39** by Mori is shown in Scheme 57 [87]. Enzymatic acetylation of (±)-2,3-epoxynerol (**A**) with vinyl acetate and lipase PS gave **B** together with **C**. The acetate **B** was converted to a multi-gram quantity of (*S*)-**39** according to Oliver [86].

9-Isopropyl-4,6-dimethyltricyclo[4.4.$0^{5,9}$]dec-3-en-2-one (**40**) with unknown absolute configuration was isolated as a putative male pheromone of a Brazilian predatory stink bug (*Tynacantha marginata*), and its enantiomers were synthesized by Kuwahara as shown in Scheme 58 [88]. The triene **A** gave **B** and **C** by an intramolecular Diels-Alder reaction, and they could be converted to the enantiomers of **40**.

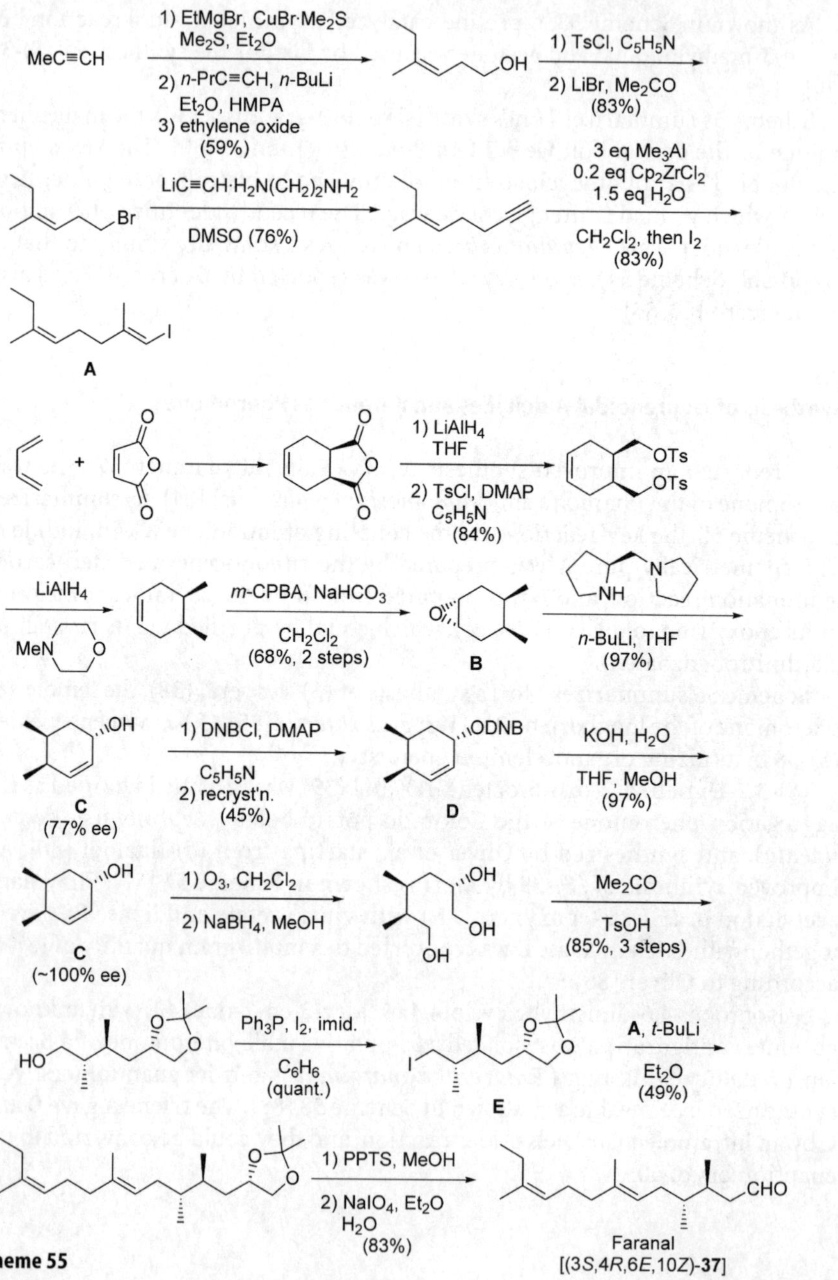

MeC≡CH
1) EtMgBr, CuBr·Me2S
Me2S, Et2O
2) n-PrC≡CH, n-BuLi
Et2O, HMPA
3) ethylene oxide
(59%)
OH
1) TsCl, C5H5N
2) LiBr, Me2CO
(83%)
Br
LiC≡CH·H2N(CH2)2NH2
DMSO (76%)
3 eq Me3Al
0.2 eq Cp2ZrCl2
1.5 eq H2O
CH2Cl2, then I2
(83%)
I
A
O
O
O
1) LiAlH4
THF
2) TsCl, DMAP
C5H5N
(84%)
OTs
OTs
LiAlH4
MeN
m-CPBA, NaHCO3
CH2Cl2
(68%, 2 steps)
B
NH
n-BuLi, THF
(97%)
OH
C
(77% ee)
1) DNBCl, DMAP
C5H5N
2) recryst'n.
(45%)
ODNB
D
KOH, H2O
THF, MeOH
(97%)
C
(~100% ee)
1) O3, CH2Cl2
2) NaBH4, MeOH
Me2CO
TsOH
(85%, 3 steps)
HO
Ph3P, I2, imid.
C6H6
(quant.)
E
A, t-BuLi
Et2O
(49%)
1) PPTS, MeOH
2) NaIO4, Et2O
H2O
(83%)
CHO
Faranal
[(3S,4R,6E,10Z)-37]

Scheme 55

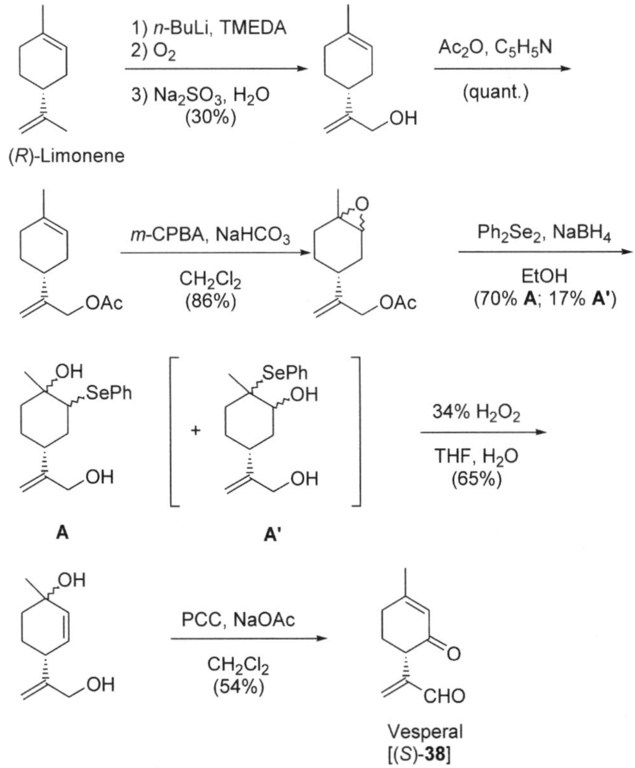

Scheme 56

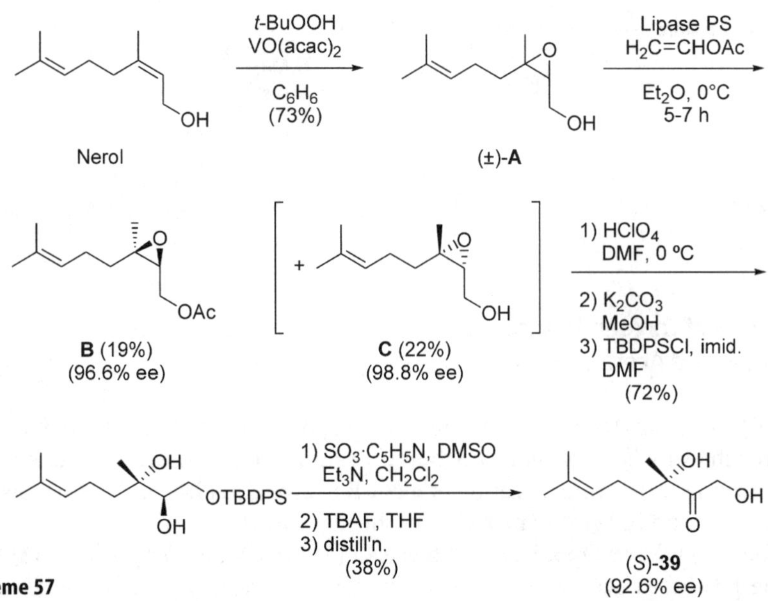

Scheme 57

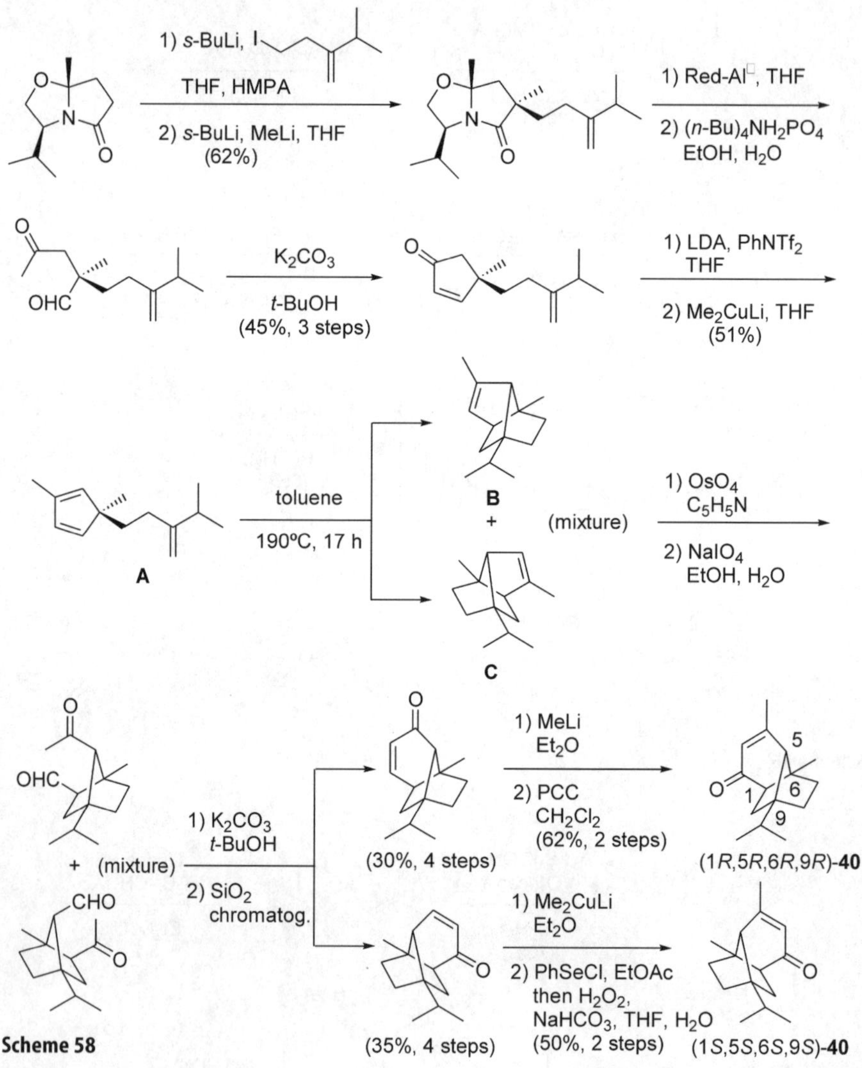

Scheme 58

10 Synthesis of Oxygen Heterocycles (Excluding Epoxides, Hemiacetals, Lactones and Acetals) as Pheromones

(2*R*,4*R*)-Supellapyrone (**41**) is the female sex pheromone of the brown banded cockroach (*Supella longipalpa*). Scheme 59 summarizes Mori's synthesis of **41** [89]. Enzymatic desymmetrization of *meso*-**B** to give (2*R*,4*S*)-**C** and Reformatsky-type cyclization of **D** to furnish **E** were the key steps.

Scheme 60 shows Oppolzer's synthesis of (2*S*,3*R*,1′*S*,2′*S*)-serricorole (**42**), the female-produced sex pheromone components of the cigarette beetle (*Lasio-*

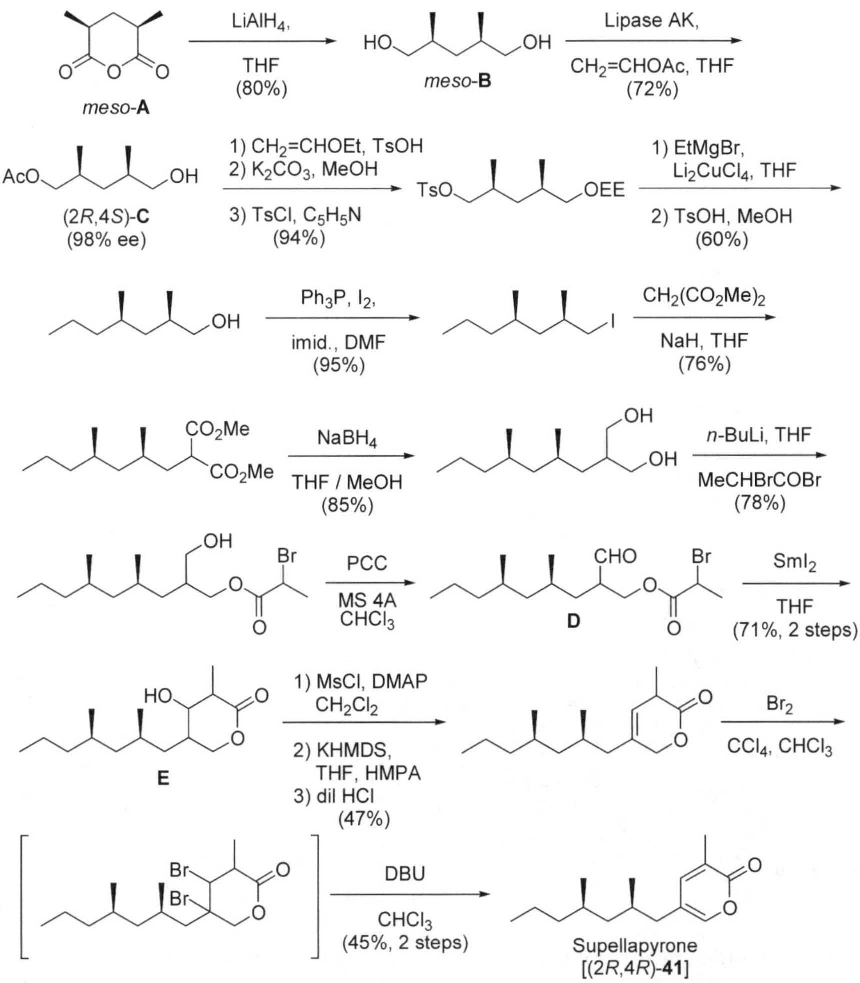

Scheme 59

derma serricorne) [90]. Two building blocks **A** and **B** were prepared enantioselectively, and coupled together to give **C**. Cyclization of **D** to give TBS-protected **42** was effected with titanium tetrachloride in a good yield, while the use of lithium hexamethyldisilazide for this cyclization gave the product in a poor yield [91].

(2*S*,3*R*,1′*R*)-Stegobinone (**43**) is the female sex pheromone of the drugstore beetle (*Stegobium paniceum*). The pure pheromone is crystalline. Due to the easy epimerization at C-1′, it was difficult to synthesize crystalline **43**. The first synthesis of crystalline **43** was achieved by Matteson as shown in Scheme 61 [92]. They employed highly stereoselective organoborane chemistry to secure **43**. The building blocks **A** and **B** were coupled to give **C**, which afforded **43** via **D**, **E** and **F**.

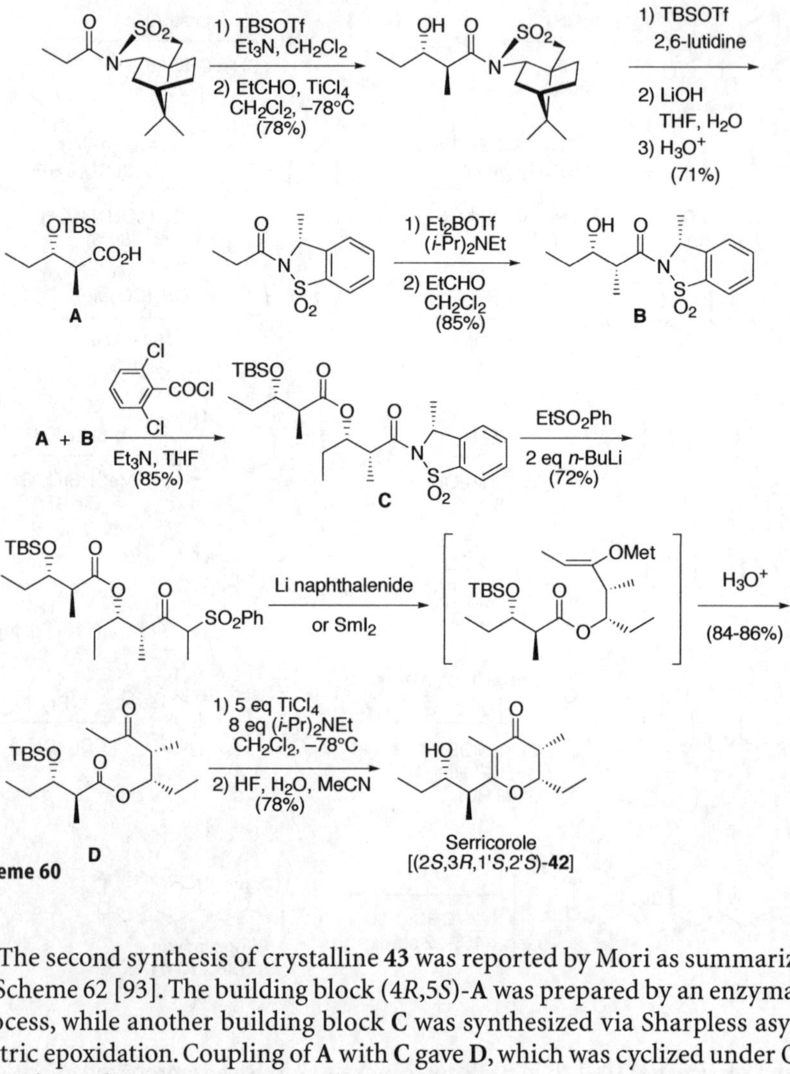

Scheme 60

The second synthesis of crystalline **43** was reported by Mori as summarized in Scheme 62 [93]. The building block (4*R*,5*S*)-**A** was prepared by an enzymatic process, while another building block **C** was synthesized via Sharpless asymmetric epoxidation. Coupling of **A** with **C** gave **D**, which was cyclized under Oppolzer's conditions to give crystalline **E**. When **E** was oxidized with Dess-Martin periodinane or tetra(*n*-propyl)ammonium perruthenate or Jones chromic acid, crystalline **43** was obtained. Swern oxidation or oxidation with 2,2,6,6-tetramethylpiperidin-1-oxyl of **E** afforded only oily materials. Accordingly, oxidation of **E** to **43** must be executed extremely carefully. A synthesis of oily **43** was reported by Gil [94].

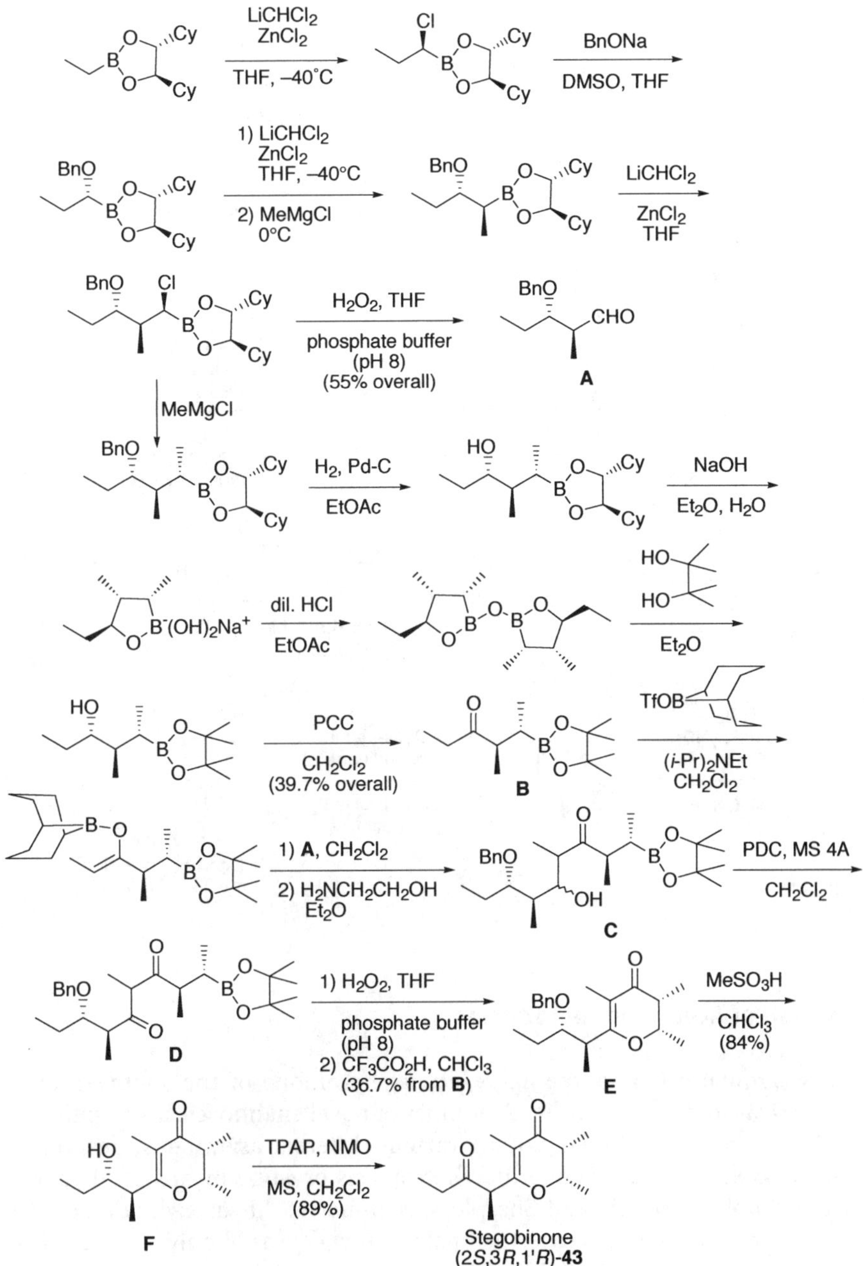
LiCHCl2
ZnCl2
THF, –40°C
Cl
BnONa
DMSO, THF
BnO
1) LiCHCl2
ZnCl2
THF, –40°C
2) MeMgCl
0°C
LiCHCl2
ZnCl2
THF
H2O2, THF
phosphate buffer
(pH 8)
(55% overall)
CHO
A
MeMgCl
H2, Pd-C
EtOAc
HO
NaOH
Et2O, H2O
B-(OH)2Na+
dil. HCl
EtOAc
Et2O
PCC
CH2Cl2
(39.7% overall)
B
TfOB
(i-Pr)2NEt
CH2Cl2
1) A, CH2Cl2
2) H2NCH2CH2OH
Et2O
OH
C
PDC, MS 4A
CH2Cl2
D
1) H2O2, THF
phosphate buffer
(pH 8)
2) CF3CO2H, CHCl3
(36.7% from B)
E
MeSO3H
CHCl3
(84%)
F
TPAP, NMO
MS, CH2Cl2
(89%)
Stegobinone
(2S,3R,1'R)-43
mp 51.5-52.5°C

Scheme 61

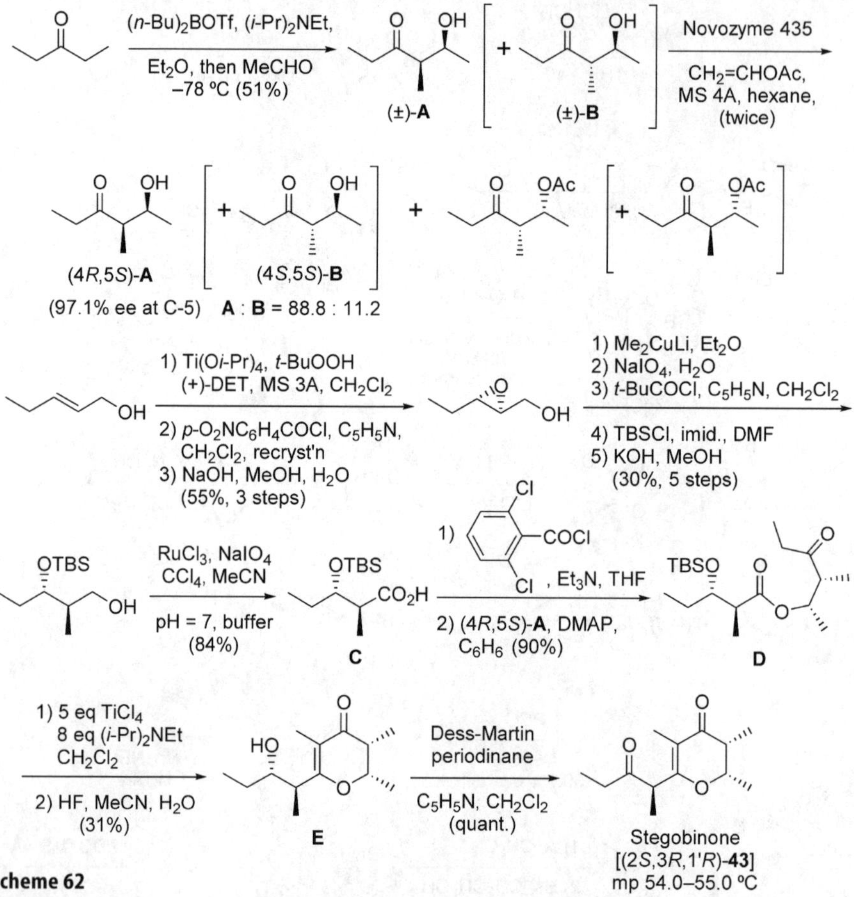

Scheme 62

11 Synthesis of Acetals as Pheromones

(1*S*,5*R*)-Frontalin (**44**) is the aggregation pheromone of the southern pine beetle (*Dendroctonus frontalis*). A number of novel enantioselective syntheses of **44** were published by employing various different asymmetric reactions. Scheme 63 summarizes Sonderquist's synthesis of **44** by means of Miyaura-Suzuki coupling (**A**→**B**) and Sharpless asymmetric dihydroxylation (**B**→**C**) [95]. Unfortunately in this case, the final product **44** was of only 35% ee.

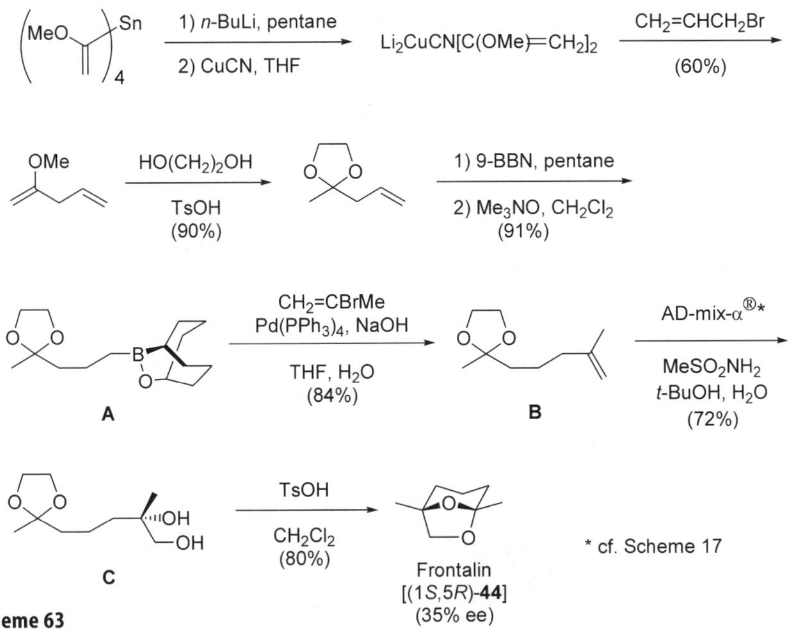

Scheme 63

By using bacterial epoxide hydrolase of *Rhodococcus equi* IFO 3730, Faber synthesized **44** as shown in Scheme 64 [96]. The enzymatic step (**A**→**B**), however, proceeded in a low yield.

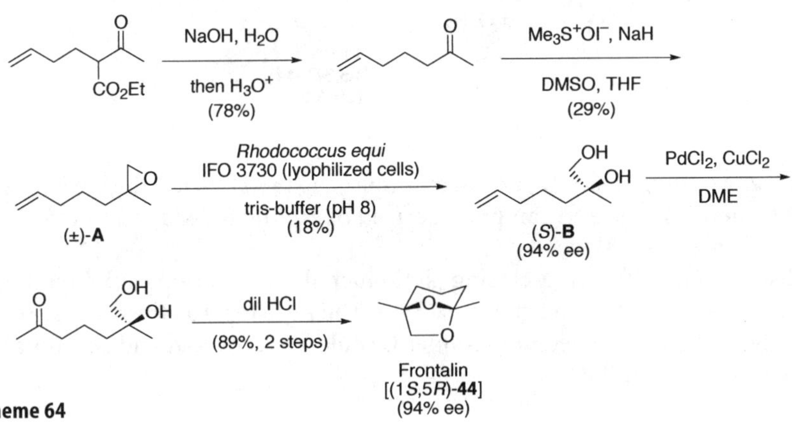

Scheme 64

Scheme 65 summarizes Mori's synthesis of **44** [97]. Reduction of keto ester **A** with baker's yeast gave hydroxy ester **B** of about 98% ee. Methylation of the dianion derived from **B** diastereoselectively gave **C**, which was converted to **44**. This process enabled the preparation of about 10 g of (1*S*,5*R*)-**44**.

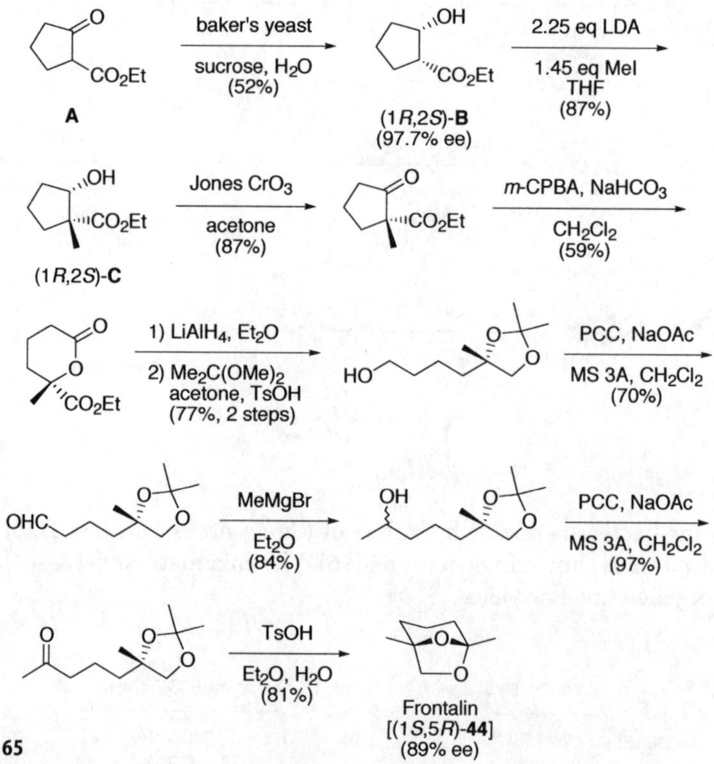

Scheme 65

Langlois synthesized (1*S*,5*R*)-**44** by employing [3+2] cycloaddition between methyl methacrylate and camphor-derived oxazoline *N*-oxide (**A**) as the key-step (Scheme 66) [98].

Grubbs applied his ring-closing olefin metathesis reaction to the synthesis of (1*S*,5*R*)-**44** as shown in Scheme 67 [99]. The key-step was the cyclization of **A** to give **C**. The unreacted *anti*-isomer **B** could be recovered and equilibrated to a mixture of **A** and **B**.

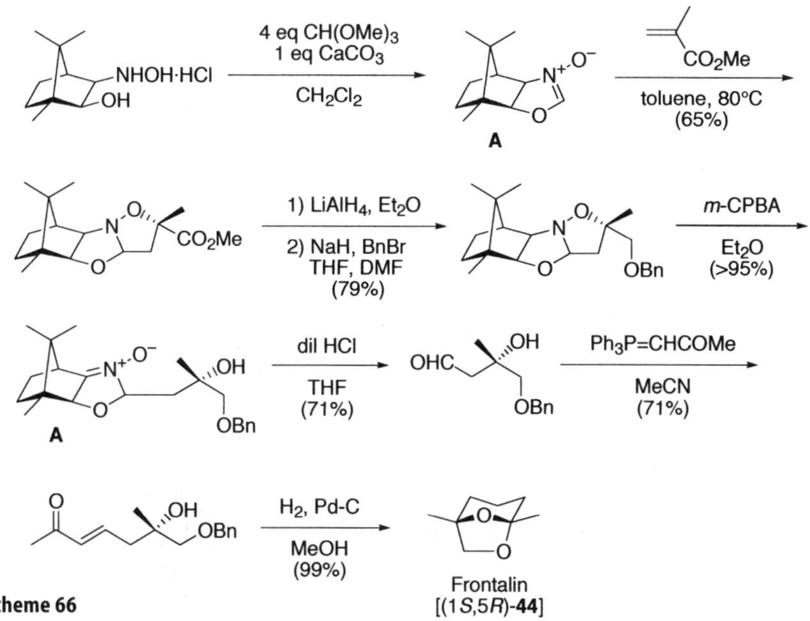

Scheme 66

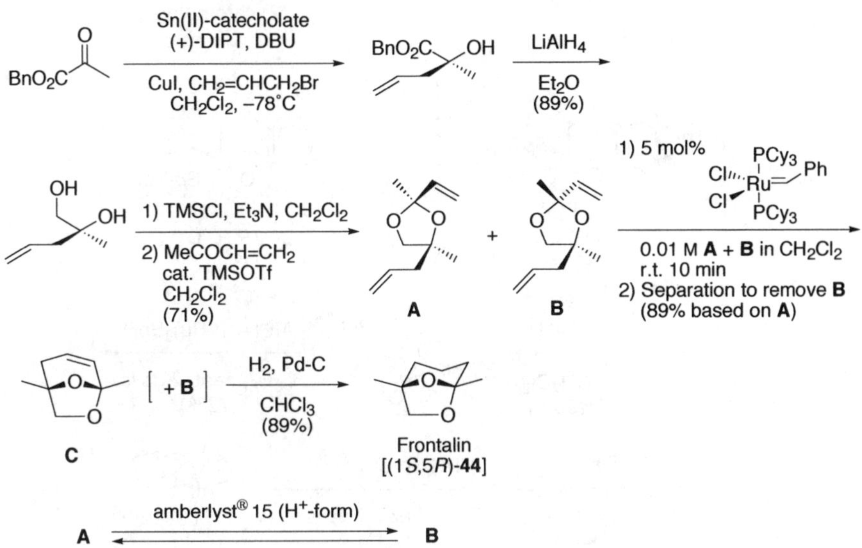

Scheme 67

(1*R*,5*S*,7*R*)-3,4-Dehydro-*exo*-brevicomin (**45**) was isolated from the urine of the male mouse (*Mus musculus*) as a pheromone component. Two new syntheses of this pheromone were reported in 1999. Maignan synthesized **45** employing asymmetric heterocycloaddition induced by a chiral sulfoxide (**A**+**B**→**C**) as the key-step (Scheme 68) [100].

Scheme 69 summarizes Mori's synthesis of **45**, in which was employed Sharpless asymmetric dihydroxylation (**A**→**B**) as the key-step [101].

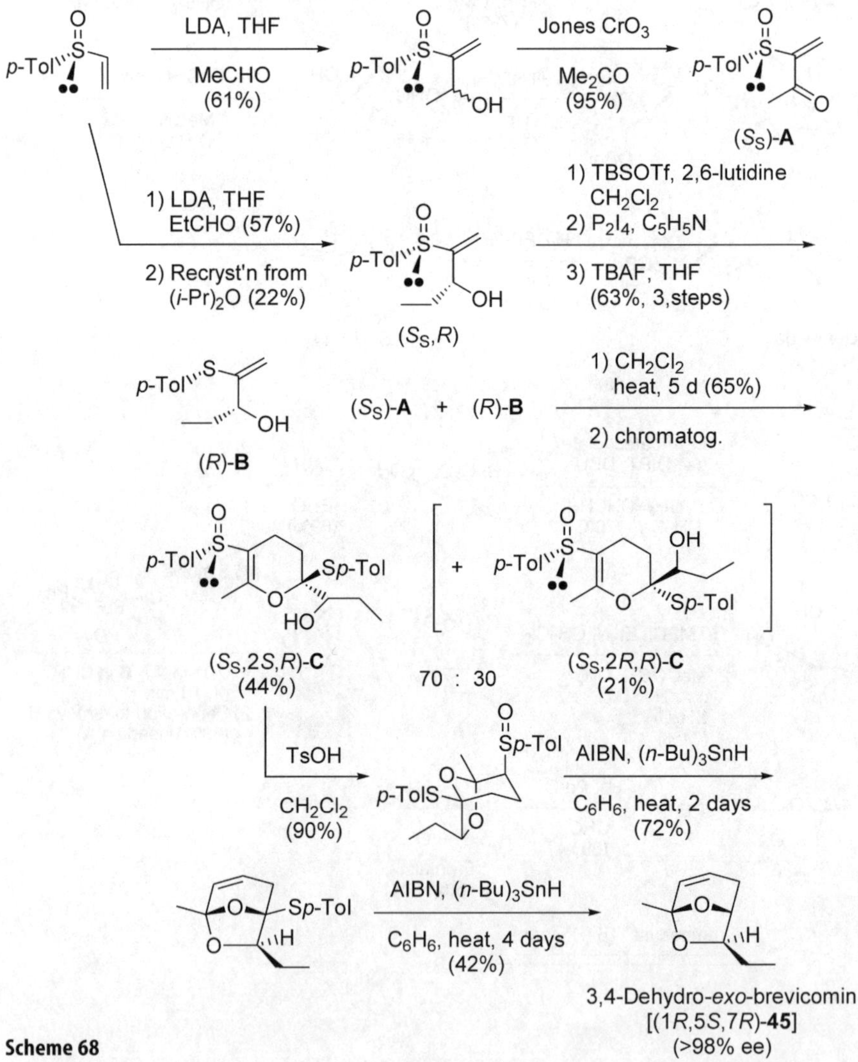

Scheme 68

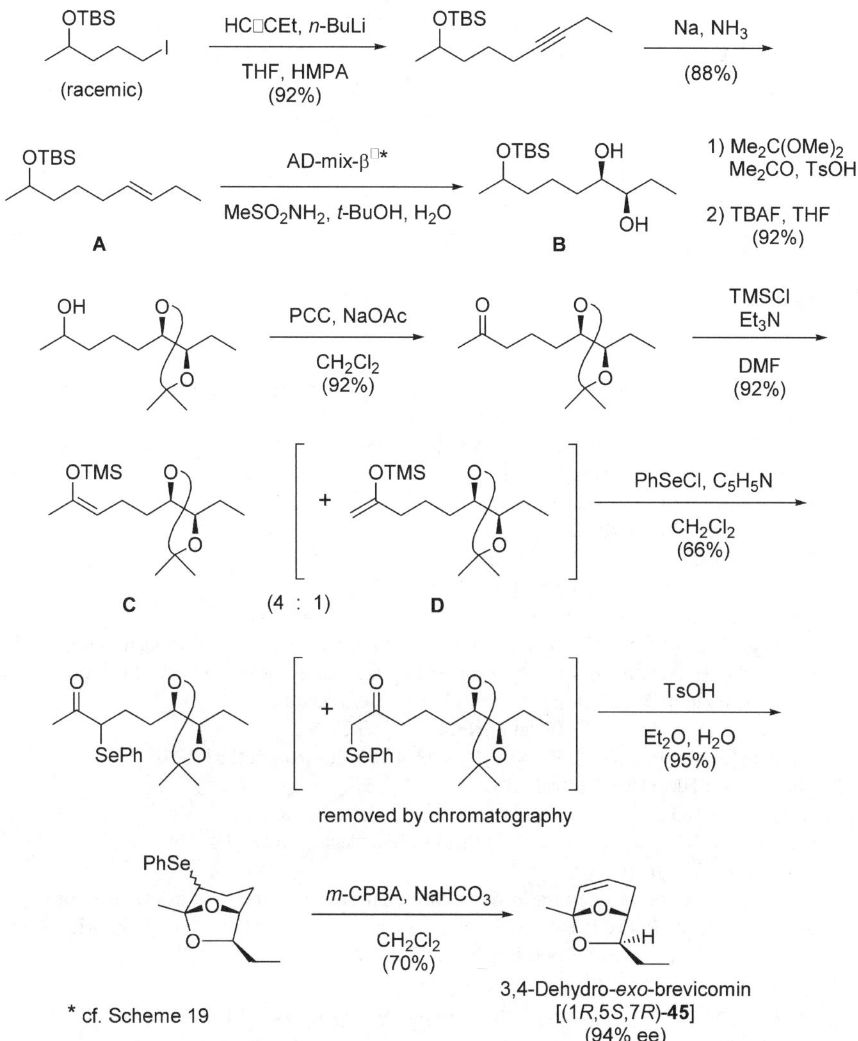

Scheme 69

12
Conclusion

It is now clear that pure pheromones can be synthesized in quantity. The problem is how to prepare them simply and efficiently. New synthetic methodologies are always welcome to improve the existing syntheses. Organoborane reactions and organotransition metal chemistry contributed much to improve the efficiency of carbon-carbon bond formation, while asymmetric epoxidations and dihydroxylations as well as enzymatic reactions greatly improved the enantiomeric purity of synthetic pheromones.

By synthesizing pure enantiomers of pheromones, various stereochemistry-pheromone activity relationships could be clarified. For example, in the case of sulcatol (6-methyl-5-hepten-2-ol), both the enantiomers are necessary for bioactivity. For other relationships, please refer to [4–9].

Acknowledgements I thank my junior colleagues especially Dr. T. Tashiro for the help in preparing the manuscript.

References

1. Mori K (1989) Tetrahedron 45:3233
2. Mori K (1981) The synthesis of insect pheromones. In: ApSimon J (ed) The total synthesis of natural products, vol 4. Wiley, New York, p 1
3. Mori K (1992) The synthesis of insect pheromones. In: ApSimon J (ed) The total synthesis of natural products, vol 9. Wiley, New York, p 1
4. Mori K (1997) Chem Commun 1153
5. Mori K (1998) Eur J Org Chem 1479
6. Mori K (2000) Acc Chem Res 33:102
7. Mori K (1996) Biosci Biotechnol Biochem 60:1925
8. Mori K (1998) Chirality 10:578
9. Mori K (2002) Chirality in the natural world: chemical communications. In: Lough WJ, Wainer JW (eds) Chirality in natural and applied science. Blackwell, Oxford, p 241
10. Chong JM, Heuft MA, Rabbat P (2000) J Org Chem 65:5837
11. Buck M, Chong JM (2001) Tetrahedron Lett 42:5825
12. Daeuble JF, McGettigan C, Stryker JM (1990) Tetrahedron Lett 31:2397
13. Choi J, Yoon NM (1996) Tetrahedron Lett 37:1057
14. Harada K, Urabe H, Sato F (1995) Tetrahedron Lett 36:3203
15. Hungerford NL, Kitching W (1998) J Chem Soc, Perkin Trans 1 1839
16. Mori K (1995) Synlett 1097
17. Mori K (2000) Chemoenzymatic synthesis of pheromones, terpenes, and other bioregulators. In: Patel RN (ed) Stereoselective biocatalysis. Marcel Dekker, New York, p 59
18. Shirai Y, Seki M, Mori K (1999) Eur J Org Chem 3139
19. Hutzinger MW, Oehlschlager AC (1995) J Org Chem 60:4595
20. Svatoš A, Kalinová B, Hoskovec M, Kindl J, Hovorka O, Hrdý I (1999) Tetrahedron Lett 40:7011
21. Cabezas JA, Oehlschlager AC (1999) Synthesis 107
22. Negishi E, Yoshida T, Abramovitch A, Lew G, Williams RM (1991) Tetrahedron 47:343
23. Uenishi J, Kawahama R, Izaki Y, Yonemitsu O (2000) Tetrahedron 56:3493
24. Fletcher MT, McGrath MJ, König WA, Moore CJ, Cribb BW, Allsopp PG, Kitching W (2001) Chem Commun 885. See also (2003) J Org Chem 68:3739
25. Alami M, Gueugnot S, Domingues E, Linstrumelle G (1995) Tetrahedron 51:1209
26. Hall DR, Beevor PS, Campion DG, Chamberlain DJ, Cork A, White RD, Almestar A, Henneberry TJ (1992) Tetrahdron Lett 33:4811
27. Keinan E, Sinha SC, Sinha-Bagchi A, Wang Z-M, Zhang X-L, Sharpless KB (1992) Tetrahedron Lett 33:6411
28. Ko SY (1994) Tetrahedron Lett 35:3601
29. Sinha-Bagchi AS, Sinha SC, Keinan E (1995) Tetrahedron: Asymmetry 6:2889
30. Marshall JA, Jablonowski JA, Jiang H (1999) J Org Chem 64:2152

31. Hu S, Jayaraman S, Oehlschlager AC (1999) J Org Chem 64:3719
32. Brevet J-L, Mori K (1992) Synthesis 1007
33. Muto S (2003) Ph. D. thesis, University of Tokyo
34. Oliver JE, Waters RM (1995) J Chem Ecol 21:199
35. Oliver JE, Waters RM, Harrison DJ (1996) J Chem Ecol 22:287
36. Zhang Z-B. Wang Z-M, Wang Y-X, Liu H-Q, Lei G-X, Shi M (2000) J Chem Soc Perkin Trans 1 53
37. Muto S, Mori K (2003) Biosci Biotechnol Biochem 67:1559
38. Wakamura S, Arakaki N, Yamamoto M, Hiradate S, Yasui H, Yasuda T, Ando T (2001) Tetrahedron Lett 42:687
39. Muto S, Mori K (2001) Eur J Org Chem 4635
40. Fernandes RA, Kumar P (2002) Tetrahedron 58:6685
41. Gries R, Holden D, Gries G, Wilmalaratne PDC, Slessor KN, Saunders C (1997) Naturwissenschaften 84:219
42. Yamamoto M, Yamazawa H, Nakajima N, Ando T (1999) Eur J Org Chem 1503
43. Razkin Lizarraga J, Mori K (2001) Nat Prod Lett 15:89
44. Muto S, Mori K (2003) Eur J Org Chem 1300
45. Hanaki N, Ishihara K, Kaino M, Naruse Y, Yamamoto H (1996) Tetrahedron 52:7297
46. Chow S, Kitching W (2002) Tetrahedron: Asymmetry 13:779
47. Hamon DPG, Massy-Westropp RA, Newton JL (1990) Tetrahedron: Asymmetry 1:771
48. Bueno AB, Carr?eno MC, Ruans JLG, Hamdouchi C (1995) Tetrahedron: Asymmetry 6:1237
49. Ceccherelli P, Curini M, Epifano F, Marcotullio MC, Roseti O (1996) J Org Chem 61:2882
50. Hamon DP, Tuck KL (2000) Tetrahedron 56:4829
51. Mori K, Amaike M (1994) J Chem Soc Perkin Trans 1 2727
52. Mori K, Furuuchi T, Kiyota H (1994) Liebigs Ann Chem 953
53. Mori K, Furuuchi T, Matsuyama K (1995) Liebigs Ann 2093
54. Kurosawa S, Takenaka M, Dunkelblum E, Mendel Z (2000) ChemBioChem 1:56
55. Matteson DS, Singh RP, Schafman B, Yang J-I (1998) J Org Chem 63:4466
56. Mori K, Takenaka M (1998) Eur J Org Chem 2181
57. Enders D, Ridder A (2000) Synthesis 1848
58. Epstein OL, Kulinkovich OG (2001) Tetrahedron Lett 42:3757
59. Takanashi S, Mori K (1997) Liebigs Ann/Recueil 825
60. Mateus CR, Feltrin MP, Costa AM, Coelho F, Almeida WP (2001) Tetrahedron 57:6901
61. Mori K, Amaike M, Watanabe H (1993) Liebigs Ann Chem 1287
62. Amaike M, Mori K (1995) Liebigs Ann 1451
63. Sugai T, Ohsawa S, Yamada H, Ohta H (1990) Synthesis 1112
64. Fukusaki E, Senda S, Nakazono Y, Omata T (1991) Tetrahedron 47:6223
65. Yamamoto Y, Sakamoto A, Nishioka T, Oda J, Fukagawa Y (1991) J Org Chem 56:1112
66. Couladouros E, Mihou AP (1999) Tetrahedron Lett 40:4861
67. Fürstner A, Langemann K (1996) J Org Chem 61:3942
68. Ramachandran PV, Reddy MVR, Brown HC (2000) Tetrahedron Lett 41:583
69. Shibata C, Mori K (2004) Eur J Org Chem 1083
70. Tashiro T, Bando M, Mori K (2000) Synthesis 1852
71. Kurosawa S, Mori K (2000) Eur J Org Chem 955
72. Tebayashi S, Hirai N, Matsuyama S, Suzuki T, Nakakita H, Memoto T, Nakanishi H (1998) J Stored Prod Res 34:99
73. Kurosawa S, Bando M, Mori K (2001) Eur J Org Chem 4395
74. Tashiro T, Kurosawa S, Mori K (2004) Biosci Biotechnol Biochem 68:663
75. Riediker M, Steiner E (1990) Helv Chim Acta 73:353

76. Brown HC, Randard RS (1990) Tetrahedron 46:4463
77. Terada M, Mikami K (1995) J Chem Soc Chem Commun 2391
78. Draillard K, Lebreton J, Villiéras J (1999) Tetrahedron: Asymmetry 10:4281
79. List B, Pojarliev P, Castello C (2001) Org Lett 3:573
80. de March P, Figueredo M, Font J, Raya J, Alvarez-Larena A, Piniella JF (2003) J Org Chem 68:2437
81. Boyer F-D, Ducrot P-H (1999) Eur J Org Chem 1201
82. Petschen I, Parilla A, Bosch MP, Amela C, Botar AA, Camps F, Guerrero A (1999) Chem Eur J 5:3299
83. Petschen J, Bosch MP, Guerrero A (2000) Tetrahedron: Asymmetry 11:1691
84. Mori K, Murata N (1995) Liebigs Ann 2089
85. Domon K, Mori K (2000) Eur J Org Chem 3783
86. Oliver JE, Dickens JC, Glass TE (2002) Tetrahedron Lett 43:2641
87. Tashiro T, Yamauchi T, Imamura Y, Mori K (to be published)
88. Kuwahara S, Ishikawa J, Leal WS, Hamade S, Kodama O (2000) Synthesis 1930
89. Fujita K, Mori K (2001) Eur J Org Chem 493
90. Oppolzer W, Rodriguez I (1993) Helv Chim Acta 76:1275
91. Ebata T, Mori K (1987) Agric Biol Chem 51:2925
92. Matteson DS, Man H-W, Ho OC (1996) J Am Chem Soc 118:4560
93. Mori K, Sano S, Yokoyama Y, Bando M, Kido M (1998) Eur J Org Chem 1135
94. Gil P, Razkin J, González A (1998) Synthesis 386
95. Santiago B, Soderquist JA (1992) J Org Chem 57:5844
96. Kroutil W, Osprian I, Mischitz M, Faber K (1997) Synthesis 156
97. Nishimura Y, Mori K (1998) Eur J Org Chem 233
98. Kouklovsky C, Dirat O, Berranger T, Langlois Y, Tran-Huu-Dan ME, Riche C (1998) J Org Chem 63:5123
99. Scholl M, Grubbs RH (1999) Tetrahedron Lett 40:1425
100. Hayes P, Maignan C (1999) Tetrahedron: Asymmetry 10:1425
101. Tashiro T, Mori K (1999) Eur J Org Chem 2167

Topics in Current Chemistry (2004) 239: 51–96
DOI 10.1007/b95449

Lepidopteran Sex Pheromones

Tetsu Ando (✉) · Shin-ichi Inomata · Masanobu Yamamoto

Tokyo University of Agriculture and Technology, Graduate School of Bio-Applications and Systems Engineering (BASE), Koganei, 184-8588 Tokyo, Japan
antetsu@cc.tuat.ac.jp

Abstract As a consequence of the diversity of Lepidoptera, including 150,000 described species, interesting species-specific sex pheromone systems are exhibited in this insect group. The quite varied pheromones, which have been identified from female moths of nearly 530 species from around the world, are classified into groups of Type I (75%), Type II (15%), and miscellaneous (10%), according to their chemical structures. Additionally, many pheromones produced by male moths and butterflies have been known. While new sex pheromones from about 70 lepidopteran species have been reported in the last five years utilizing GC-EAD, GC-MS, LC, and NMR, our information about the pheromones is still rudimentary, and these kinds of semiochemicals remain an exciting research target for natural product chemistry. In addition to the overview of their chemical structures, this chapter deals with current methods for their identification. Furthermore, an actual application of the synthetic pheromones for pest control is briefly introduced.

Keywords Pheromones · Female moth · Male attractants · Chemical ecology · Mating disruption

List of Abbreviations

DMDS	Dimethyl disulfide
EAG	Electroantennogram
EI	Electron impact ionization
FID	Flame ionization detector
GC-EAD	Gas chromatography combined with an EAG detector
GC-MS	Gas chromatography combined with mass spectrometry
HPLC	High performance liquid chromatography
KI	Kovats retention index
LC-MS	Liquid chromatography combined with mass spectrometry
MPLC	Medium pressure liquid chromatography
MTAD	4-Methyl-1,2,4-triazoline-3,5-dione
RID	Refractive index detector
SPME	Solid phase micro-extraction
TIC	Total ion chromatogram

1 Introduction

Lepidoptera is the second largest insect group and includes nearly 150,000 described species, which have evolved over 100 million years since the Mesozoic era. For the birth of a new species, it must be isolated from other species by some factor to prevent inter-species crossing. The sex pheromone, which is secreted by the adult (usually by a female moth and sometimes by a male moth or butterfly) for the benefit of a specific partner, plays an important role in reproductive isolation. Therefore, it is no wonder that the chemical structures of the species-specific pheromones exhibit considerable differences.

As a consequence of the large number of pest species which comprise this order, enormous efforts have been expended, especially since the identification of bombykol from the silkworm female moth (*Bombyx mori*) in the late 1950s [1], on studies of lepidopteran sex pheromones. Namely, in addition to the early studies done with harmful pests such as the cabbage looper (*Trichoplusia ni*) [2] and the gypsy moth (*Lymantria dispar*) [3], the chemical structures of female sex pheromones with a long-range attractive activity have been determined for nearly 530 moth species from across the world to date. Furthermore, field evaluations of the identified sex pheromones and related compounds have revealed attractants of males for another 1250 moth species. Additionally, many male sex pheromones produced by moths and butterflies have been reported. Considering the species diversity within the Lepidoptera, it should be mentioned that our information is still rudimentary. Other natural products, however, have not been investigated on so many different species as the lepidopteran sex pheromone have. In the last 5 years, new pheromones from about 70 lepidopteran species and attractants for another 30 species have been reported. Despite the plethora of research, these semiochemicals remain an exciting research target for organic chemistry, chemical ecology, and applied entomology. The structures of the known pheromones and attractants are listed in books [4], and recently on inter-net web sites, "Pherolist" and "The pherobase", which are complied and maintained by Arn et al. [5], and by A. M. El-Sayed [6], respectively. The Chemical Ecology Laboratory of Tokyo University of Agriculture and Technology also proposes a similar list in addition to other information such as Japanese common names of insects and their distribution in Japan [7]. This chapter summarizes characteristics of the chemical structures of lepidopteran sex pheromones, recent research into their identification, and their application to pest control.

2 Structural Diversity of Pheromone Components

2.1 Taxonomy of Moths and Overview of Their Female Sex Pheromones

The Lepidoptera are currently divided into about 120 families in 46 superfamilies [8] as mainly indicated in Fig. 1, in which the more highly evolved groups occupy the upper branches. Except for minor species in Micropterigoidea, Eriocranioidea, Nepticuloidea and another ten primitive superfamilies, nearly 99% of the species belong to Ditrysia possessing two separated genital openings, copulatory orifice and ovipore. Ditrysia includes two superfamilies of butterflies, Hesperioidea and Papilionoidea, and about 31 superfamilies of moths. For each superfamily of moths, the total number of species whose female sex pheromones or male attractants are reported has been counted and listed in Fig. 1. Since Gelechioidea, Tortricoidea, Pyraloidea, and Noctuoidea comprise major groups and include many harmful pest insects, their pheromones have been extensively studied.

Among the diverse pheromones, primary alcohols and their derivatives (mainly acetates and aldehydes) with a long straight chain (C_{10}–C_{18}) have most commonly been detected in lepidopteran insects. The chemicals in this most predominant group (Type I compounds, ☆ and ★ shown in Fig. 1) comprise about 75% of the known pheromones and were recorded widely from almost all the superfamilies of Ditrysia. For example, the following notorious pest insects are classified into the above superfamilies and their pheromones consist of Type I compounds; the potato tuberworm (*Phthorimaea operculella*, Gelechiidae) in Gelechioidea, the smaller tea tortrix (*Adoxophyes honmai*, Tortricidae: Tortricinae) and the codling moth (*Cydia pomonella*, Tortricidae: Olethreutinae) in Tortricoidea, the rice stem borer (*Chilo suppressalis*, Pyralidae: Crambinae) and the European corn borer (*Ostrinia nubilalis*, Pyralidae: Pyraustinae) in Pyraloidea, *T. ni* (Noctuidae: Plusiinae) and the tobacco budworm moth (*Heliothis virescens,* Noctuidae: Heliothinae) in Noctuoidea. For the former three

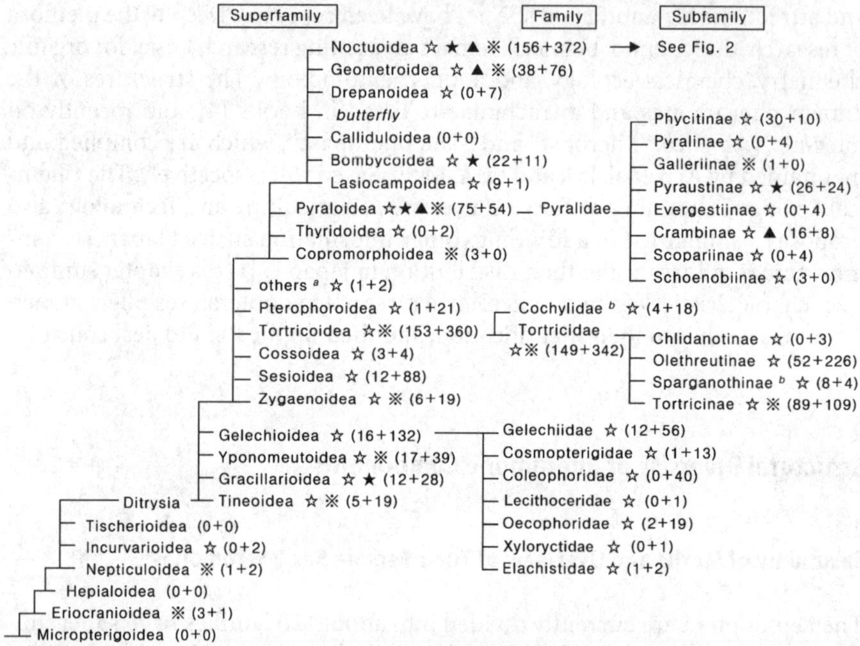

Fig. 1 Taxonomy and pheromone studies of lepidopteran insects. The more highly evolved superfamily of moths is arranged at the upper position. The numbers before and after + in parenthesis of each group indicates the total number of species whose female sex pheromone and male attractant have been reported respectively. Each mark after the group name indicates that some species within the group produces a pheromone component of Type I (☆) with a common functional group, ★ with a novel functional group), Type II (▲), or others (※). Male sex pheromones of moths and butterflies (Hesperioidea and Papilionoidea) are excluded. *a*: Galacticoidea (0+0), Choreutoidea ☆(1+1), Schreckensteinoidea ☆(0+1), Epermeinioidea (0+0), Alucitoidea (0+0); *b*: groups included in Tortricinae in the classification by Kristensen [8]

superfamilies, further information about their families and subfamilies is shown in Fig. 1. Only a few compounds not belonging to Type I have been identified from these insect groups. In contrast, Type I compounds are not always found as a major class in the females of the Noctuoidea. The Noctuoidea superfamily is subdivided into five families; i.e., Noctuidae, Lymantriidae and so on. Noctuidae, which is the largest family in Lepidoptera, can be further subclassified into about 15 subfamilies. While this taxonomy is still open to dispute, pheromone components identified from the species in Trifinae, which appear to be more highly evolved than the species in Quadrifinae, are composed of Type I chemicals as shown in Fig. 2.

Polyunsaturated hydrocarbons and the epoxy derivatives with a longer straight chain (C_{17}–C_{23}) comprise a second major group [9], the Type II pheromones (▲ in Fig. 1). They lack a functional group at the terminal position,

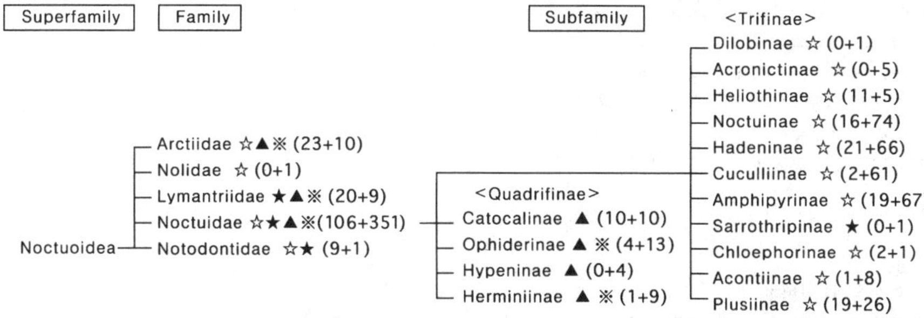

Fig. 2 Taxonomy and pheromone studies of insects in the superfamily of Noctuoidea. The numbers before and after + in parenthesis of each group indicates the total number of species whose female sex pheromone and male attractant have been reported respectively. Each mark after the group name indicates that some species within the group produces a pheromone component of Type I (☆ with a common functional group, ★ with a novel functional group), Type II (▲), or others (※)

represent about 15% of the known lepidopteran pheromones, and have been identified from insects in highly evolved groups such as Geometroidea and Noctuoidea (Fig. 1). Geometridae, in the superfamily of Geometroidea, is second only to the Noctuidae in terms of numbers. In the Noctuoidea, the Type II pheromones are produced by some groups which scarcely secrete Type I pheromones; i.e., Lymantriidae, Arctiidae, and Quadrifinae in Noctuidae (Fig. 2). Other pheromone components that exhibit structural features different compared to those of Types I and II have been mainly reported from those insects in micro-lepidopteran groups and Noctuoidea (Fig. 1). Especially, structural variety of the pheromones is noteworthy in Lymantriidae.

The variation of the chemical structures in both Type I and Type II compounds results from differences in both biosynthetic enzyme systems and their starting material. This topic, however, has been judiciously described in the chapter by R. Jurenka. While the grouping employed in this chapter is based on biosynthetic origin, some chemicals were involuntarily classified considering their functional groups more sizably than the origin. Incidentally, taxonomic information is important for insect pheromone research. In this chapter, the family name (with the common suffix -idae) is associated with the species name. For those species whose family name is not listed in Figs. 1 and 2, the superfamily name (with the common suffix -oidea) is associated. The subfamily name (with the common suffix -inae) is also described for the species in Tortricidae, Pyralidae, and Noctuidae.

2.2
Type I Compounds with a Common Functional Group

Typical Type I pheromone components (☆) that exhibit a common functional group (hydroxyl, acetoxyl, or formyl group) at the terminal position are shown

Table 1 Type I lepidopteran female pheromones

(A) Compounds with a common functional group (☆)

Formulae and abbreviations of representative compounds

E10,Z12-16:OH (bombykol)

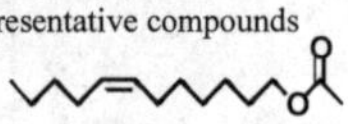
Z7-12:OAc

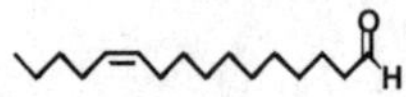
Z11-16:Ald

(B) Compounds with a novel functional group (★)

Compound class / Formula and abbreviation	Super-family / Family	Identified component	Insect species	Year [Ref.] [a]
Acetylenic derivative	Pyraloidea			
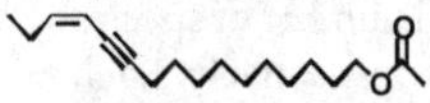	Pyralidae	≡11-16:Ald	*Desmia funeralis*	02 [49]
	Noctuoidea			
	Notodontidae	≡11,Z13-16:OAc	*Thaumetopoea pityocampa*	81 [45]
≡11,Z13-16:OAc			*T. jordana*	93 [46]
			T. wilkinsoni	93 [46]
		≡11,Z13-16:OH	*T. jordana*	93 [46]
		≡11,Z13-16:Ald	*T. jordana*	93 [46]
Nitrate ester	Gracillarioidea			
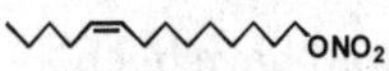	Bucculatrigidae	Z9-14:ONO_2	*Bucculatrix thurberiella*	92 [50]
Z9-14:ONO_2		Z8-13:ONO_2		
Ester with miscellaneous acid	Bombycoidea			
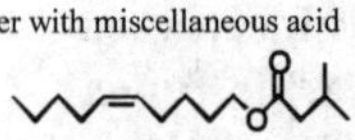	Saturniidae	Z5-10:OisoVal	*Nudaurelia cytherea*	73 [51]
	Noctuoidea			
Z5-10:OisoVal	Lymantriidae	Z7-18:OMe2-But	*Euproctis similis*	94 [52]
		Z7-18:OisoVal	*E. similes*	94 [52]
		Z7-18:OBut	*E. similes*	94 [52]
Z7-18:OMe2-But		Z7-18:OisoBut	*E. similes*	94 [52]
		Z7,Z13,Z16,Z19-22:OisoBut [b]	*E. chrysorrhoea*	91 [53]
		Z11,Z14,Z17-20:OisoBut [b]	*E. pulverea*	01 [54]
		Z11,Z14,Z17-20:OMe4-Val [b]	*E. pulverea*	01 [54]

[a] Publication year and [reference].

[b] Related to Type II compounds based on an aspect of their predicted biosynthetic pathway. Abbreviations: OisoBut=isobutyryl ester, and OMe4-Val-4-methylvaleryl ester.

in Table 1(A); i.e., (10*E*,12*Z*)-10,12-hexadecadien-1-ol (E10Z12–16:OH, bombykol) identified from *B. mori* (Bombycoidea: Bombycidae), (*Z*)-7-dodecenyl acetate (Z7–12:OAc) initially identified from *T. ni* [2] and subsequently found in another 25 species, as well as (*Z*)-11-hexadecenal (Z11–16:Ald) initially identified from *H. virescens* [10] and currently known to be present in another 40 species. Chemicals in this class are characterized by a C_{10}–C_{18} straight chain and 0–3 double-bonds. Since the chemical structures are simple, Type I compounds are rarely named but usually abbreviated as follows; Z=(*Z*)-double bond, E=(*E*)-double bond, the number before the hyphen=position of the double bond, number after the hyphen=the carbon number of the straight chain, OAc=acetate, OH=alcohol, and Ald=aldehyde. Some saturated compounds, which can act as a synergist for the unsaturated component, have also been identified.

The straight chain is dominantly formed with even-numbered carbons because the Type I compounds are derived de novo via general saturated fatty acids such as palmitic acid (16:Acid) and stearic acid (18:Acid). However, exceptions do exist as pheromone components with an odd number chain have been determined from six species: E4,Z7–13:OAc and E4,Z7,Z10–13:OAc from *P. operculella*, [11], E4–13:OAc from the tomato pinworm (*Keiferia lycopersicella*, Gelechiidae) [12], E3–13:OAc from the tobacco stem borer (*Scrobipalpa heliopa*, Gelechiidae) [13], Z8–13:OAc and Z10–15:OAc from the sugarcane stalk borer (*Chilo auricilius*, Pyralidae: Crambinae) [14], E8,Z10–15:OAc and E9–15:OAc from the cranberry fruitworm (*Acrobasis vaccinii*, Pyralidae: Phycitinae) [15], and Z11–17:OAc from the cabbage armyworm (*Mamestra brassicae*, Noctuidae: Hadeninae) [16].

Figure 3A shows the location of the double-bond relative to a functional group of the Type I monoenyl components. The unsaturation is observed at more than 40% of the possible positions. Double bonds at odd-numbered positions are more predominant than those at even-numbered positions, and an unsaturation at the 11-position occurs in all even-numbered chain skeletons. Compounds monounsaturated at the 4- or 6-positions have not been identified, although an odd-numbered chain component, E4–13:OAc, has been identified. Currently, it is a rare occurrence to find a component unsaturated at a new position, as double bonds at the 10-position in a C_{12} chain and the 3-position

(A)

Chain length	Double bond position, counting from the functional group															
	2	3	4	5	6	7	8	9	10	11	12	13	14	15	16	17
C-10				▼		▼			—	—	—	—	—	—	—	—
C-12		▼		▼		▼	▼	▼	▼	▼	—	—	—	—	—	—
C-14		▼		▼		▼	▼	▼	▼	▼	▼		—	—	—	—
C-16						▼		▼	▼	▼					—	—
C-18	▼									▼		▼				

(B)

Chain length	Double bond position (ω), counting from the terminal methyl group															
	16	15	14	13	12	11	10	9	8	7	6	5	4	3	2	1
C-10	—	—	—	—	—	—	—	—				▼		▼		
C-12	—	—	—	—	—	—		▼		▼		▼	▼	▼	▼	▼
C-14	—	—	—	—		▼		▼		▼	▼	▼	▼	▼	▼	
C-16	—	—						▼		▼	▼	▼				
C-18	▼					▼				▼		▼				

Fig. 3A,B Double-bond positions of monoenyl components in Type I pheromones; **A** counting from the functional group; **B** counting from the terminal methyl group. See text for components with an odd numbered chain (C_{13}, C_{15}, and C_{17})

in a C_{14} chain were recognized in the last decade for the following pheromones: E10–12:OAc from the spotted tentiform leafminer (*Phyllonorycter blancardella*, Tineoidea: Gracillariidae) [17], Z3–14:OAc from the carpenterworm moth (*Holcocerus insularis*, Cossoidea: Cossidae) [18], and E3–14:OAc from the South American potato tuber moth (*Symmetrischema tangolias*, Gelechioidea: Gelechiidae) [19]. In the future, however, it is clear that compounds containing a double bond at either the 4- or 6-positions will be discovered as a real pheromone component, because E4–12:OAc and Z6–14:OAc have been shown to function as an attractant in some species. Indeed, the 10–12 and 3–14 compounds had been recognized as a sex attractant of some species prior to their detection in the female moths.

Although these monoenyl compounds are produced by a desaturation of a saturated fatty acyl intermediate in the pheromone gland, each double-bond position does not directly indicate occurrence of the corresponding biosynthetic step. As confirmed by experiments with labeled precursors in Plusiinae species [20, 21], 7-dodecenyl (7–12), 9-tetradecenyl (9–14), and 11-hexadecenyl (11–16) compounds, which include the double bond at the same ω5-position counted from the terminal methyl group, are derived from an 11–16 acyl compound (see Fig. 6). This monoenyl intermediate is produced by Δ11-deasturation of 16:Acid, and shortened into C_{14}, C_{12}, and further C_{10} compounds by a β-oxidation reaction. Given this, it is possible that 5–12, 7–14, and 9–16 compounds with a double bond at the same ω7-position are formed via Δ9-desaturation of 16:Acid or Δ11-desaturation of 18:Acid. Figure 3B shows the double-bond positions of the monoenyl components counted from the terminal methyl group. Even though several double bonds at the same ω-position are introduced by a common desaturation enzyme, this figure also indicates the necessity of many enzymes with a different target position. The double bond of the monoenyl compounds tends to be located to the side of the terminal methyl group, and compounds unsaturated at ω8- or ω10-position have not been found. With respect to configuration of the double bonds, (*Z*)-isomers have been more frequently identified than (*E*)-isomers, although females of some species secrete both isomers. In this case, their mixing ratio is an important factor for male attraction.

In addition to the monoenyl components, a number of dienyl and trienyl Type I compounds with an even-numbered carbon chain skeleton have been identified as follows: 1,3-diene with a C_{10} chain, 1,3- and 1,7-dienes with a C_{12} chain, 1,3-, 1,4-, 1,5-, and 1,6-dienes with a C_{14} or C_{16} chain, and 1,3-, 1,4-, 1, 11-, and 1,12-dienes with a C_{18} chain. Figure 4 shows double-bond positions of these polyenyl compounds. Among them, variation of the double-bond positions in the compounds with a conjugated 1,3-diene system is remarkable. Some polyenes have the unsaturation at 4-, 6-, ω8-, and ω10-positions. These variations, not observed in the monoenyl compounds, strongly indicate that insects evolved multiple desaturases possessing differing restricted target sites and substrate specificity. The number of possible combinations involving two or three double bonds is enormous, as new pheromone components with a

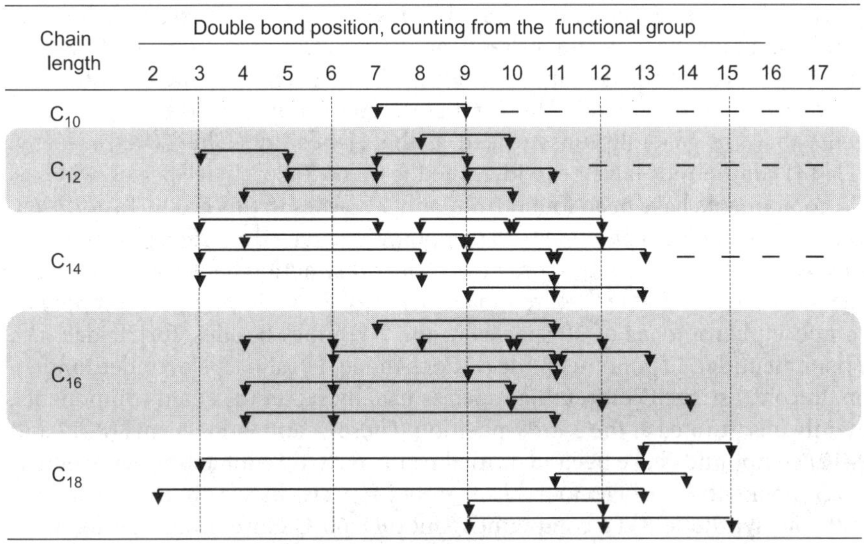

Fig. 4 Double-bond positions of dienyl and trienyl components in Type I pheromones. See text for components with an odd numbered chain (C_{13} and C_{15})

polyene system at novel positions are routinely discovered. The following six compounds containing new unsaturated positions have been reported since 1995: E3,Z5–12:OAc from the Brazilian apple leafroller (*Phtheochroa cranaodes*, Tortricoidea: Cochylidae) [22], E3,Z7–14:OAc from *S. tangolias* [19], E3,Z8–14:OAc and E3,Z8,Z11-14:OAc from the tomato leafminer (*Scrobipalpuloides absoluta*, Gelechiidae) [23–26], E4,Z6–16:OAc from the persimmon fruit moth (*Stathmopoda masinissa*, Gelechioidea: Oecophoridae) [27], and E11,E14–18: Ald from the tea cluster (*Andraca bipunctata*, Bombycidae) [28]. In addition to Type II pheromone components, Z9,Z12–18:Ald and Z9,Z12,Z15–18:Ald have been identified from the castor semilooper (*Achaea janata*, Noctuidae: Catocalinae) [29], the saltmarsh caterpillar moth (*Estigmene acrea*, Arctiidae) [30], and the fall webworm moth (*Hyphantria cunea*, Arctiidae) [31, 32]. These unsaturated skeletons are most likely derived from dietary linolic and linolenic acids.

2.3 Common Structures of Type I Compounds in Some Insect Groups

It is expected that taxonomically related species which have developed from a common ancestor also exhibit similarity at the pheromone level. An interesting typical representative can be observed in the family Tortricidae. This family includes a multitude of agricultural pest insects such as leaf-rollers and fruit-borers, whose sex pheromones have been exhaustively investigated [33]. Among the four subfamilies in Tortricidae, Tortricinae and Olethreutinae are the major two, and the pheromone components have been reported from 89 species in Tortricinae including *A. honmai* and the redbanded leafroller (*Argyrotaenia velutinana*) and 52 species in Olethreutinae including *C. pomonella* and the Oriental fruit moth (*Grapholita molesta*). The Tortricinae species usually secret C_{14} components that are generally unsaturated at the 11-position. The 11-tetradecenyl (11–14) compounds have been identified from 66 Tortricinae species, whereas C_{12} components have been found from only 7 species in this group. In addition, field tests using the synthetic 11–14 compounds attracted another 72 Tortricinae species. This 11–14 structure provides a clear distinction between the Tortricinae pheromones and the Olethreutinae pheromones, while the 11–14 compounds compose pheromones of 30 species in the 7 families besides Tortricidae; i.e., Yponomeutidae (Yponomeutoidea), Cochylidae, Pyralidae, Noctuidae, and so on. In contrast, the Olethreutinae species usually secret C_{12} chain components, mainly unsaturated at the 8- or 9-position. These 8- and 9-dodecenyl (8–12 and 9–12) compounds have been identified from 26 Olethreutinae species, whereas C_{14} components have been found from only 10 species in this group. In field tests with the synthetic 8–12 compounds, another 60 Olethreutinae species were attracted. The 8–12 compounds are quite specific to Olethreutinae and have yet to be found in other families. On the dienyl derivatives, 11,13–14 and 8,10–12 compounds have exclusively been identified from three Tortricinae species and 14 Olethreutinae species, respectively. While some Tortricinae species produce aldehyde components, Olethreutinae pheromones are always composed of alcohols or acetates except for E8,E10–12:Ald of *Cydia trasias* [34].

In clearwing moths in Sesiidae (Sesioidea) which exhibit wasp-mimicry diurnally fly, the males are not captured by a light-trap but attracted by female pheromones. Following research with peachtree borers (*Synanthedon exitiosa* and *S. pictipes*) [35], sex pheromones of another 12 Sesiidae species have been identified, these being composed of 3,13- and 2,13-octadecadienyl (3,13–18 and 2,13–18) compounds. These dienyl structures have also been assigned for the pheromones of three species in Tineidae (Tineoidea) and Cossidae (Cossoidea); any other components of the Sesiidae pheromones, however, are not known except for Z13–18:OAc, a minor pheromone component of the currant borer (*Synanthedon tipuliformis*) [36]. Field trials with synthetic lures found new attraction of many Sesiidae species [37]. In particular, attraction of 76 species by 3,13–18 compounds has been reported.

Identification of 9,12-tetradecadienyl (9,12–14) compounds began with studies on two cosmopolitan pests of stored products, the almond moth (*Cadra cautella*, Pyralidae: Phycitinae) and the Indian meal moth (*Plodia interpunctella*, Phycitinae) [38, 39]. This 9,12–14 structure has been reported from another 13 Pyralidae (only in Phycitinae) species and 11 Noctuidae species (9 species in Amphipyrinae, and 1 species each in Hadeninae and Plusiinae). These two families, however, are not closely related. Most likely, the females classified in distant groups happened to produce the same chemical in the train of their perpetual evolution of modifying the original systems for pheromone biosynthesis. The 5,7-dodecadienyl (5,7–12) structure is a carbon skeleton common to pheromone components of eight Lasiocampidae (Lasiocampoidea) species such as the pine moth (*Dendrolimus spectabilis*) [40], indicating that the conjugated dienyl system is peculiar to this family. The same dienyl structure, however, has recently been found in addition to monoenyl 7–12 compounds, in two Noctuidae species (*Thysanopluia intermixta* and *T. orichalcea*, Plusiinae) [41] (see Fig. 6). Biosynthetic experiments with *T. intermixta* revealed that the 5,7-dienyl structure was produced by Δ5-desaturation of the 7–12 acyl intermediate [42]. Instead of the 7–12 compounds, 5–12 compounds have been identified from Lasiocampidae species such as *D. punctatus* [43] and *Gastropacha quercifolia* [44], suggesting that dienyl structure of the Lasiocampidae pheromone might be produced via a 5–12 acyl or analogous intermediate in a manner different to the Plusiinae species.

2.4
Type I Compounds with Unique Structures

Some lepidopteran species produce unique compounds (★) shown in Table 1B, which have been classified into the Type I grouping because their structures are analogous to the common components in this class. From the processionary moth (*Thaumetopoea pityocampa*, Notodontidae) and two other species in the same genus, 13-en-11-ynyl compounds (≡11, Z13–16:OH and its derivatives) have been identified [45, 46]. Since the corresponding (*Z,Z*)-11,13-dienes are known as pheromone components of three *Thaumetopoea* species [46, 47], the triple bond at the 11-position was expected to be introduced by further desaturation of the double bond. This assumption has been confirmed by experiments with deuterium-labeled precursors [48]. Recently, a monoynyl compound (≡11–16:Ald) has been identified form the grape leaffolder (*Desmia funeralis*, Pyralidae: Pyraustinae), a species simultaneously secretes monoenyl and dienyl analogs (Z11–16:Ald and Z11,Z13–16:Ald) [49]. Other acetylene compounds have yet to be identified from female moths; consequently it would be valuable to discover pheromone components with a triple bond introduced at a position other than the 11-position.

Although almost all of the esters with a long chain alcohol are acetates, some other esters are exceptionally produced by certain female moths. Nitrate esters are quite singularly known as the pheromone of a lone micro-lepidopteran species in Bucculatrigidae (Gracillarioidea) [50]. Furthermore, they were the first naturally occurring nitrate esters identified. Esters with a C_4–C_6 acid have been identified from one Saturniidae (Bombycoidea) species [51] and three Lymantriidae species [52]. The monoenyl 7–18 chain skeleton of the Lymantriidae pheromone is worthy of note, because 7–18 compounds with a common functional group have not been found (see Fig. 3). The polyenyl 7,13,16,19–22 and 11,14,17–20 compounds containing an isobutyryl (isoBut) or 4-methylvaleryl (Me4-Val) ester function [53, 54], which are produced by Lymantriidae species and might be derived from linolenic acid via chain elongation, are closely correlated to the Type II pheromones based on their biosynthetic origin.

2.5
Structures of Pheromones in Type II

Hill et al. identified the first pheromone component of the Type II group, *cis*-9,10-epoxy-(*Z*,*Z*)-3,6-henicosadiene (Z3,Z6,epo9-21:H), from *E. acrea* (Arctiidae) in 1981 [30]. Since then, the compounds presented in Table 2 have been found from 65 species, although almost always within the families of Geometridae (30 species), Noctuidae (12 species), Lymantriidae (7 species), and Arctiidae (16 species) (▲ in Fig. 1) [4–7, 9]. Exceptionally, Z3,Z6,Z9-23:H has identified from one Pyralidae species, the tomato fruit borer (*Neoleucinodes elegantalis*, Crambinae) [55]. In this chapter, these Type II chemicals are abbreviated similarly to those of Type I; 'epo' and 'H' representing the *cis*-epoxy ring and the absence of a terminal functional group, respectively. A majority of the Type II compounds are dienes and trienes, which are biosynthesized from linolic and linolenic acids [20], respectively, and their monoepoxy derivatives. Therefore, the double bonds with the *Z* configuration and *cis*-epoxy ring are fixed at the 3-, 6-, and 9-positions.

Interestingly, the *N. elegantalis* female also secretes a Type I compound (E11-16:OH) like other Pyralidae species [55]. Only this species constructs a unique hybrid pheromone system that is composed of the Type I component biosynthesized de novo via a saturated fatty acid and the Type II component derived from dietary linolenic acid. It is noteworthy that no combination of a compound in the de novo type and an epoxy compound has been reported. Although *E. acrea* and *H. cunea* produce epoxy Type II components in addition to C_{18} polyunsaturated aldehydes that are classified as Type I, according to the terminal functional group [30–32, 56], aldehydes are the most likely compounds from dietary acids, as already mentioned above. No epoxy derivatives of a Type I compound that includes an epoxy ring and a terminal functional group as a hybrid compound of Types I and II (such as E10,epo12–16:OH and epo7–12:OAc) have been found in a pheromone gland. Epoxy pheromones have been identified from highly evolved species, indicating that the epoxydase is a relatively new enzyme in the evolution of pheromone biosynthesis.

Table 2A shows the frequency with which each compound has been identified from the species in the above families. All known pheromones are C_{17}–C_{23} straight chain compounds. While odd-numbered compounds are dominant, some C_{18} and C_{20} pheromone components are known. Since it has been reported that one Geometridae species was specifically attracted by a lure containing Z3,Z6,Z9–22:H [57], some natural C_{22} pheromone components will undoubtedly be identified from female moths someday. This table highlights the tendency of pheromones from Geometridae species to be shorter in length than those of Arctiidae species. Surveying the chemical groups, many pheromones are constructed with trienes and their monoepoxides rather than dienes and their monoepoxides. The 9,10-epoxides have been most frequently found in insects among the epoxydienes, in particular at this point in time, all the known Arctiidae pheromones are 9,10-epoxides.

Table 2 Type II lepidopteran female pheromones (▲)

(A) Trienes, dienes, and their monoepoxy derivatives

Formulae and abbreviations of representative compounds

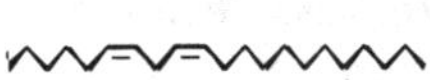

Z3,Z6-21:H

Z6,epo9-18:H

Z3,Z6,Z9-17:H

epo3,Z6,Z9-19:H

Number of lepidopteran species whose sex pheromones were found to be the indicated component

Chain length	Diene Z6,Z9	Epoxymonoene epo6,Z9	Epoxymonoene Z6,epo9	Triene Z3,Z6,Z9	Epoxydiene epo3,Z6,Z9	Epoxydiene Z3,epo6,Z9	Epoxydiene Z3,Z6,epo9
17	no		no	G(6)	G(6)	no	no
18	no	no	G(1)	G(2)	no	G(1)	G(6)
19	G(2)	G(2)	no	G(10)	G(3)	G(4)	L(1)
20	G(1), A(1)	no	no	G(1), N(6)	no	no	N(1), A(2)
21	G(1), N(2), A(3)	no	N(1), A(1)	G(1), N(10), L(1), A(10)	no	N(1), L(1)	N(3), A(12)
22	no	no	no	no	no	no	no
23	no	no	no	P(1), A(2)	no	no	no
Total	G(4), N(2), A(4)	G(2)	N(1), G(1), A(1)	P(1), G(20), N(16), L(1), A(12)	G(9)	G(5), N(1), L(1)	G(6), N(4), L(1), A(14)

P: Pyralidae, G: Geometridae, N: Noctuidae, L: Lymantriidae, A: Arctiidae.

(B) Further modified compounds [a]

Compound class / Formula and abbreviation	Super-family / Family	Identified component	Insect species	Year [Ref.] [b]
Polyene unsaturated at 1-, 4-, or 11-position	Geometroidea			
	Geometridae	1,Z3,Z6,Z9-19:H	*Operophtera brumata*	82 [58, 59]
			O. bruceata	87 [60]
1,Z3,Z6,Z9-19:H		1,Z3,Z6,Z9-21:H	*Epirrita autumnata*	95 [61]
		Z3,Z6,Z9,Z11-19:H	*Alsophila pometaria*	84 [65]
		Z3,Z6,Z9,E11-19:H	*A. pometaria*	84 [65]
		E4,Z6,Z9-19:H	*Bupalus piniaria*	98 [9]
	Noctuoidae			
	Arctiidae	1,Z3,Z6,Z9-21:H	*Arctia villica*	84 [62]
			Pareuchaetes pseudoinsulata	93 [63]
			Utetheisa ornatrix	82 [64]
Epoxytriene	Noctuoidae			
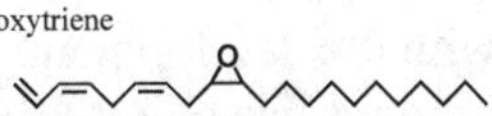	Arctiidae	1,Z3,Z6,epo9-20:H	*Hyphantria cunea*	89 [56]
		1,Z3,Z6,epo9-21:H	*H. cunea*	89 [56]
1,Z3,Z6,epo9-21:H				
trans-Epoxide	Geometroidea			
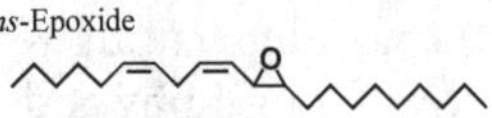	Geometridae	*t*-epo4,Z6,Z9-19:H	*B. piniaria*	98 [9]
	Noctuoidae			
Z6,Z9,*t*-epo11-21:H (posticlure)	Lymantriidae	Z6,Z9,*t*-epo11-21:H (11*S*,12*S*)	*Orgyia postica*	01 [66]
Diepoxy derivative	Noctuoidae			
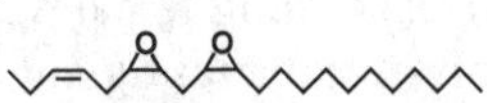	Lymantriidae	Z3,epo6,epo9-21:H (3*R*,4*S*,6*S*,7*R*)	*Leucoma salicis*	97 [68]
Z3,epo6,epo9-21:H (leucomalure)		epo3,epo6,Z9-21:H (3*S*,4*R*,6*S*,7*R*)	*Perina nuda*	02 [70]

[a] Keto derivatives such as ket3,Z6,Z9–19:H and Z6,ket9–21:H are listed in Table 3.

[b] Publication year and [reference].

In spite of the large number of species in the above families, the structural diversity of the compounds listed in Table 2A is quite limited. Some females have modified these compounds with additional desaturation and/or epoxidation steps as shown in Table 2B. The C_{19} tetraene with an extra double bond at the terminal position (1,Z3,Z6,Z9–19:H) is a well-known compound which was identified from the winter moth (*Operophtera brumata*) during the first study on Geometridae pheromones [58, 59]. This tetraene is also a pheromone component of another *Operophtera* species [60], and a longer chain derivative (1,Z3,Z6,Z9–21:H) has been found in one Geometridae species [61] and three Arctiidae species [62–64]. Other polyenes unsaturated at the 11-positon of a 3,6,9-triene and the 4-position of a 6,9-diene have been reported from Geometridae species, including Z3,Z6,Z9,E11–19:H [65], Z3,Z6,Z9,Z11–19:H [65], and E4,Z6,Z9–19:H [9]. These additional double bonds are conjugated with the original polyene system, and an *E* configuration is noteworthily introduced. With the epoxidation of these polyenes, novel components are created such as 9,10-epoxides derived from 1,Z3,Z6,Z9-tetraenes (1,Z3,Z6,epo9–20:H and 1,Z3,Z6,epo9–21:H) [56] and a 4,5-epoxide derived from E4,Z6,Z9-diene (*t*-epo4,Z6,Z9–19:H) [9]. The 4,5-epoxide has a *trans*-epoxy ring. Another pheromone compound with a *trans*-epoxy ring, Z6,Z9,*t*-epo11–21:H from the tussock moth in Lymantriidae [66, 67], could be biosynthesized from the corresponding triene (Z6,Z9,E11–21:H). This intermediate, however, has not been found in insects. On the other hand, extra epoxidation of epoxydienes produces diepoxy derivatives such as the 6,7–9,10-diepoxide (Z3,epo6,epo9–21:H) [68] and 3,4–6,7-diepoxide (epo3,epo6,Z9–21:H) [69, 70] of Lymantriidae species. The latter diepoxide is a minor component and has been shown to have an important synergistic effect on the attractive activity of the major component (Z3,epo6,Z9–21:H). Any 3,4–9,10-diepoxides and triepoxy derivatives have so far eluded identification.

The epoxy compounds include chiral centers and the absolute configuration used to be defined by biological evaluation of both enantiomers prepared via chiral synthesis or resolution. A substantial supply of optically pure samples is essential for pheromone research and the current notable syntheses are reviewed in the chapter by K. Mori in this volume. Recently, with the development of chiral chromatographic techniques, some epoxy pheromones have directly been defined by GC or HPLC. The kinds of available chiral columns and their resolution capabilities are described later. Table 3 shows the absolute configuration of some natural pheromones as analyzed by chiral chromatography and the attractive activity of synthetic epoxides in the field [67, 71–76]. Interestingly, males were not always effectively attracted by the synthetic epoxide despite exhibiting the same configuration as produced by the corresponding females. This insufficient coincidence as evidenced by the case of the Japanese giant looper (*Ascotis selenaria cretacea*) and the fruit-piercing moth (*Oraesia excavata*) indicates that determination of the stereochemistry of natural pheromones cannot be accomplished only by field evaluation. Combining the data from ten other species, which were tentatively determined by field tests of synthetic epox-

Table 3 Stereochemistry of natural pheromones containing an epoxy ring (main components) and field attractancy of the synthetic racemate

Species	Main pheromonal component	Stereochemistry of the epoxy ring (main component)				[Reference]
		Natural configuration	Method for chiral analysis	Field attraction		
				Optimum isomer [a]	Racemate [b]	
Ascotis selenaria cretacea [c]	epo3,Z6,Z9-19:H	racemate	GC [f], HPLC [g]	3*R*,4*S*	weak	[71]
Biston robustum [c]	epo6,Z9-19:H	6*S*,7*R*	HPLC [h]	6*S*,7*R* + 6*R*,7*S* (9:1)	none	[72]
Semiothisa clathrata [c]	epo3,Z6,Z9-17:H	3*R*,4*S*	GC [i]	3*R*,4*S*	none	[73]
Colotois pennaria [c]	Z3,epo6,Z9-19:H	6*R*,7*S*	GC [i]	6*R*,7*S*	weak	[74]
Erannis defoliaria [c]	Z3,epo6,Z9-19:H	6*S*,7*R*	GC [i]	6*S*,7*R*	same	[74]
Hemerophila atrilineata [c]	Z3,Z6,epo9-18:H	9*S*,10*R*	HPLC [h,j]	9*S*,10*R*	same	[75]
Oraesia excavata [d]	Z6,epo9-21:H	9*S*,10*R*	HPLC [h]	racemate	best	[76]
Orgyia postica [e]	Z6,Z9,*t*-epo11-21:H	11*S*,12*S*	HPLC [h]	11*S*,12*S*	same	[67]

[a] Optimum configuration of the main component to attract male moths in the field.
[b] Activity of racemate in a field test, comparison to lure baited with the optimum isomer.
[c] Geometridae.
[d] Noctuidae: Ophiderinae.
[e] Lymantriidae.
[f] Chiraldex A-PH column.
[g] Chiralpak AS column.
[h] Chiralpak AD column.
[i] Custom-made column with a 1:1 mixture modified cyclodextrin and OV-1701.
[j] Chiralcel OJ-R.

ides, the overview of their stereochemistry indicates that the *S*,*R* configuration has been assigned for the epoxy ring more frequently than the *R*,*S* configuration and that no 9,10-epoxides with 9*R*,10*S* configuration have been found. The female of *A. s. cretacea* exceptionally produce a racemic pheromone, while the male is attracted more by an optically pure sample with a 3*R*,4*S* configuration [71].

2.6 Structures of Other Female Pheromones

Miscellaneous pheromone components, which are classified as neither Type I nor II, are listed in Table 4. Secondary alcohols with a C_7 or C_9 chain have been identified from some primitive species; i.e., saturated and monoenyl alcohols from three species in Eriocraniidae (Eriocranioidea) [77, 78] and dienyl alcohols from one species in Nepticulidae (Nepticuloidea) [79]. Their hydroxyl group is always located at the 2-position, and this is illustrated in the table with the abbreviation 2-OH in place of the OH which was used for the Type I pheromones. Furthermore, esters between a long chain acid and a rather small alcohol have been characterized from species in Tineoidea and Zygaenoidea. Phero-

Table 4 Lepidopteran female pheromones of other groups (※)

Compound class Formula and abbreviation	Super-family Family	Identified comp.	Insect species	Year [Ref.] [a]
Secondary alcohol	Ereiocranioidea			
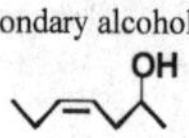	Eriocraniidae	Z4-7:2-OH (*R*)	*Eriocrania cicatricella*	95 [77]
		7:2-OH (*R*)	*E. cicatricella*	95 [77]
Z4-7:2-OH		Z6-9:2-OH (*S*)	*E. sangii*	96 [78]
			E. semipurpurella	96 [78]
	Nepticuloidea			
	Nepticulidae	Z6,8-9:2-OH (*S*)	*Stigmella malella*	95 [79]
		E6,8-9:2-OH (*S*)	*S. malella*	95 [79]
Ester of a long chain acid	Tineoidea			
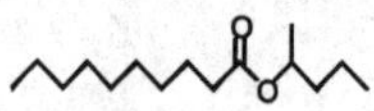	Psychidae	(*R*)-1-Methylbutyl 10:Ate	*Thyridopteryx ephemeraeformis*	83 [80]
			Oiketicus kirbyi	94 [81]
1-Methylbutyl 10:Ate		Isopropyl 8:Ate	*Megalophanes viciella*	00 [82]
	Zygaenoidea			
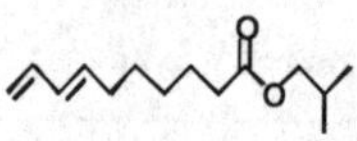	Limacodidae	Isobutyl E7,9-10:Ate	*Darna bradleyi*	00 [83]
		Methyl E7,9-10:Ate	*D. bradleyi*	00 [83]
		(*E*)-2-Hexenyl E7,9-10:Ate	*D. trima*	00 [83]
Isobutyl E7,9-10:Ate		(*S*)-2-Methylbutyl E7,9-10:Ate	*D. trima*	00 [83]
	Zygaenidae	(*S*)-*sec*-Butyl Z7-14:Ate	*Harrisina brillians*	82 [84]
			Theresimima ampellophaga	98 [85]
Unsaturated ketone	Copromorphoidea			
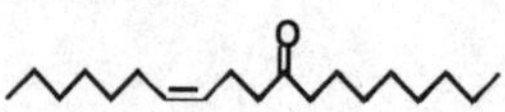	Carposinidae	Z7,ket11-19:H	*Carposina niponensis*	77 [86]
		(= ket9,Z12-19:H)	*Heterocrossa rubophaga*	00 [87]
Z7,ket11-19:H		Z7,ket11-20:H	*C. niponensis*	77 [86]
	Geometroidea Geometridae	ket3,Z6,Z9-19:H [b]	*Peribatodes rhomboidaria*	85 [88]
	Noctuoidea	Z6,ket11-21:H	*Orgyia pseudotsugata*	75 [89]
	Lymantriidae		*O. thyellina*	99 [90]
			O. postica	01 [91]
		1,Z6,ket11-21:H	*O. pseudotsugata*	78 [92]
		Z6,E8,ket11-21:H	*O. pseudotsugata*	97 [93]
		Z6,ket9-21:H [b]	*O. thyellina*	99 [90]
		Z6,Z9,ket11-21:H [b]	*O. leucostigma*	03 [94]
Epoxyalkane	Noctuoidea			
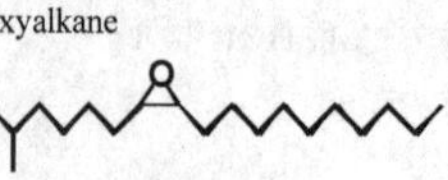	Lymantriidae	Me2,epo7-18:H (7*R*,8*S*)	*Lymantria dispar*	70 [3]
			L. fumida	99 [95]
			L. monacha	75 [96]
Me2,epo7-18:H (disparlure)		epo7-18:H (7*R*,8*S*)	*L. monacha*	01 [97]
		Me2,epo7-20:H	*L. xylina*	99 [98]
Methyl-branched hydrocarbon	Yponomeutoidea			
	Lyonetiidae	Me5,Me9-17:H (5*S*,9*S*)	*Leucoptera malifoliella*	87 [99]
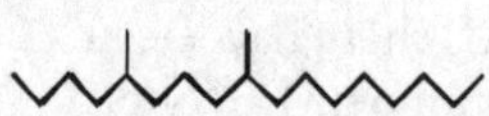		Me5,Me9-16:H	*L. malifoliella*	87 [99]
			Perileucoptera coffeella	88 [100]
Me5,Me9-17:H		Me5,Me9-15:H	*P. coffeella*	88 [100]
		Me5,Me9-18:H	*L. malifoliella*	90 [101]
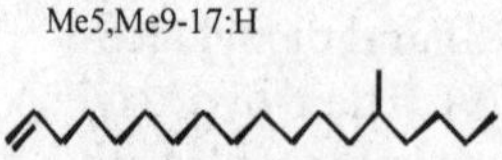		1,Me14-18:H (*S*)	*L. clerkella*	84 [102]
		1,Me10,Me14-18:H	*L. prunifoliella*	97 [103]
1,Me14-18:H				

[a] Publication year and [reference].

[b] Related to Type II compounds based on an aspect of their predicted biosynthetic pathway.

Table 4 (continued)

Compound class Formula and abbreviation	Super-family Family	Identified comp.	Insect species	Year [Ref.] [a]
	Geometroidea			
	Geometridae	Me2,Me5-17:H (*R*), (*S*)	*Lambdina fiscellaria*	91 [104]
		Me5,Me11-17:H (5*R*,11*S*)	*L. fiscellaria*	91 [104]
		Me7-17:H (*S*)	*L. athasaria*	94 [105]
			L. fiscellaria	93 [106]
			L. pellucidaria	98 [107]
		Me7,Me11-17:H (meso)	*L. athasaria*	94 [105]
			L. pellucidaria	98 [107]
		Me3,Me13-17:H (3*S*,13*R*)	*Nepytia freemani*	93 [108]
	Noctuoidea			
	Noctuidae	Z6,Me13-21:H	*Scoliopteryx libatrix*	00 [109]
	Lymantriidae	Me2,Z7-18:H	*Lymantria fumida*	99 [95]
			L. lucescens	02 [110]
			L. monacha	01 [97]
			L. serva	02 [110]
	Arctiidae	Me2-17:H	*Holomelina aurantiaca*	71 [111]
			H. ferruginosa etc.	71 [111]
			Pyrrharctia isabella	71 [111]
Methyl-branched alkenyl acetate	Tortricoidae			
	Tortricidae	Me10-12:OAc	*Adoxophyes honmai*	79 [129]

monal components of three Psychidae species in the former superfamily are esters of C_8 or C_{10} saturated acids [80–82] and those of two Limacodidae and two Zygaenidae species in the latter superfamily are esters of C_{10} or C_{14} unsaturated acids [83–85]. The abbreviations in Table 4 are made with respect to the acid moiety. Unsaturated ketones with a C_{19}–C_{21} chain have been identified from two species in Carposinidae (Copromorphoidea) [86, 87], one species in Geometridae [88], and five species in Lymantriidae [89–94]. Their chemical structures have been abbreviated corresponding to the Type II pheromones without a terminal functional group, and supplying 'ket' to indicate the position of the carbonyl group. Among the known eight unsaturated ketones, six compounds possess the carbonyl group at the 11-position. Double-bond positions in the other two compounds (ket3,Z6,Z9–19:H and Z6,ket9–21:H) suggest a relationship with the Type II pheromones. The dienyl 3-ketone and monoenyl 9-ketone seem to be biosynthesized by an opening of the epoxy ring of epo3,Z6,Z9–19:H and Z6,epo9–21:H, respectively, with successive oxidation of the produced alcohol. Similar speculation is possible for the biosynthesis of Z6,Z9,ket11–21:H from Z6,Z9,*t*-epo11–21:H, which is identified from the species in *Orgyia* in Lymantriidae and classified into the Type II pheromone.

Some lepidopteran species secret methyl-branched chemicals for their sexual communication. These have been abbreviated with 'Me' to indicate the position of the methyl group. Disparlure (Me2,epo7–18:H) is a well-known pheromone identified from *Lymantria dispar* [3] and two other species in the same genus, *L. fumida* [95] and *L. monacha* [96]. *L. monacha* also secrets an unbranched derivative (epo7–18:H) [97], and a C_{20} analog (Me2,epo7–20:H) has recently been found in females of another related species, *L. xylina* [98]. In addition to the epoxides, six methyl-branched hydrocarbons with a C_{15}–C_{18} main chain have been identified from four micro-lepidopteran species in Lyonetiidae (Yponomeutoidea) [99–103]. These Lyonetiidae pheromones possess one methyl group at the common 5-position, because the 14-position of 1,Me14–18:H and 1,Me10,Me14–18:H is represented as the 5-positon counting from the opposite terminal methyl group. Furthermore, other hydrocarbons have been discovered form macro-lepidopteran species in the rather evolved families, such as Geometridae [104–108], Noctuidae [109], Lymantriidae [95, 97, 110], and Arctiidae [111]. Me2,Z7–18:H, a pheromone component of *Lymantria* spp., is also a biosynthetic precursor of disparlure [112]. In Arctiidae, Me2–17:H was identified from female moths of four *Holomelina* spp. (*H. fragilis*, *H. immaculata*, *H. lamae*, and *H. nigricans*) in addition to two species listed in Table 4 [111].

2.7 Male Sex Pheromones

Males of many lepidopteran species, both moths and butterflies, produce sex pheromones in scent-organs located on the abdomen, thorax, legs, and wings. The organs vary from simple scales to complex eversible structures and have many descriptive terms such as androconial scales, scent fans, costal hairs, brushes, hair-pencils, and coremata [113]. As summarized in the review of Birch et al. [113], their chemical structures are quite different from that of the female pheromones, as shown in Fig. 5, and their species-specificity is low. Among moths, benzaldehyde, benzyl alcohol, and 2-phenyl ethanol have been identified from several Noctuidae species, and danaidal (**1a**) and hydroxydanaidal (**1b**) have been found in the coremata of several Arctiidae species. While no female pheromones of Hepialidae species in Hepialoidea, a primitive group in

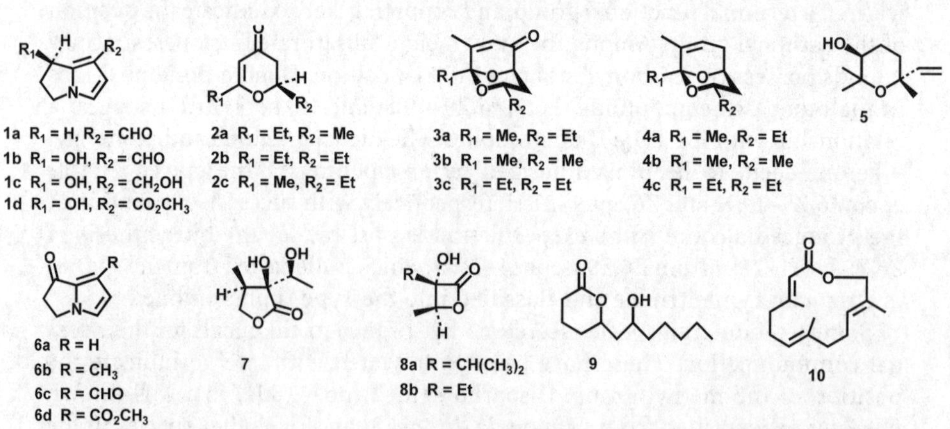

Fig. 5 Representative sex pheromones produced by male moths and butterflies

Lepidoptera (see Fig. 1), have been reported, novel dihydropyrans (**2a–c**) and bicyclic acetals (**3a–c** and **4a–c**) have been identified from the males of *Hepialus hecta* in this family [114]. Three of them, **2a**, **3a**, and **4a**, are major components in its hindleg extract, and compounds **2c** and **3b** have been identified from the males of two other Hepialidae species, *H. californicus* [115] and *Endoclita excrescens* [116], respectively. In addition to these sex-specific compounds, one polyunsaturated hydrocarbon (Z3,Z6,Z9–21:H), which is classified as a Type II female pheromone, has been found in an abdominal tip extract of the males of *Anticarsia gemmatalis* (Noctuidae: Catocalinae) [117]. The finding is noteworthy, because the females of this species produce a sex pheromone composed of Z3,Z6,Z9–21:H and Z3,Z6,Z9–20:H, and the males are effectively attracted to their 3:5 mixture [118]. The effect of the hydrocarbon on the females has not been examined, but a single dose of Z3,Z6,Z9–21:H at a relatively high release rate attracted the males, indicating an unknown male-male interaction.

The male pheromones stimulate male-searching behaviors of females, such as walking and close-range attraction, and/or evoke acceptance by females [113]. Experimental confirmation of these functions, however, is more difficult than of those of female pheromones with a long-range attractive activity, and the structure determinations of male pheromones have been sometimes reported without details of their biological activities. In the case of *E. excrescens*, field observations of mating behavior confirmed a female flight straight toward a male in a swing-like pendulum flight (calling by the male) and indicated that the female was attracted to the male by visual cues from a long distance and by the male scent from a short distance in the final encounter [119]. As an exceptional instance, a sticky trap baited with a mixture of tetrahydropyran (**5**) and vanillin, male pheromone components of *Tirathaba mundella* (Pyralidae: Galleriinae), captured the females released in a screen cage, suggesting the activity for long-range female attraction [120]. Since female antennae response to male pheromones as well as male antennae are stimulated by female pheromones, the GC-EAD technique (see below) is also useful for the analysis of male pheromones [120, 121].

From the hair-pencils of butterflies in Danainae and Ithomiinae (Papilionoidea: Nymphalidae), a wider variety of pyrrolizines (**1a–d**, and **6a–d**) have been identified than from Arctiidae moths. These compounds are biosynthesized from pyrrolizidine alkaloids, which are included in host plants fed by the larvae and protect them from the attacks of other herbivores [122]. In addition to novel lactones (**7**, **8a**, and **8b**) derived from an acid part of the alkaloids, many volatiles of more than 100 compounds (aromatics, terpenoids, hydrocarbons, and others) constitute scent bouquets of the male butterflies [123]. For example, the hair-pencil of *Idea leuconoe* (Danainae) which is distributed in South-East Asia contained 16 compounds (**6b**, **8a**, **8b**, **9**, and others), and a mixture of the major volatiles applied to a butterfly dummy elicited an abdomen-curling acceptance posture in the females as a crude extract of the male hair-pencils did [124]. A chiral GC analysis revealed the absolute configuration of the lactone components utilizing authentic samples synthesized by asymmetric dihydroxylation or hydrogenation. The β-lactones (**8a** and **8b**) of *I. leuconoe* have an *S,S* configuration, which, interestingly, suggests their biosynthetic origins, while the natural δ-lactone **9** is a mixture of all enantiomers [125]. Recently, a new macrocyclic sesquiterpene (niaviolide, **10**) and its 10,11-monoepoxide were identified from *Amauris niavius* living in African forests [126]. These male pheromones possess a unique 13-membered macrolide ring, originating from an α,ω-oxidation pattern of the sesquiterpene backbone.

2.8 Reproductive Isolation by Sex Pheromones

Contrary to the structure similarity of the pheromones secreted by taxonomical related moths, some differences are necessary for their sexual communication systems to play an important role in their reproductive isolation. In addition to further modifications of the various structures, diversity of the lepidopteran sex pheromones is generated by blending multiple components. Innumerable pheromone blends are based not only on combinations of different components but also on variations in the mixing ratio. A pioneer study with *Adoxophyes* spp. (Tortricidae: Tortricinae) had already proposed this concept in the early 1970s. While the smaller tea tortrix (*A. honmai*) and the Japanese summerfruit tortrix (*A. orana fasciata*) had been considered to be variant strains with different host preferences in the same species, Tamaki et al. found that females of the former pest insect in the tea garden secreted Z9–14:OAc and Z11–14:OAc in a ratio of 7:4 but females of the latter defoliator of apple trees secreted them in a ratio of 13:4 [127, 128]. Furthermore, two other components (E11–14:OAc and Me10–12:OAc) were subsequently identified from the former species [129].

Since then, many researches have revealed the reproductive isolation by sex pheromones of sibling species such as *Ostrinia* spp. (Pyralidae: Pyraustinae). In this genus, 20 species are reported in the world including the European corn borer (*O. nubilalis*), which is known to be polymorphic with respect to a ratio of the pheromone components (Z11–14:OAc + E11–14:OAc); i.e., 97:3 of the Z-strain and 4:96 of the E-strain [130, 131]. In addition to the Asian corn borer (*O. furnacalis*) emanating the positional isomers (Z12–14:OAc + E12–14:OAc, 1:1–3:2) [132, 133], another six *Ostrinia* spp. are distributed throughout Japan, these being differentiated by morphological data, mitochondrial gene analysis, and host plants. Recently, sex pheromones of all of them have been identified as follows; *O. latipennis* (E11–14:OH) [134], *O. palustralis* (Z11–14:OAc + E11–14:OAc, 1:99) [135], *O. orientalis* (Z11–14:OAc + E11–14:OAc) [134], *O. scapulalis* (Z11–14:OAc + E11–14:OAc, 97:3) [136], *O. zaguliaevi* (Z11–14:OAc + E11–14:OAc + Z9–14:OAc, 50:5:45) [137], and *O. zealis* (Z11–14:OAc + E11–14:OAc + Z9–14:OAc, 8:22:70) [138]. This important factor for their reproductive isolation proposes one cue toward understanding their evolutional relationships. While the 12-monoenyl components produced only in *O. furnacalis* suggest the occurrence of a specific desaturation enzyme specific to this species, two genes of a Δ11-desaturase for the 11-monoenes and of a Δ14-desaturase for the 12-monoenes have recently been characterized from both *O. furnacalis* and *O. nubilalis*, indicating that the two enzymes of these species utilize different transcriptional mechanisms [139].

Another example of the diversity is observed among the insects in Plusiinae, which is a well assembled subfamily in Noctuidae and includes 400 species across the world. After the identification of pheromone components from *Trichoplusia ni*, sex pheromones of 19 species and sex attractants of 24 species have been reported in Plusiinae [4–7, 140–146]. Many of them are composed of Z7–12:OAc as a common component which is further blended with other minor components. Whereas some species produce more than ten components, only two or three compounds are essential for male attraction. Based on the chemical structures of these primary components, Plusiinae pheromones have been classified into the following 8 groups (A–H, Fig. 6); A: 7–12 compounds only, B: 7–12 and 5–10 compounds, C: 7–12 and 9–14 compounds, D: 7–12 and 5–12 compounds, E: 7–12 and 9–12 compounds, F: 7–12 and 5,7–12 compounds, G: 5–10 and 7–10 compounds, and H: 5–12 and 7–14 compounds. The components in groups A–C are biosynthesized via Δ11-desaturation of 16:Acid, while another desaturase is necessary for the biosynthesis of components in the other groups. If we hypothesize that a primitive species of Plusiinae utilized Z7–12:OAc for its sexual communication, it is conceivable that the β-oxidation enzyme of a newly evolved species gained an ability to produce Z5–10:OAc and Z9–14:OAc. After that, in addition to the Δ11-desaturase for the compounds unsaturated at the ω5-position, another desaturase was created to produce compounds unsaturated at the ω3- or ω7-position during the long history of this subfamily.

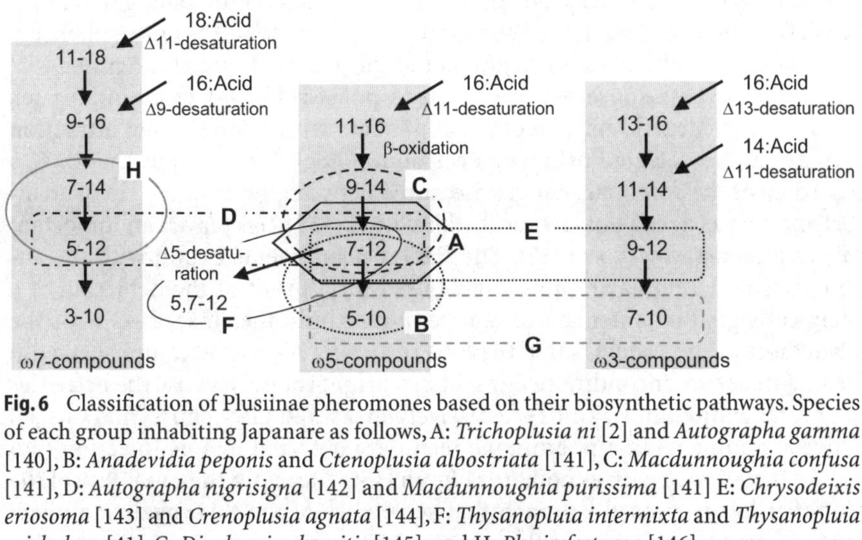

Fig. 6 Classification of Plusiinae pheromones based on their biosynthetic pathways. Species of each group inhabiting Japanare as follows, A: *Trichoplusia ni* [2] and *Autographa gamma* [140], B: *Anadevidia peponis* and *Ctenoplusia albostriata* [141], C: *Macdunnoughia confusa* [141], D: *Autographa nigrisigna* [142] and *Macdunnoughia purissima* [141] E: *Chrysodeixis eriosoma* [143] and *Crenoplusia agnata* [144], F: *Thysanopluia intermixta* and *Thysanopluia orichalcea* [41], G: *Diachrysia chrysitis* [145], and H: *Plusia festucae* [146]

3 Current Techniques for Structure Determination

3.1 Collection of Pheromones and Evaluation of Their Activity

The sex pheromone glands of female moths are commonly found as modified intersegmental membranes between the 8th and 9th abdominal segments [147]. After removing the glands, the female pheromones are usually extracted with a limited amount of a non-polar solvent, such as *n*-hexane or *n*-pentane, for several minutes to hold down the contamination with lipids and unknown materials. The collection of pheromones actually emitted from calling females, however, has been one of the key issues in the dynamic research of semiochemicals. The airborne, volatile pheromones have been trapped on Porapak Q, Tenax, activated carbon, glass wool, glass beads, or glass capillary tubes [148, 149]. These methods have the advantage of allowing continuous collection of pheromones; however, they have not been widely applied for the identification of lepidopteran pheromones because of the weak absorption of the lipophilic volatiles, which are released by female moths at a quite low level.

Recently, in addition to these collection methods, a technique for SPME of airborne volatiles has been developed [149] and used for the collection of lepidopteran female pheromones as follows: Z10–14:OAc and others of *Phyllonorycter sylvella* (Gracillariidae) [150], E10–12:OAc and others of *P. blancardella* [17], Z8–14:OAc and others of *P. heegerella* [151], Z11–16:OAc and others of *Sesamia nonagrioides* (Noctuidae) [152], and E3,Z5–12:OAc of *Phtheochroa cranaodes* (Tortricidae) [153]. The pheromones of the former three leafminer species have been newly identified after collection using a SPME fiber that was placed a few mm from the protruded abdominal gland of the calling female for 2–3 h. For the collection of the *S. nonagrioides* pheromone, the entire fiber surface was gently rubbed on the tegument of the glandular area for 5 min.

In addition to the observation of male responses in a flask or a wind tunnel, the activity of female pheromone components with a long-range attraction activity can be detected using an electrophysiological technique. The EAG, a recording of the potential changes measured between the base and tip of an insect antenna as a result of chemical stimulation [154], has played an important role as a bioassay system [155]. The EAG measurements begun with the antennae of male moths are now widely used for a proof of the detection of a chemical signal by antennae of female moths and other insects. One of the advantages of the system is that the EAG recording is easily accomplished with an antenna of an unconditioned insect in a bright room. Because the activities of several compounds can be successively measured after short intervals, all active components in a pheromone blend are detected by one injection into GC-EAD equipped with a capillary column [156]. Recently, the capability of the amplifier has increased so much that sensitivity of the EAD against a pheromone component is higher than that of FID, as shown in Fig. 7. On the GC-

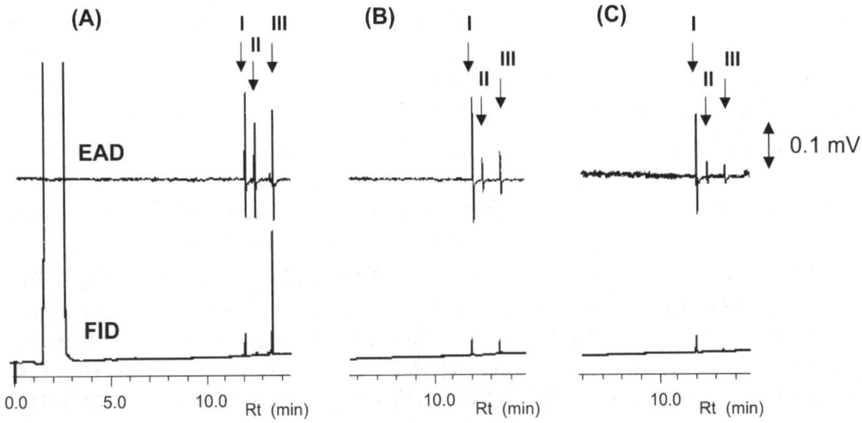

Fig. 7A–C GC-EAD measurements of synthetic Type II pheromones [Z6,epo9–18:H (I) and Z3,Z6,epo9–18:H (II)] and a homologue [Z3,Z6,epo9–19:H (III)] by a male antenna of *Hemerophila artilineata* (Geometridae): **A** I (10 ng), II (1 ng), and III (100 ng); **B** I (10 ng), II (0.1 ng), and III (10 ng); **C** I (10 ng), II (0.01 ng), and III (1 ng). GC is equipped with a DB-23 column (0.25 mm ID × 30 m), the end of which is divided into two lines to detect simultaneously by FID and EAD, with a column temperature program of 50 °C for 2 min, 10 °C/min up to 160 °C and then 4 °C/min up to 220 °C

EAD analysis with synthetic compounds, the male antenna of a Geometridae species, *Hemerophila artilineata*, was stimulated even by 0.01 ng of the main sex pheromone (Z3,Z6,epo9–18:H) of this species [75], which was scarcely detected by FID. When researching species that have a very low pheromone content, the chromatographic behaviors of EAG-active components provide valuable information for the structure elucidation. Actually, without using spectral methods, a new pheromonal compound (E8,Z10–14:Ald) of a leafminer moth, *Cameraria ohridella* (Gracillariidae), was identified in pg quantities by the GC-EAD technique [157]. Figure 7 also indicates that the minor component (Z6,epo9–18:H) was reproducibly detected by three different antennae and that a C_{19} homologous compound (Z3,Z6,epo9–19:H) was detected by the antenna, but its activity was 100 times weaker than that of the real pheromone component.

3.2 GC-MS Analysis

3.2.1 Identification of Type I Components Without Derivatization

The extract, which is prepared with pheromoneglands severed exclusively from abdomens, can be directly analyzed by GC-MSwithout any purification. EI at 70 eV is widely used for this analysis, and are liable spectrum is recorded with a sample at least at a level of several ng. The gland of a large insect such as a Noctuidae species contains around 10–100 ng of the sex pheromone [21], and, thus, the injection of an extra ctincluding pheromone equivalent to one female is sufficient to measure the fullmass spectrum of a main component and, further, to obtain some information on minor components. Although a quantitative correlation between the size of a female moth and its pheromone titer is not always observed, the pheromone content in micro-lepidopteran species, such as leafminer moths in the superfamilies of Gracillarioidea and Yponomeutoidea, is very limited, sometimes lower than a 1-pg order [25, 158, 159]. In this case, it is necessary to have a large number of females for the analysis.

Since the molecular ions (M^+) of C_{10}–C_{18} monoenyl compounds in Type I are hardly detected by EI, their occurrence in the extract is examined by systematic analysis of the GC-MS data using mass chromatograms for the fragment ions at the highest *m/z* values; i.e., $[M\text{-}H_2O]^+$ for alcohols (*m/z* 138, 166, 194, 222, and 250) andaldehydes (*m/z* 136, 164, 192, 220, and 248) and $[M\text{-}AcOH]^+$ for acetates (*m/z* 138, 166, 194, 222, and 250) with a C_{10}–C_{18} even-numbered chain.While the alcohols and acetates with the same chain length show the same fragment ion, the ion at *m/z* 61differentiates these functional derivatives. Dienyl and trienyl compounds with a C_{10}–C_{18} chain are easily recognized by the mass chromatograms of their M^+; i.e., *m/z* 154, 182, 210, 238, and 266 for dienyl alcohols, *m/z* 152, 180, 208, 236, and 264 for dienyl aldehydes and *m/z* 196, 224, 252, 280, and 308 for dienyl acetates with an even-numbered chain.

The positional isomers of monoenyl compounds show similar EI-mass spectra, but the spectra can differentiate the relative intensities of some fragment ions [160]. While these subtle differences are not rationally explained, it is possible to estimate the double-bond position of a natural pheromone by calculating the similarity of the spectra published for a series of synthetic monoenyl compounds [146, 161]. Recently, direct determination by the spectra has been attempted operating chemical ionizationion-trap mass spectrometry [162]. A cetonitrile used as a reagent gas creates $[C_3H_4N]^+$. This reagent ion reacts with carbon–carbon double bonds and generates fragmentions that are characteristic of the double-bond position under the condition that no ions below *m/z* 50 are present in the trap during the reaction period.

Type I compounds with a conjugated diene system are distinguishable from unconjugated ones by their high abundances of M^+ and long retention times (Rts) on a polarcapillary GC column. The conjugated dienes unsaturated at a side of theterminal methyl group [ωnω(n+2)-diene,n=1–5] show characteristic fragment ions, $[CH_2CH{=}CHCH{=}CHC_{n-1}H_{2n-1}]^+$ and $[CH_2{=}CHCH{=}CHC_{n-1}H_{2n-1}]^+$, reflecting the position of unsaturations [163]. Namely, the positional isomers of hexadecadienyl compounds are distinguished with the following fragmentions; *m/z* 67 and 54 of 13, 15–16 compounds (ω1,ω3-dienes), *m/z* 81 and 68 of 12,14–16 compounds (ω2,ω4-diene), *m/z* 95 and 82 of 11,13–16 compounds (ω3,ω5-diene), *m/z* 109 and 96 of 10,12–16 compounds (ω4,ω6-diene), and *m/z* 123 and 110 of 9,11-16 compounds (ω5,ω7-diene). These diagnosticions, particularly the characteristic base peaks at *m/z* 67 (ω1,ω3-dienes), 81 (ω2,ω4-diene), and 95 (ω3,ω5-diene), have been observed to be unrelated to the chain length and functional group, as shown in Table 5.

Table 5 Base peaks in the EI-mass spectra of Type I compounds including a conjugated diene system [163, 164][a]

Compounds	Base peak (*m/z*)						
	3,5-diene [b]	4,6-diene [b]	5,7-diene [c]	6,8-diene - ω4,ω6-diene [d]	ω3,ω5-diene [e]	ω2,ω4-diene [e]	ω1,ω3-diene [e]
Alcohol	67	79	79	67	95 (B)	81 (C)	67 (D)
Acetate	79	79	79	67	95 (B)	81 (C)	67 (D)
Aldehyde	—	84 (A)	79 or 80	67	95 (B)	81 (C)	67 (D)

[a] The following structures (**A–D**) are proposals for the formation of these base peaks.

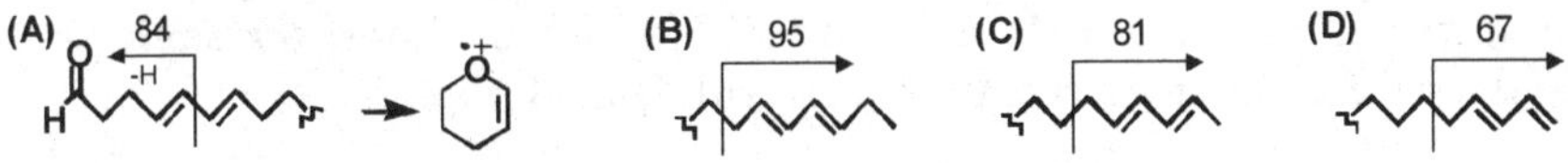

[b] C_{16} dienyl compounds.
[c] C_{16} and C_{12} dienyl compounds. C_{16} 5,7-dienal: *m/z* 79 (78%) and 80 (100%), C_{12} 5,7-dienal: *m/z* 79 (100%) and 80 (79%).
[d] C_{16} 6,8-,7,9-, 8,10-, and 10,12-dienyl compounds, C_{14} 7,9- and 8,10-dienyl compounds, and C_{12} 6,8-dienyl compounds.
[e] C_{16} 11,13-, 12,14-, and 13,15-dienyl compounds, C_{14} 9,11-, 10,12-, and 11,13-dienyl compounds, and C_{12} 7,9-, 8,10-, and 9,11-dienyl compounds.

On the other hand, the mass spectra of the conjugated dienes unsaturated at a side of the functional group are too similar to differentiate each positional isomer by the spectral patterns. 4,6-Dienal, however, shows a characteristic base peakat *m/z* 84, while the base peaks of the corresponding alcohol and acetate have been recorded at *m/z* 79 (Table 5). The ion at *m/z* 84 is expected to have a stable 2,3-dihydropyranyl structure ($[C_5H_8O]^+$), which might be formed by a cyclization and hydrogen rearrangement after cleavage of the bond between the 5- and 6-positions [164]. On the basis of this notable base peak, a 4,6–16 structure was recently assigned for a pheromone component of *Stathmopoda masinissa* [27]. A homologous ion at *m/z* 98 ($[C_6H_{10}O]^+$) is diagnostic for 5,7-dienal.

3.2.2 Identification of Type I Components by Derivatization

When a sufficient amount of sample is available (ca. 1 μg), monoenyl compounds can be analyzed by micro-ozonolysis with and without a solvent [146, 165]. Ozonides, directly injected into GC-MS, are reductively decomposed into two aldehydes by heat. Besides this chemical reaction, the double-bond position is easily and high-sensitively confirmed by making an adduct with DMDS, which gives rise to M^+ and two characteristic fragment ions produced by cleavage of the bond between the sulfur-substituted carbons on the EI-MS analysis [165, 166], as illustrated with the mass spectrum of DMDS adduct of Z7–12:OAc in Fig. 8A. The fragment ion sat *m/z* 117 ($[C_5H_{10}SMe]^+$), 143 (M-AcOH-117), and 203 (M-117) reveal the original double bond at the 7-position. Figure 9 shows a TIC and mass chromatograms monitoring M^+ and the above fragment ions to analyze the DMDS adducts of monoene acetates in the gland extract of *Anadevidia peponis* [141]. This measurement revealed the existence of DMDS adducts derived from Z5–10:OAc (**I** with ions at *m/z* 117, 175, and 292 [M^+]), Z5–12:OAc(**II** with ions at *m/z* 145, 175, and 320 [M^+]), Z7–12:OAc (**III** with ions at *m/z* 117, 203, and 320 [M^+]), 11–12:OAc (**IV** with ions at *m/z* 259 and 320 [M^+]), Z9–14:OAc (**V** with ions at *m/z* 117, 231, and 348 [M^+]), and Z11–16:OAc (**VI** with ions at *m/z* 117, 259, and 376 [M^+]). Peak areas of **I**–**VI** in the TIC indicate a mixingratio of 5:1:10:0.5:0.4:0.1, which is almost the same as the analytical data of an extract untreated with DMDS.

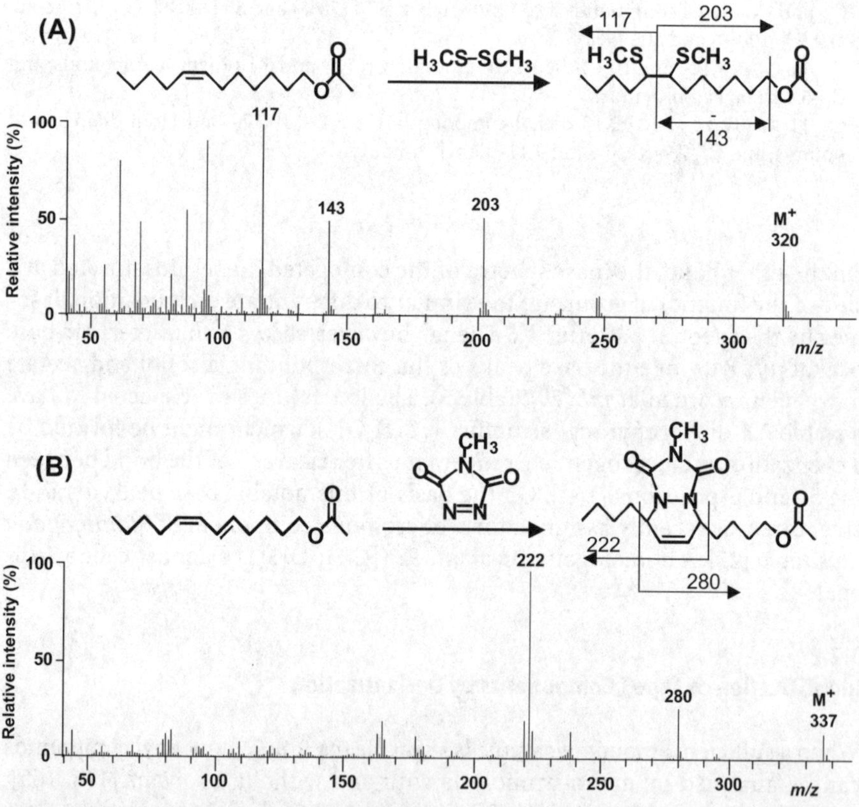

Fig. 8A,B Derivatization of Type I pheromones: **A** Z7–12:OAc with dimethyl disulfide (DMDS) and mass spectra of the DMDS adduct; **B** E5,Z7–12:OAc with 4-methyl-1,2,4-triazoline-3,5-dione (MTAD) and mass spectra of the MTAD adduct

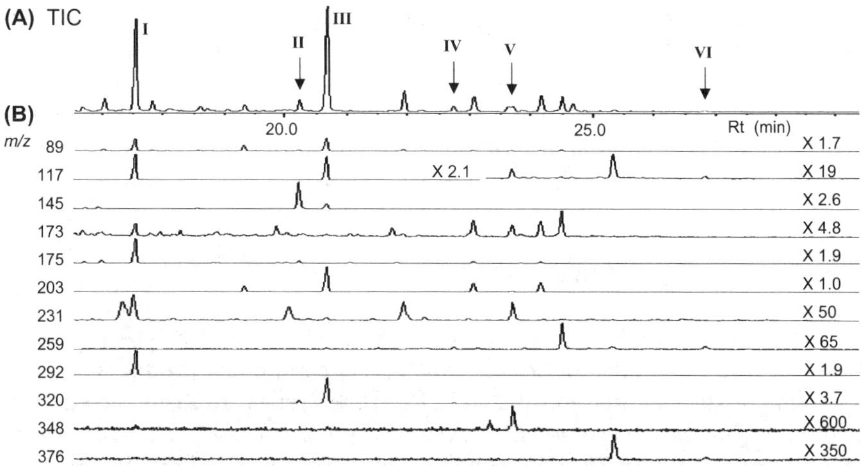

Fig. 9A,B GC-MS analysis of the pheromone extract of *Anadevidia peponis* (Noctuidae, 1 FE) treated with DMDS; **A** TIC; **B** mass chromatograms [141]. The mass chromatograms, which are multiplied by indicated factors, monitor the M^+ of DMDS adducts derived from C_{10} to C_{16} monoenyl acetates (*m/z* 292, 320, 348, and 376) and some diagnostic fragment ions (*m/z* 89, 117, 145, 173, 175, 203, 231, and 259) to determine their double-bond position. Peaks **I–VI** indicate the DMDS adducts of the following components in the pheromone gland; Z5-10:OAc (**I**), Z5-12:OAc (**II**), Z7-12:OAc (**III**), 11-12:OAc (**IV**), Z9-14:OAc (**V**), and Z11-16:OAc (**VI**)

In the case of dienyl compounds, derivatization with DMDS leads to either linear or cyclic polythioethers, depending on the number of methylene groups separating the two double bonds [165, 167]. When the two double bonds are not conjugated, some fragment ions reflecting the original positions of the double bonds are observed in the mass spectra of the DMDS adducts. The double-bond positions of new natural pheromones, 3,7-diene of *Symmetrischema tangolias* [19] and 11,14-diene of *Andraca bipunctata* [28], have been identified by GC-MS analysis of their adducts. The 3,8,11-trienyl structure of *Scrobipalpuloides absoluta* has also been successfully determined by the DMDS derivatization [24]. However, the intensities of the diagnostic ions are sometimes lower and the spectra are more complicated than those of monoenyl compounds. This derivatization proceeds without stereoselectivity and produces multiple diastereomeric adducts which cause less sensitivity. Diimide reduction of a polyene proceeds without selectivity for the position and configuration of double bonds and yields a mixture of monoenyl compounds under a restricted condition. All unsaturated positions of polyenyl pheromones, 3,8,11-triene of *S. absoluta* [23] and 11,14,17-triene of *Euproctis pulverea* [54], have been determined by GC-MS analysis of the partially reduced products after derivatization with DMDS or ozonolysis.

For the components with a conjugated diene system, derivatization with MTAD is useful to determine the unsaturated positions [168]. Adducts of the dienophile MTAD, which is formed with conjugated dienes via Diels-Alder cycloaddition, give mass spectra including abundant fragmentions that are diagnostic for the unsaturated positions in the parent compounds, as illustrated with a mass spectrum of the MTAD adduct of E5,Z7–12:OAc in Fig. 8B. Fragment ions at *m/z* 222 and 280 reveal the original 5,7-diene system. By GC-MS analysis of the MTAD adduct of a pheromone component secreted by *Hemileuca eglanterina* (Bombycoidea: Saturniidae), the 10,12-dienyl structure was revealed [169].

3.2.3 Identification of Type II Components

GC-MS is a useful tool to elucidate chemical structures for the Type II pheromones as well [9]. On the EI measurement, every component produces M^+ and some diagnostic fragmentions listed in Fig. 10. Polyenes are easily recognized by their high abundance of ions, for example, at *m/z* 81 (base peak, $[C_2H_5(CH{=}CH)_2]^+$) and 110 for Z6,Z9-dienes and *m/z* 79 (basepeak, $[H(CH{=}CH)_3]^+$), 108, and M-56 for Z3,Z6,Z9-trienes. These two monoepoxides derived from Z6,Z9-dienes show similar mass spectra but are distinguishable by comparing the intensities of several ions, which are produced by fragmentations around the epoxy ring [170]. The three monoepoxides derived from Z3,Z6,Z9-trienes show rather different spectra because a homoconjugated diene system remained in the 3,4-epoxy-6,9-dienes and 9,10-epoxy-3,6-dienes. These two positional isomers show a base peak at *m/z* 79 and other characteristic fragment ions (*m/z*

A

B

C

D

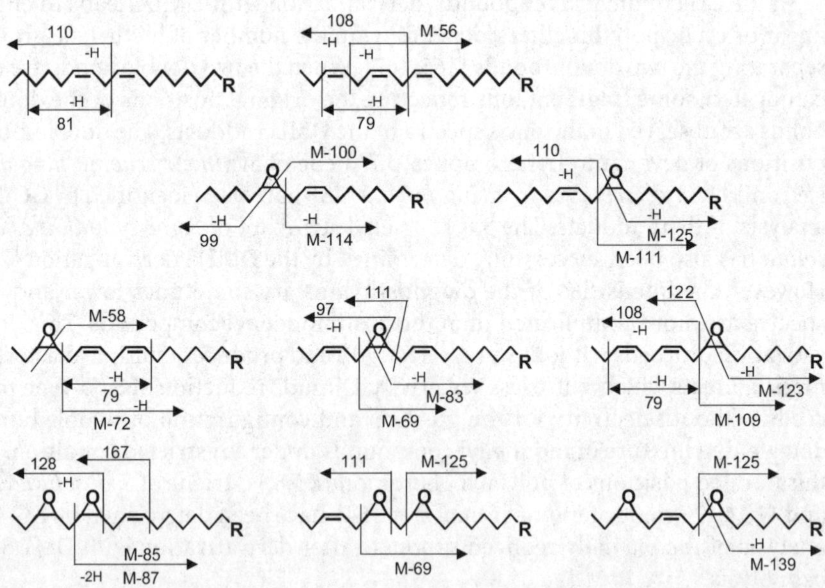

Fig. 10A–D Diagnostic fragment ions for the GC-MS analysis of Type II pheromones: **A** Z6,Z9-dienes and Z3,Z6,Z9-trienes; **B** monoepoxy derivatives of Z3,Z6-dienes; **C** monoepoxy derivatives of Z3,Z6,Z9-trienes; **D** diepoxy derivatives of Z3,Z6,Z9-trienes

M-72 for the 3,4-epoxides and *m/z* 108 for the 9,10-epoxides), which include the diene part. 6,7-Epoxy-3,9-dienes show a basepeak at *m/z* 67 and different diagnostic ions produced by fragmentation around the epoxy ring [171]. Each diepoxide derived from Z3,Z6,Z9-trienes loses the homoconjugated diene system; as a result, characteristic ions with high abundance have not been detected for each positional isomer. However, some diagnostic ions whose formation is easily interpreted by the fragmentation, such as cleavage α to the epoxy ring, are useful for the structure determination [69].

The origins of the above diagnostic ions, which have been estimated by comparing the mass spectra of a series of synthetic compounds with a different chain length [170, 171], contribute to the analysis of natural pheromones. Moreover, the validity of this estimation is confirmed by comparing the spectra of differently unsaturated homologues. Figure 11 shows the mass spectra of four 9,10-epoxides with a different number of double bonds. Epoxides with a saturated chain produce two highly abundant ions after cleavage α to the epoxy ring, for example, fragment ions at *m/z* 155 and 197 of epo9–21:H. The same fragmentations are expected for epoxides with an unsaturated chain to produce homologous ions, such as at *m/z* 153 of Z6,epo9–21:H, 151 of Z3,Z6,epo9–21:H, and 149 of 1,Z3,Z6,epo9–21:H, in addition to their common ion at *m/z* 197. These ions have actually been detected, but their intensities decrease according to the increment of the unsaturation degree. On the contrary, the cleavage of the epoxy ring is probably accelerated by homoconjugation with a double bond, and characteristic ions, such as *m/z* 124 and 183 of Z6,epo9–21:H, are remarkably detected. The intensities of the corresponding ions at *m/z* 126 and 183 of epo9–21:H are very small, indicating the nonexistence of a double bond. Double bonds in Z3,Z6,epo9–21:H and 1,Z3,Z6,epo9–21:H cause the ions produced by the epoxy ring cleavage (*m/z* 122 and 183 of Z3,Z6,epo9–21:H and *m/z* 120 and 183 of 1,Z3,Z6,epo9–21:H) to be distinct, but their relative intensity is still small because of the strong base peaks (*m/z* 79 and 106). The base peak at *m/z* 106($[H(CH{=}CH)_4H]^+$) of1,Z3,Z6,epo9–21:H is a counter part of the*m/z* 197 ion, which is produced by the cleavage α to the epoxy ring. The homologous ions at *m/z* 108 of Z3,Z6,epo9–21:H and 110 of Z6,epo9–21:H are diagnostic, but the intensity of the *m/z* 112 ion of epo9–21:H is worthless for determining the structure.

On a capillary GC analysis, the separation of positional isomers of epoxy compounds is generally well accomplished by a high polar column, such as DB-23, rather than by a low polar column, such as DB-1. For the positional isomers, a different elution order depending on the kinds of column has not been reported. In the case of two mono epoxides derived from Z6,Z9-dienes, 6,7-epoxides elute slightly faster than 9,10-epoxides [72, 170], but the separation is insufficient even on the high polar column. Three monoepoxides derived from Z3,Z6,Z9-trienes elute in the order of 6,7-, 3,4-, and 9,10-epoxides [9]. The former two isomers are sufficiently separated on the high polar column, while the elution of the latter two isomers overlaps [71]. For each positional isomer of diepoxides derived from the Z3,Z6,Z9-trienes, two diastereomeric

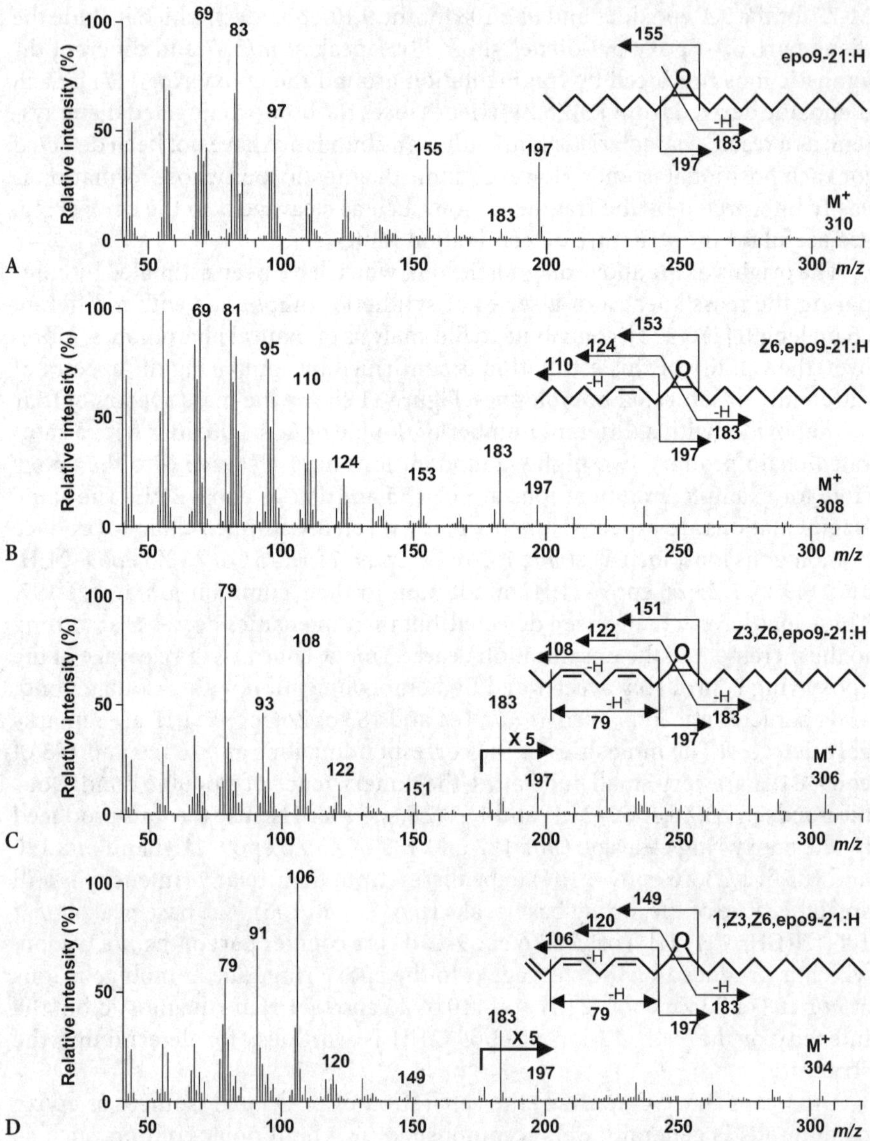

Fig. 11A–D Mass spectra of C_{21} 9,10-epoxy compounds unsaturated with a different degree (0–3): A epo9–21:H; B Z6,epo9–21:H; C Z3,Z6,epo9–21:H; D 1,Z3,Z6,epo9–21:H

configurations are possible and the diastereomers of each positional isomer are separable on a GC column. The isomers with a configuration orienting two epoxy rings to the opposite face [(S^*,R^*,R^*,S^*)-isomers, i.e., 3,4–6,7-diepoxides with a 3*S*,4*R*,6*R*,7*S* configuration, 6,7–9,10-diepoxides with a 6*S*,7*R*,9*R*,10*S* configuration, 3,4–9,10-diepoxides with a 3*S*,4*R*,9*R*,10*S* configuration, and their antipodes] elute faster than the isomers with a configuration orienting two epoxy rings to the same face [(S^*,R^*,S^*,R^*)-isomers]. The high polar column demonstrates higher ability for the separation of the diastereomers than the low polar column. Nevertheless, 3,4–6,7-diepoxides and 6,7–9,10-diepoxides with the S^*,R^*,R^*,S^* configuration showed almost the same Rts, and it was difficult to separate all six isomers of the diepoxide on GC [69]. In all cases of Type II epoxy pheromones, GC analysis was insufficient to determine which isomer is present in a pheromone gland. Mass spectral data, particularly information on the diagnostic ions, are necessary for determining the structure.

To accomplish unambiguously the structure determination of epoxymonoenes, DMDS derivatization has been applied [72]. The DMDS adducts of 6,7-epoxides show M^+ and fragmentions at *m/z* 187 ($[C_9H_{16}OSMe]^+$) and M-187 indicating the original double bond at the 9-position and *m/z* 99 ($[C_6H_{11}O]^+$) indicating the 6,7-epoxy ring. On the other hand, the adducts of 9,10-epoxides show M^+ and fragment ions at *m/z* 131 ($[C_6H_{12}SMe]^+$) and M-131 indicating the original double bond at the 6-position and *m/z* M-219 ($[M\text{-}C_9H_{17}(SMe)_2]^+$) indicating the 9,10-epoxy ring. The structure of the main component of the *Biston robustum* pheromone (epo6,Z9–19:H) was confirmed by this derivatization [72]. In addition to this direct derivatization, a double-bond position of the epoxymonoenes can be determined by GC-MS analysis of DMDS adducts of the hydroxyl derivatives produced by their $LiAlH_4$ reduction [70]. Whereas a reasonable DMDS derivatization of epoxydienes was not accomplished, epoxydienes can be partially reduced with diimide, and the produced epoxymonoenes have been used for this derivatization. The structure of the main component of the *Perina nuda* pheromone (Z3,epo6,Z9–21:H) was confirmed by the DMDS experiment after successive reduction with diimide and $LiAlH_4$ [70]. The three diepoxymonoenes can also be converted into the corresponding DMDS adducts, which show characteristic ions from the fragmentation between the two thiomethyl groups reflecting the position of an original double bond [69].

3.2.4
Identification of Methyl-Branched Components

By cleavage at the branched position, monomethylalkanes (R_1–$CHCH_3$–R_2) give characteristic secondary fragment ions, $[R_1\text{–}CHCH_3]^+$ (or $[R_1\text{–}CHCH_2]^+$) and $[R_2\text{–}CHCH_3]^+$(or $[R_2\text{–}CHCH_2]^+$). Although the relative intensity of their M^+ and these diagnostic ions are rather small under the usual EI measurement, they are valuable for the structure determination. Dimethylalkanes ($R_1\text{–}CHCH_3\text{–}(CH_2)_n\text{–}CHCH_3\text{–}R_2$) are also expected to predominantly produce similar secondary ions. The mass spectra of some pheromone components with two methyl groups identified from Lyonetiidae and Geometridae species [105–107] indicate selective cleavages at the branched positions to produce the fragment ions shown in Table 6. In addition to the secondary ions such as $[R_1\text{–}CHCH_3]^+$ at *m/z* A (or $[R_1\text{–}CHCH_2]^+$ at *m/z* A-1) and $[R_1\text{–}CHCH_3\text{–}(CH_2)_n\text{–}CHCH_3]^+$ at *m/z* M-R_2 (or $[R_1\text{–}CHCH_3\text{–}(CH_2)_n\text{–}CHCH_2]^+$ at *m/z* M-R_2-1), primary ions composed of their counter parts, such as $[(CH_2)_n\text{–}CHCH_3\text{–}R_2]^+$ at *m/z* M-A and $[R_2]^+$, are

Table 6 Diagnostic fragment ions for some dimethyl-branched pheromone components

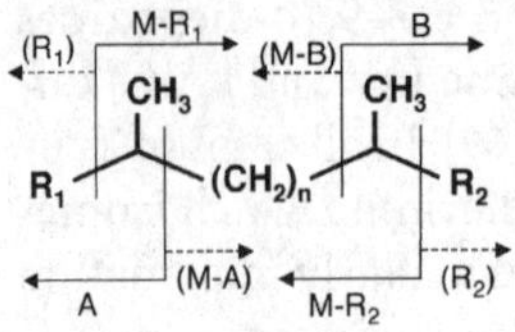

				Fragment ion (*m/z*)			
	R_1	R_2	n	M-R_1	A	M-R_2	B
Me5,Me9-17:H	C_4H_9	C_8H_{17}	3	211	85	155	141
Me5,Me11-17:H	C_4H_9	C_6H_{13}	5	211	85	183	113
Me7,Me11-17:H	C_6H_{13}	C_6H_{13}	5	183	113	183	113

M = 268, A = R_1-$CHCH_3$, B = R_2-$CHCH_3$.

also informative. Since many branched hydrocarbons have been identified from insect cuticle [172] and silk lipids of a spiderweb [173], several dimethylalkanes with a different number of methylene units between methyl groups (n=0–7) have been synthesized, and analysis of their mass spectra has indicated the effect of the methylene units on the fragmentation, particularly the ratio between $[R_1{-}CHCH_3]^+$ and $[R_1{-}CHCH_2]^+$ [174]. EI-MS analysis of methyl-branched compounds synthesized more systematically will more obviously generalize this fragment pattern, which will be helpful for the structure determination of a new pheromone component in this class.

Elution patterns from capillary GC for many methyl-branched alkenes have been reported [172, 173], and their KI values have been predicted from the chemical structures [175, 176]. The elution order of monomethylalkanes can be estimated by the KI value reported for several positional isomers with a C_{31} backbone; namely, the highest KI value of Me3–31:H decreases when the branched methyl moves to the middle of the chain. Moreover, the KI values of polymethylalkanes are interestingly calculated by a semi-empirical topological index (I_{ET}) on the basis of the chain length, the number of methyl groups, and their positions [176]. Namely, $I_{ET}=\Sigma(C_i+\delta_i)$, where C_i is the value attributed to each carbonatom (i) in the molecule and δ_i is the sum of the logarithms of the values for each adjacent carbon atom ($\log C_1+\log C_2+\log C_3+\log C_4$; the values of C_i for primary, secondary, tertiary, and quaternary carbon atoms are 1.0, 0.9, 0.8,and 0.7, respectively). For the alkane with two methyl groups at the n_1-position and n_2-position, the optimum I_{ET}($I_{ET(opt)}$) and KI values are calculated as follows:

$$I_{ET(opt)} = I_{ET} - 1/3\,(n_1+1)+(n_2+8)$$

$$KI = -39.5251 + 123.1610\,I_{ET(opt)}$$

Using these equations, the KI values for four known pheromone components are estimated to be 1807.7 for Me2,Me5–17:H, 1790.6 for Me5,Me9–17:H, 1788.6 for Me5,Me11–17:H, and 1783.5 for Me7, Me11–17:H. These positional isomers separately elute from a capillary GC column in a coincidental order with these KI values [105, 106], indicating the usefulness of $I_{ET(opt)}$ for the pheromone research.

3.3
LC and LC-MS Analysis

3.3.1
Separation with a chiral Columns

While lepidopteran sex pheromones are volatile and GC is commonly used for their analysis and separation, HPLC is suitable for the quantitative analysis of pheromone components with a conjugated diene system, such as bombykol. The UV detector can selectively record them with high sensitivity. Furthermore, HPLC showed its effectiveness as a purification tool on a biosynthetic experiment with a radiolabeled compound [177]. For the measurement of radioactivity incorporated into the pheromone by liquid scintillation counting, a purification method ensuring perfect recovery is necessary. Both normal-phase and reversed-phase conditions are available for the HPLC of usual lipophilic pheromone components with a C_{10}–C_{23} chain. Particularly, separation depending on the chain length and the number of double bonds (sometimes, the configuration of the double bonds [164]) can be achieved by the latter chromatography. By end absorption around UV 210 nm, ca. 1 µg of monoenyl acetates and aldehydes in the Type I class and homoconjugated dienyl compounds in the Type II class can be detected. An RID records every type of compound, also at the level of 1 µg. Figure 12 shows the chromatograms of Type II pheromone components produced by an Arctiidae species, *Spilosoma imparilis* [178], indicating that all of the five components are easily separated by an ODS column. After this separation, the epoxy components were used for a subsequent analysis with a chiral column to determine their absolute configuration.

LC is also useful for the preparation of Type II pheromones. Monoepoxy compounds in this class have been systematically synthesized by the oxidation

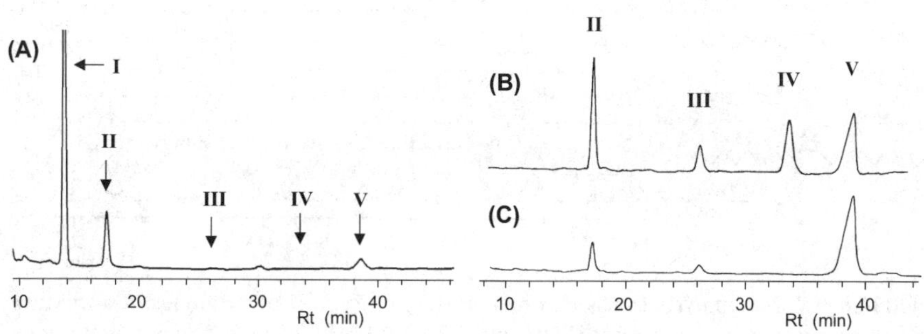

Fig. 12A–C Separation of Type II pheromone components by HPLC with an ODS column (4.6 mm ID X 25 cm): **A** a crude pheromone extract of *Spilosoma imparilis* (Arctiidae, 2 FE) including 1,Z3,Z6,epo9-21:H (**I**), Z3,Z6,epo9-21:H (**II**), Z3,Z6,epo9-23:H (**III**), Z6,epo9-23:H (**IV**), and Z3,Z6,Z9-21:H (**V**) detected by UV 215 nm; **B** a mixture of synthetic standards (ca. 1 µg each of **II–V**) detected by RID; **C** the same synthetic mixture detected by UV 215 nm. The solvent system is 3.5% water in MeOH (1.0 ml/min)

of (Z3,Z6,Z9)-trienes and (Z6,Z9)-dienes with 3-chloroperoxybenzoic acid. This epoxidation proceeds without regio-selectivity and mixtures of three epoxydienes and two epoxymonoenes are produced from the trienes and dienes, respectively. When the epoxydiene mixture is injected into a Lobar column of MPLC, the positional isomers separately elute in the order of 6,7-, 9,10-, and 3,4-epoxydienes [179]. This MPLC separation is better than that of HPLC with an ODS column, which performs the elution in a different order of 9,10-, 3,4-, and 6,7-epoxydienes (see Fig. 13). In contrast, two positional isomers of the epoxymonoenes are barely separated by repeated MPLC [170], but the reversed-phase HPLC accomplishes their separation more sufficiently. The 9,10-epoxymonoene elutes faster than the 6,7-epoxymonoene on the reversed-phase HPLC in the opposite order to the normal-phase MPLC. LC is also useful for the preparation of diepoxy pheromones derived from the (Z3,Z6,Z9)-trienes [69].

Currently, LC-MS is widely used for the analysis of polar compounds, such as medicinal metabolites and bioactive peptides, since the interface has been improved and several new ionization methods have been developed. The sensitivity and reproducibility are sufficient for a daily quantitative analysis. The usefulness of the LC-MS has been demonstrated for studies on Type II pheromones using a time-of-flight MS with electrospray ionization (ESI) [180]. Each epoxydiene derived from the (Z3,Z6,Z9)-triene shows three ion series of $[M+NH_4]^+$, $[M+H]^+$, and $[M\text{-}OH]^+$ with high resolution and good sensitivity, indicating its molecular formula. In addition to these, characteristic fragment

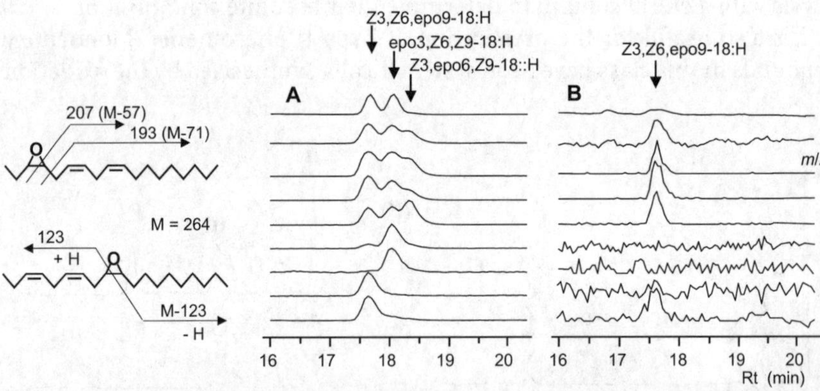

Fig. 13A,B LC-MS analysis (mass chromatograms) of Type II pheromones: **A** a mixture including monoepoxy derivatives (**I–III**) of Z3,Z6,Z9–18:H (500 ng each); **B** a pheromone extract of the *Hemerophila artilineata* (10 FE) [180]. Their $[M+NH_4]^+$ at *m/z* 282, $[M+H]^+$ at *m/z* 265, and $[M\text{-}OH]^+$ at *m/z* 247 were monitored, as were the diagnostic fragment ions at *m/z* 207 (M-57) and 193 (M-71) for epo3,Z6,Z9–18:H (**I**) and 141 (M-123) and 123 for Z3,Z6, epo9–18:H (**III**). No diagnostic ions for Z3,epo6,Z9–18:H (**II**) were detected. The LC/ESI-TOF-MS analysis was conducted using an ODS-2 column (2.1 mm ID × 150 cm) eluted with a 10 min linear gradient of 20–5% water in MeOH (0.2 ml/min)

ions at *m/z* M-57 and M-71 for the 3,4-epoxides and at *m/z* M-123 and 123 for the 9,10-epoxides have been detected, while the 6,7-epoxides do not produce fragment ions that reflect their structures. Monitoring these diagnostic ions on the LC-MS analysis of a gland extract, the structure of a main pheromone component of *Hemerophila artilineata* was confirmed, as shown in Fig. 13 [180].

3.3.2 Enantiomeric Separation with Chiral Columns

A chiral GC column is able to separate enantiomers of epoxy pheromones in the Type II class, but the applications are very limited as follows; a custom-made column packed with a β-cyclodextrin derivative as a liquid phase for the stereochemical identification of natural 3,4- and 6,7-epoxydienes [73, 74] and a commercialized column of an α-cyclodextrin type (Chiraldex A-PH) for the 3,4-epoxydiene [71] (See Table 3). The resolution abilities of chiral HPLC columns have been examined in detail, as shown in Table 7 and Fig. 14 [75, 76, 179]. The Chiralpak AD column operated under a normal-phase condition separates well two enantiomers of 9,10-epoxydienes, 6,7-epoxymonoenes and 9,10-epoxymonoenes. Another normal-phase column, the Chiralpak AS column, is suitable for the resolution of the 3,4-epoxydienes. The Chiralcel OJ-R column operated under a reversed-phase condition sufficiently accomplishes enantiomeric separation of the 6,7-epoxydienes and 6,7-epoxymonoenes.

The stereochemistry of each enantiomer separated by the chiral HPLC has been studied after methanolysis of the epoxy ring. Examining the ^{1}H NMR data of esters of the produced methoxyalcohols with *(S)*- and *(R)*-α-methoxy-α-(trifluoromethyl) phenylacetic acid by a modified Mosher's method [181], it has been indicated that the earlier eluting parent epoxides are (3*S*,4*R*)-, (6*S*,7*R*)-, and (9*R*,10*S*)-isomers (Table 7) [75, 76, 179]. The above three chiral HPLC columns show different resolution abilities but a different elution order is not observed. The resolution profile by the reversed-phase OJ-R column has been generalized with molecular shapes of the epoxy compounds considering the

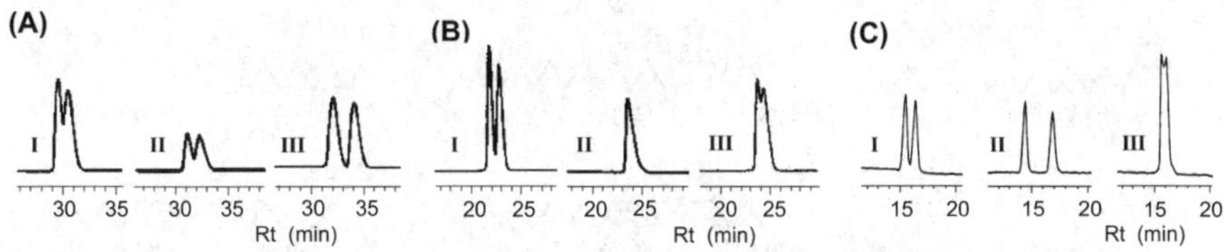

Fig. 14A–C Chromatography of the racemic monoepoxy derivatives (I–III) of Z3,Z6,Z9–18: on chiral HPLC columns: **A** Chiralpak AD; **B** Chiralpak AS; **C** Chiralcel OJ-R. The solvent system for the former two normal-phase columns is 0.1% 2-propanol in *n*-hexane (0.45 ml/min), and that of the third column is 15% water in MeOH (0.45 ml/min). Homoconjugated dienes, epo3,Z6,Z9–18:H (I) and Z3,Z6,epo9–18:H (III), were detected by UV (215 nm), and Z3,epo6,Z9–18:H (II) was detected by RID. The earlier eluting isomers have a 3*S*,4*R*, 6*S*,7*R*, or 9*R*,10*S* configuration

Table 7 Enantiomeric separation of monoepoxides derived from Z3,Z6,Z9-trienes and Z6,Z9-dienes with aC_{17}–C_{23} straight chain on chiral HPLC columns [75, 76, 179]

Compound	Separation factor (α)			Configuration of the earlier eluting isomer
	Normal-phase[a]		Reversed-phase[b]	
	Chiralpak AD[c]	Chiralpak AS[c]	Chiralcel OJ-R[d]	
Epoxydiene				
3,4-Epoxide	1.03–1.08	1.08–1.11	1.07–1.10	(+)-3*S*,4*R*
6,7-Epoxide	1.08–1.09	1.00	1.17–1.26	(+)-6*S*,7*R*
9,10-Epoxide	1.11–1.12	1.00–1.06	1.03–1.08[e]	(+)-9*R*,10*S*
Epoxymonoene				
6,7-Epoxide	1.17–1.21	1.00	1.10–1.28	(–)-6*S*,7*R*
9,10-Epoxide	1.14–1.18	1.00	1.00–1.02	(–)-9*R*,10*S*

[a] Solvent: 0.1% 2-propanol in *n*-hexane.
[b] Solvent: 15% water in MeOH for C_{17}–C_{19} compounds, and 10% water in MeOH for C_{20}–C_{23} compounds.
[c] Column: 0.46 cm ID×50 cm.
[d] Column: 0.46 cmID×15 cm.
[e] Values for C_{17}–C_{20} compounds. Enantiomers of C_{21}–C_{23} compounds are not separable.

lengths of two chains connected to the epoxy ring [75]. Figure 15 shows the stereoisomers of the C_{17}–C_{23} epoxymonoenes and epoxydienes eluted faster than their antipodes, in addition to those of *cis*-epoxides with a saturated C_{20} chain (epoxyicosanes). The shorter-Rt isomers of 3,4-, 6,7-, and 9,10-epoxyicosanes with an *S*,*R* configuration have small alkyl groups on the left side and large alkyl groups on the right side in this figure, where the epoxy oxygen atom is set upward and behind the plane of this paper. It has been speculated that the enantiomer of a di-substituted epoxyalkene with this configuration regularly

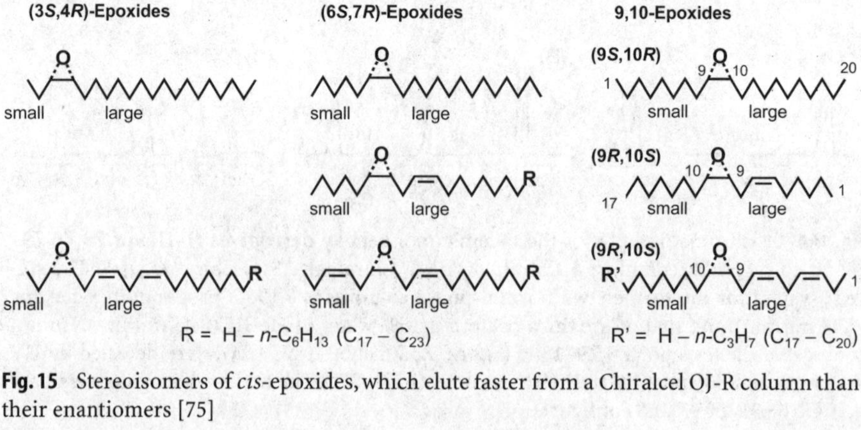

Fig. 15 Stereoisomers of *cis*-epoxides, which elute faster from a Chiralcel OJ-R column than their enantiomers [75]

elutes faster from the OJ-R column. The (*S*,*R*)-isomers of 3,4- and 6,7-epoxydienes with a C_{17} to C_{23} chain, which also possess this configuration, invariably have shorter Rts. In the case of 9,10-epoxydienes, double bonds have an influence on the elution order, and the (*R*,*S*)-isomers with a C_{18} to C_{20} chain elute faster than the antipodes, as in the case of the C_{17} chain compounds. Two homoconjugated double bonds prevent folding of the unsaturated C_8 chain, which might be larger than a saturated C_8 to C_{10} chain.

Utilizing the Chiralpak AD column, the absolute configurations of the *cis*-epoxy rings in the natural epoxydienes [71, 75] and epoxymonoenes [72, 76] secreted by Geometridae and Noctuidae species were determined, as shown in Table 3. Additionally, the stereochemistry of a *cis*-epoxy ring in disparlure (Me2,epo7–18:H of *Lymantria dispar*) [112], two *cis*-epoxy rings in a diepoxy pheromone component (epo3,epo6,Z9–21:H of *Perina nuda*) [70], and a *trans*-epoxy ring in posticlure (Z6,Z9,*t*-epo11–21:H of *Orgyia postica*) [67] were defined by the same column. Besides analysis with a UV detector or RID, quantitative GC analyses of the chiral HPLC eluants have confirmed some assignments of the stereochemistry [67, 70, 72, 76].

3.4 NMR Analysis

^{1}H NMR measurements of natural pheromones produced by lepidopteran insects have been rarely achieved because of their quantitative restriction and inapplicability to samples with impurities, which are rather easily overcome in GC-MS measurements. Configuration of the double bond and epoxy ring in Types I and II pheromones, however, cannot be determined by the mass spectra. The coupling constant between two olefinic protons with an *E* configuration (~15 Hz) is bigger than the coupling constant in those with a *Z* configuration (~10 Hz), and the constant between two protons on a *cis*-epoxy ring (~5 Hz) is bigger than that on a *trans*-epoxy ring (~2.5 Hz). The *Z*-double bond of the *Thaumetopoea pityocampa* pheromone (≡11,Z13–16:OAc) [45] and the *trans*-epoxy ring of the *Orgyia postica* pheromone (Z6,Z9,*t*-epo11–21:H, posticlure) [66, 67] were determined by FT-NMR analysis of the natural products. The pheromone titers of these species are rather high, and about 10 μg of the components was used for the NMR analysis. Because the quantity is rather few, the ^{1}H NMR spectra of the *T. pityocampa* pheromone at 250 MHz was recorded with a 60-μl microcell in the early 1980s. The current measurement of posticlure at 600 MHz has been accomplished with a usual 5-mm OD tube using a field gradient system and the ^{1}H signal correlations in the COSY spectrum, which clarify the peaks due to contaminated impurities, have revealed the novel structure.

Signals of two olefinic protons on a C=C bond, which are isolated from the other double bond and the functional group in a long chain, appear at almost the same chemical shift, and their coupling constant is not clear because of f ailing the first-order approximation. Instead of that, the chemical shift values of olefinic and allylic protons indicate the double-bond configuration; i.e., ^{1}H at ~5.35 and ~2.0 ppm for the *Z*-double bond, and ^{1}H at ~5.4 and ~1.95 ppm for the *E*-double bond. Additionally, allylic carbon signals differentiate the configuration; i.e., ^{13}C at ~27 ppm for the *Z*-double bond and ^{13}C at ~32.5 ppm for the *E*-double bond. Although ^{13}C NMR is not available for the identification of natural pheromones produced by moths, this spectrometer is useful for the structure confirmation of synthetic pheromones. Some carbons of geometrical and positional isomers, which are located in a slightly different environment, afford sufficient separation of their signals. The ^{13}C signal assignments of dienyl and trienyl Type I pheromones [182, 183] and typical Type II compounds [170, 171] were completed by 2D-NMR techniques.

Accumulated ^{13}C NMR data support the proposal that the chemical shift for an olefinic carbon of monoenyl compounds can be calculated with ethylene as a reference [184]. A similar method was applied to the polyenyl Type I components with a conjugated system to estimate ^{13}C chemical shift parameters. Using 1,3-butadien for the base values (C^1 117.6 and C^2 137.8 ppm), the chemical shift values of the outer carbon and the inner carbon of 1,4-dialkyl conjugated dienes with the *E,E* configuration were separately calculated using different substituent parameters and the correction factors for other geometrical isomers with *Z,E*, *E,Z*, and *Z,Z* configurations [182]. Based on the data of (*E*)-1,3,5-hexatriene (C^1 117.7, C^2 136.9, and C^3 133.7 ppm), the substituent parameters of three carbons in 1,6-dialkyl conjugated trienes were estimated [183]. ^{13}C NMR measurements of the conjugated polyenyl compounds can be expected to give some clues as to their unsaturated positions, geometry, and purity.

4 Applications of Synthetic Pheromones for Pest Control

Female pheromones strongly attract male moths in a field; as a result, their use as a monitoring tool has been developed in integrated pest management (IPM) programs. The synthetic pheromone (0.1–1 mg) is usually immersed in a small rubber septum (8 mm OD), and a trap baited with this lure can effectively catch males for at least one month. Mass trapping of insects is one concept of the direct application for the control of pest species. Besides the potential for protecting stored products, there has been no success with attempts to diminish outdoor populations of lepidopteran insects, probably because of the multiple-coupling ability of surviving males. On the other hand, disruption of the chemical communication between female and male moths has been achieved by permeating a field with a synthetic pheromone [185]. After the registrations of 'Nomate PBW,' consisting of Z7,Z11–16:OAc and Z7,E11–16:OAc, for the control of the pink bollworm moth (*Pectinophora gossypiella,* Gelechiidae) in the USA (1976) and 'Hamaki-con,' containing Z11–14:OAc, for the simultaneous control of several leafroller species in Tortricidae in Japan (1983), other disruptants have been developed for more than 20 lepidopteran species. The use of a reservoir-type formulation, such as 'Hamaki-con,' is supplanting the use of sprays, such as 'Nomate PBW.' Mostly, the reservoir-type formulation makes use of a polyethylene tube (20 cm long), which encloses ca. 100 mg of the synthetic pheromone and releases it gradually over two to three months. The polyethylene dispensers are settled in a field at a density of 50–500 tubes ha^{-1}.

Different from the use of ordinal insecticides, this disruption method has high target selectivity and, as would be desired, ensures the survival of natural enemies. The sex pheromone, which shows no toxicity to mammals, is an ideal insect-behavior regulator (IBR). Table 8 shows the application areas of main mating disruptants for lepidopteran insects. In addition to the use of the synthetic pheromone of *P. gossypiella* in large cotton fields, many disruptants are

Table 8 Utilization of representative mating disruptans in the world[a]

Crop	Pest insect species (common name)	Disruptant (mixing ratio)	Country	Applied field (ha)	
				1977	2002
Cotton	*Pectinophora gossypiella* (pink bollworm moth)	Z7,Z11–16:OAc + Z7,E11–16:OAc (1:1) (gossyplure)	USA	30,000	40,000
			Egypt	328,000	–
			Israel	8,000	5,000
			Brazil	–	5,000
Apple and pear	*Cydia pomonella* (codling moth)	E8,E10–12:OH (codlelure)	USA	13,200	63,000
			Italy	6,800	14,000
			South Africa	5,200	13,000
Peach and nectarine	*Grapholita molesta* (Oriental fruit moth)	Z8–12:OAc + others	USA	7,200	24,000
			Australia	1,200	3,000
			Italy	–	11,000
Grape[b]	*Lobesia botrana* (European grapevine moth)	E7,Z9–12:OAc	EU	25,000	73,000
Japanese plum	*Synanthedon hector* (cherry treeborer)	Z3,Z13–18:OAc + E3,Z13–18:OAc (1:1)	Japan	4,000	3,800
Tomato	*Keiferia lycopersicella* (tomato pinworm moth)	E4–13:OAc	Mexico	3,200	10,000
Tea	*Adoxophyes honmai* (smaller tea tortix) *Homona magnanima* (Oriental tea tortrix)	Z11–14:OAc	Japan	400	500
Forest	*Lymantria dispar* (gypsy moth)	Me2,epo7–18:H (disparlure)	USA	10,000	150,000

[a] Information supplied from Shin-Etsu Chemical Co.Ltd. in Japan.
[b] Including European grape moth (*Eupoecilla ambiguella*, Z9–12:OAc + 12:OAc).

being used on about 500,000 ha worldwide in order to prevent damage from pests to fruits, vegetables, rice, and tea. Most of them are chemicals of the Type I pheromones. As an exception, the epoxy pheromone of *L. dispar* (disparlure) has been widely applied in forests in the USA. While disruptants, including Type II pheromones, have not been commercially available, an interesting disruption by an epoxyalkenyl pheromone was reported for *A. s. cretacea* [186]. Mating of the tethered females is completely inhibited in a field permeated with a mixture of the epoxydienes derived from Z3,Z6,Z9–19:H, although the female moths produce only one epoxydiene (epo3,Z6,Z9–19:H). Preparation of the mixture via the unselective epoxidation of the triene is much easier than the selective synthesis of one positional isomer, indicating a possible application of the Type II chemicals for actual pest control. Generally, a synthetic material that mimics a natural pheromone blend is an optimal lure for male attraction. For the mating disruption, it is unclear whether or not the natural pheromone blend is the best. The best disruptants have been selected after many trials and errors.

While details of the mechanism of mating disruption are not clear, it is expected that males are meaninglessly excited by the permeated pheromone and that females are masked by a higher level of the synthetic pheromone or by a modification of the natural mixing ratio. In a tea garden and an apple orchard, 'Hamaki-con' including only one chemical (Z11–14:OAc) effectively suppressed the mating of several leaf-roller moths, which universally secrete this component. Particularly, it targets *A. honmai*, a harmful defoliator of tea leaves whose pheromone is a mixture of Z9,14:OAc, Z11–14:OAc, E11–14:OAc, and Me10–12:OAc in a ratio of 63:31:4:2 [129]. The responses of *A. honmai* males have been examined in a wind tunnel containing a lure baited with the synthetic pheromone at the center and disruptants at the upwind end [187]. The orientation of male moths to the lure was strongly inhibited by permeation with Z11–14:OAc alone as well as with all four components in a natural mixing ratio; however, different patterns of movement were observed. Every male in the tunnel with the four components moved to the upwind side, but most of the males in the tunnel with the single component stayed on the downwind side, where they were released. This result indicates that the permeation of Z11–14:OAc alone prevents upwind flights of the males even though the synthetic lure is present in the tunnel.

Sexual communication is one of the most essential actions of insects for the reproduction of the next generation. When males do not respond to the sex pheromone, they miss a chance to reproduce. Therefore, it is presumed that it is not easy to alter a communication system and that insects rarely develop resistance to a mating disruptant, as shown by the case of *P. gossypiella*, which has been controlled by a disruption method for more than 20 years without any resistance. However, in the mid-1990s, a resistant population was recognized for *A. honmai* at a tea garden in Japan. The area had been treated with 'Hamaki-con' every year since the early 1980s. Recently, this problem was resolved by the introduction of a dispenser that included the four-component blend [188].

Acknowledgments The authors are grateful to Dr. F. Komai of Osaka University of Arts for helpful discussions about insect taxonomy and K. Ogawa of Shin-Etsu Chemical Co. Ltd. for information on the utilization of mating disruptants.

References

1. Butenandt A, Beckmann R, Stamm D, Hecker E (1959) Z Naturforsch 14b:283
2. Berger RS (1966) Ann Entomol Soc Am 59:767
3. Bierl BA, Beroza M, Collier CW (1970) Science 170:87
4. Arn H, Tóth M, Priesner E (1992) List of sex pheromones of Lepidoptera and related attractants. OILB-SROP/IOBC-WPRS, France, ISBN 92-9067-044-4; Arn H, Tóth M, Priesner E (1997) Techno Trans Mating Disrup 20:257
5. http://www-pherolist.slu.se/
6. http://www-pherobase.com/
7. http://www.tuat.ac.jp/~antetsu/LepiPheroList.htm
8. Kristensen NP (1999) Lepidoptera, moths and butterflies, vol 1. Evolution, systematics, and biogeography. In: Fischer M (ed) Handbook of zoology, vol IV. Arthropoda: Insecta 35:1–487
9. Millar JG (2000) Annu Rev Entomol 45:575
10. Roelofs WL, Hill AS, Cardé RT, Baker TC (1974) Life Sci 14:1555
11. Persoons CJ, Voerman S, Verwiel PEJ, Ritter FJ, Nooyen WJ, Minks AK (1976) Entomol Exp Appl 20:289
12. Charlton RE, Wyman JA, McLaughlin JR, Du J, Roelofs WL (1991) J Chem Ecol 17: 175
13. Baker R, Herbert RH, Neequaye NN, Rao KN (1985) J Chem Ecol 11:989
14. Nesbitt BF, Beevor PS, Cork A, Hall DR, David H, Nandagopal V (1986) J Chem Ecol 12:1377
15. McDonough LM, Averill AL, Davis HG, Smithhisler CL, Murray DA, Chapman PS, Voerman S, Dapsis LJ, Averill MM (1994) J Chem Ecol 20:3269
16. Novák L, Tóth M, Balla J, Szántay C (1979) Acta Chim Acad Sci Hung 102:135
17. Mozuraitis R, Borg-Karlson A, Buda V, Ivinskis P (1999) J Appl Entomol 123:603
18. Zhang J, Han Y, Meng X (2001) Z Naturforsch 56c:423
19. Griepink FC, Beek TA, Visser JH, Voerman S, Groot A (1995) J Chem Ecol 21:2003
20. Roelofs W, Bjostad L (1984) Bioorg Chem 12:279
21. Komoda M, Inomata S, Ono A, Watanabe H, Ando T (2000) Biosci Biotechnol Biochem 64:2145
22. Unelius CR, Eiras A, Witzgall P, Bengtsson M, Kovaleski A, Vilela EF, Borg-Karlson A (1996) Tetrahedron Lett 37:1505
23. Attygalle AB, Jham GN, Svatoš A, Frighetto RTS, Meinwald J, Vilela EF, Ferrara Uchoa-Fernandes MA (1995) Tetrahedron Lett 36:5471
24. Griepink FC, Beek TA, Posthumus MA, Groot A, Visser JH, Voerman S (1996) Tetrahedron Lett 37:411
25. Svatoš A, Attygalle AB, Jham GN, Frighetto RTS, Vilela EF, Saman D, Meinwald J (1996) J Chem Ecol 22:787
26. Kalinová B, Svatoš A, Kindl J, Hovorka O, Hrdý I, Kuldová J, Hoskovec M (2003) J Chem Ecol 29:387
27. Naka H, Vang LV, Inomata S, Ando T, Kimura T, Honda, Tsuchida K, Sakurai H (2004) J Chem Ecol 29:2447
28. Ho H, Tao Y, Tsai R, Wu Y, Tseng H, Chow Y (1996) J Chem Ecol 22:271

29. Krishnakumari B, Prasuna AL, Jyothi KN, Valli MY, Sighamony S, Prasad AR, Yadav JS (1998) J Entomol Res 22:197
30. Hill AS, Roelofs WL (1981) J Chem Ecol 7:655
31. Hill AS, Kovalev BG, Nikolaeva LN, Roelofs WL (1982) J Chem Ecol 8:383
32. Einhorn J, Lallemand J, Zagatti P, Gallois M, Virelizier H, Riom J, Menassieu P (1982) C R Acad Sci Ser II 294:41
33. Roelofs WL, Brown RL (1982) Ann Rev Ecol Syst 13:395
34. Fu X, Pan X, Meng X (2002) Entomol Exp Appl 103:183
35. Tumlinson JH, Yonce CE, Doolittle RE, Heath RR, Gentry CR, Mitchell ER (1974) Science 185:614
36. Szöcs G, Schwarz M, Sziraki G, Tóth M, Klun JA, Leonhardt BA (1985) Entomol Exp Appl 39:131
37. Bûda V, Mäeorg U, Karalius V, Rothschild GHL, Kolonistova S, Ivinskis P, Mozûraitis R (1993) J Chem Ecol 19:799
38. Kuwahara Y, Kitamura C, Takahashi S, Hara H, Ishii S, Fukami H (1971) Science 171: 801
39. Brady UE, Tumlinson JH, Brownlee RG, Silverstein RM (1971) Science 171:802
40. Vu MH, Ando T, Takahashi N, Tatsuki S, Yamane A, Ikeda T, Yamazaki S (1980) Agric Biol Chem 44:231
41. Ando T, Inomata S, Shimada R, Nomura M, Uehara S, Pu G (1998) J Chem Ecol 24:1105
42. Ono A, Imai T, Inomata S, Watanabe A, Ando T· (2002) Insect Biochem Mol Biol 32:701
43. Zhao C, Li Q, Guo X, Wang X (1993) Kunchong Xuebao 36:247
44. Bestmann HJ, Attygalle AB, Garbe W, Kern F, Martichonok V, Schäfer D, Vostrowsky O, Hasenfuß I (1993) Insect Biochem Mol Biol 23:793
45. Guerrero A, Camps F, Coll J, Riba M, Einhorn J, Descoins C, Lallemand JY (1981) Tetrahedron Lett 22:2013
46. Frerot B, Demolin G (1993) Boll Zool Agric Bachic II 25:33
47. Frérot B, Malosse C, Milat ML, Demolin G, Martin JC, Khemici M, Zamoun M, Gachi M (1990) J Appl Entomol 109:210
48. Villorbina G, Rodriguez S, Camps F, Fabriàs G (2003) Insect Biochem Mol Biol 33:155
49. Millar J, McElfresh JS, Marques FA (2002) J Econ Entomol 95:692
50. Hall DR, Beevor PS, Campion DG, Chamberlain DJ, Cork A, White RD, Almestar A, Henneberry TJ (1992) Tetrahedron Lett 33: 4811
51. Henderson HE, Warren FL, Augustyn OPH, Burger BV, Schneider DF, Boshoff PR, Spies HSC, Geertsema H (1973) J Insect Physiol 19:1257
52. Yasuda T, Yoshii S, Wakamura S (1994) Appl Entomol Zool 29:21
53. Leonhardt BA, Mastro VC, Schwarz M, Tang JD, Charlton RE, Pellegrini-Toole A, Warthen JD, Schwalbe CP, Cardé RT (1991) J Chem Ecol 17:897
54. Wakamura S, Arakaki N, Ono H, Yasui H (2001) Entomol Exp Appl 100:109
55. Cabrera A, Eiras AE, Gries G, Gries R, Urdaneta N, Mirás B, Badji C, Jaffe K (2001) J Chem Ecol 27:2097
56. Tóth M, Buser HR, Peña A, Arn H, Mori K, Takeuchi T, Nikolaeva LN, Kovalev BG (1989) Tetrahedron Lett 30:3405
57. Wong JW, Underhill EW, MacKenzie SL, Chisholm MD (1985) J Chem Ecol 11:727
58. Roelofs WL, Hill AS, Linn CE, Meinwald J, Jain SC, Herbert HJ, Smith RF (1982) Science 217:657
59. Bestmann HJ, Brosche T, Koschatzky KH, Michaelis K, Platz H, Roth K, Süß J, Vostrowsky O, Knauf W (1982) Tetrahedron Lett 23:4007
60. Underhill EW, Millar JG, Ring RA, Wong JW, Barton D, Giblin M (1987) J Chem Ecol 13:1319

61. Zhu J, Löfstedt C, Philipp P, Francke W, Tammaru T, Haukioja E (1995) Entomol Exp Appl 75:159
62. Einhorn J, Boniface B, Renou M, Milat M (1984) C R Acad Sci Ser 3 298:573
63. Frérot B, Malosse C, Chenon RD, Ducrot PH, Cain A (1993) C R Acad Sci Ser 2 317:1045
64. Jain SC, Dussourd DE, Conner WE, Eisner T, Guerrero A, Meinwald J (1983) J Org Chem 48:2266
65. Wong JW, Palaniswamy P, Underhill EW, Steck WF, Chisholm MD (1984) J Chem Ecol 10:463
66. Wakamura S, Arakaki N, Yamamoto M, Hiradate S, Yasui H, Yasuda T, Ando T (2001) Tetrahedron Lett 42:687
67. Wakamura S, Arakaki N, Yamamoto M, Hiradate S, Yasui H, Kinjo K, Yasuda T, Ando T (2004) J Chem Ecol 29 (in press)
68. Gries R, Holden D, Gries G, Wimalaratne PDC, Slessor KN, Saunders C (1997) Naturwissenschaften 84:219
69. Yamazawa H, Nakajima N, Wakamura S, Arakaki N, Yamamoto M, Ando T (2001) J Chem Ecol 27:2153
70. Wakamura S, Arakaki N, Yamazawa H, Nakajima N, Yamamoto M, Ando T (2002) J Chem Ecol 28:449
71. Ando T, Ohtani K, Yamamoto M, Miyamoto T, Qin X, Witjaksono (1997) J Chem Ecol 23:2413
72. Yamamoto M, Kiso M, Yamazawa H, Takeuchi J, Ando T (2000) J Chem Ecol 26:2579
73. Tóth M, Szöcs G, Francke W, Schmidt F, Philipp P, Löfstedt C, Hansson BS, Farag AI (1994) Z Naturforsch 49c:516
74. Szöcs G, Tóth M, Francke W, Schmidt F, Philipp P, König WA, Mori K, Hansson BS, Löfstedt C (1993) J Chem Ecol 19:2721
75. Pu G, Yamamoto M, Takeuchi Y, Yamazawa H, Ando T (1999) J Chem Ecol 25:1151
76. Yamamoto M, Takeuchi Y, Ohmasa Y, Yamazawa H, Ando T (1999) Biomed Chromatogr 13:410
77. Zhu J, Kozlov MV, Philipp P, Francke W, Löfstedt C (1995) J Chem Ecol 21:29
78. Kozlov MV, Zhu J, Philipp P, Francke W, Zvereva EL, Hansson BS, Löfstedt C (1996) J Chem Ecol 22:431
79. Tóth M, Szöcs G, Nieukerken EJ, Philipp P, Schmidt F, Francke W (1995) J Chem Ecol 21:13
80. Leonhardt BA, Neal JW, Klun JA, Schwarz M, Plimmer JR (1983) Science 219:314
81. Rhainds M, Gries G, Li J, Gries R, Slessor KN, Chinchilla CM, Oehlschlager AC (1994) J Chem Ecol 20:3083
82. Subchev M, Toshova T, Stanimirova L, Stan G, Embacher G, Francke W, Reckziegel A, Ferreira JT, Priesner E (2000) J Chem Ecol 26:487
83. Sasaerila Y, Gries R, Gries G, Khaskin G, King S, Boo TC (2000) J Chem Ecol 26:1969
84. Myerson J, Haddon WF, Soderstrom EL (1982) Tetrahedron Lett 23:2757
85. Subchev M, Harizanov A, Francke W, Franke S, Plass E, Reckziegel A, Schröder F, Pickett JA, Wadhams LJ, Woodcock CM (1998) J Chem Ecol 24:1141
86. Tamaki Y, Honma K, Kawasaki K (1977) Appl Entomol Zool 12:60
87. Foster SP, Thomas WP (2000) J Chem Ecol 26:2549
88. Buser HR, Guerin PM, Tóth M, Szöcs G, Schmid A, Francke W, Arn H (1985) Tetrahedron Lett 26:403
89. Smith RG, Daterman GE, Daves GD (1975) Science 188:63
90. Gries G, Clearwater J, Gries R, Khaskin G, King S, Schaefer P (1999) J Chem Ecol 25:1091
91. Chow YS, Tsai RS, Gries G, Gries R, Khaskin G (2001) Entomol Sinica 8:13
92. Smith LM, Smith RG, Loehr TM, Daves GD (1978) J Org Chem 43:2361

93. Gries G, Slessor KN, Gries R, Khaskin G, Wimalaratne PDC, Gray TG, Grant GG, Tracey AS, Hulme M (1997) J Chem Ecol 23:19
94. Grant GG, Slessor KN, Liu W, Abou-Zaid MM (2003) J Chem Ecol 29:589
95. Schaefer PW, Gries G, Gries R, Holden D (1999) J Chem Ecol 25:2305
96. Bierl BA, Beroza M, Adler VE, Kasang G, Schröter H, Schneider D (1975) Z Naturforsch 30c:672
97. Gries G, Schaefer PW, Gries R, Liška J, Gotoh T (2001) J Chem Ecol 27:1163
98. Gries G, Schaefer PW, Khaskin G, Hahn R, Gries R, Chao J (1999) J Chem Ecol 25:2535
99. Francke W, Franke S, Toth M, Szöcs G, Guerin P, Arn H (1987) Naturwissenschaften 74:143
100. Francke W, Toth M, Szöcs G, Krieg W, Ernst H, Buschmann E (1988) Z Naturforsch 43c:787
101. Riba M, Rosell JA, Eizaguirre M, Canela R, Guerrero A (1990) J Chem Ecol 16:1471
102. Sugie H, Tamaki Y, Sato R, Kumakura M (1984) Appl Entomol Zool 19:323
103. Gries R, Gries G, King GGS, Maier CT (1997) J Chem Ecol 23:1119
104. Gries G, Gries R, Borden JH, Li J, Slessor KN, King GGS, Bowers WW, West RJ, Underhill EW (1991) Naturwissenschaften 78:315
105. Gries R, Gries G, Li J, Maier CT, Lemmon CR, Slessor KN (1994) J Chem Ecol 20:2501
106. Gries G, Gries R, Krannitz SH, Li J, King GGS, Slessor KN, Borden JH, Bowers WW, West RJ, Underhill EW (1993) J Chem Ecol 19:1009
107. Maier CT, Gries R, Gries G (1998) J Chem Ecol 24:491
108. Gries G, King GGS, Gries R, Wimalaratne PDC, Gray TG, Shepherd RF, Li J, Slessor KN, Khaskin G (1993) J Chem Ecol 19:1501
109. Francke W, Plass E, Zimmermann N, Tietgen H, Tolasch T, Franke S, Subchev M, Toshova T, Pickett JA, Wadhams LJ, Woodcock CM (2000) J Chem Ecol 26:1135
110. Gries G, Schaefer PW, Gries R, Fan Y, Higashiura Y, Tanaka B (2002) J Chem Ecol 28:469
111. Roelofs WL, Cardé RT (1971) Science 171:684
112. Jurenka RA, Subchev M, Abad J, Choi M, Fabriàs G (2003) Proc Natl Acad Sci USA 100:809
113. Birch MC, Poppy GM, Baker TC (1990) Annu Rev Entomol 35:25
114. Schulz S, Francke W, König WA, Schurig V, Mori K, Kittmann R, Schneider D (1990) J Chem Ecol 16:3511
115. Kouo I, Matsumoto T, Wagner DL, Shoolery JN (1985) Tetrahedron Lett 26:563
116. Marukawa K, Mori K (2002) Eur J Org Chem 3974
117. Heath RR, Landolt PJ, Leppla NC, Dueben BD (1988) J Chem Ecol 14:1121
118. Heath RR, Tumlinson JH, Leppla NC, McLaughlin JR, Dueben B, Dundulis E, Guy RH (1983) J Chem Ecol 9:645
119. Kan E, Hidaka T, Sato T, Kitajima H (2002) Appl Entomol Zool 37:163
120. Sasaerila Y, Gries R, Gries G, Khaskin G, King S, Takács S, Hardi (2003) Chemoecology 13:89
121. Takács S, Gries G, Gries R (2001) Chemoecology 11:153
122. Schulz S (1998) Eur J Org Chem 13
123. Schulz S, Nishida R (1996) Bioorg Med Chem 4:341
124. Nishida R, Schulz S, Kim CS, Fukami H, Kuwahara Y, Honda K, Hayashi N (1996) J Chem Ecol 22:949
125. Stritzke K, Schulz S, Nishida R (2002) Eur J Org Chem 3884
126. Stritzke K, Schulz S, Boppré M (2003) Eur J Org Chem 1337
127. Tamaki Y, Noguchi H, Yushima T, Hirano C (1971) Appl Entomol Zool 6:139
128. Tamaki Y, Noguchi H, Yushima T, Hirano C, Honma K, Sugawara H (1971) Kontyu 39:338
129. Tamaki Y, Noguchi H, Sugie H, Sato R, Kariya A (1979) Appl Entomol Zool 14:101

130. Klun JA, Chapman OL, Mattes KC, Wojtkowski PW, Beroza M, Sonnet PE (1973) Science 181:661
131. Kochansky J, Cardé RT, Liebherr J, Roelofs WL (1975) J Chem Ecol 1:225
132. Ando T, Saito O, Arai K, Takahashi N (1980) Agric Biol Chem 44:2643
133. Huang Y, Takanashi T, Hoshizaki S, Tatsuki S, Honda H, Yoshiyasu Y, Ishikawa Y (1998) J Chem Ecol 24:2079
134. IshikawaY, Takanashi T, Kim C, Hoshizaki S, Tatsuki S, Huang Y (1999) Entomol Exp Appl 91:237
135. Huang Y, Tatsuki S, Kim C, Hoshizaki S, Ishikawa Y (1998) Entomol Exp Appl 86:313
136. Huang Y, Tatsuki S, Kim C, Hoshizaki S, Yoshiyasu Y, Honda H, Ishikawa Y (1997) J Chem Ecol 23:2791
137. Huang Y, Honda H, Yoshiyasu Y, Hoshizaki S, Tatsuki S, Ishikawa Y (1998) Entomol Exp Appl 89:281
138. Ishikawa Y, Takanashi T, Huang Y (1999) Chemoecol 9:25
139. Roelofs WL, Liu W, Hao G, Jiao H, Rooney AP, Linn CE (2002) Proc Natl Acad Sci USA 99:13621
140. Dunkelblum E, Gothilf S (1983) Z Naturforsch 38c:1011
141. Inomata S, Komoda M, Watanabe H, Nomura M, Ando T (2000) J Chem Ecol 26:443
142. Sugie H, Kawasaki K, Nakagaki S, Iwata N (1991) Appl Entomol Zool 26:71
143. Benn MH, Galbreath RA, Holt VA, Young H, Down G, Priesner E (1982) Z Naturforsch 37c:1130
144. Sugie H, Kawasaki K, Higuchi H, Kitamura C, Tamaki Y (1990) Appl Entomol Zool 25:467
145. Löfstedt C, Hansson BS, Tóth M, Szöcs G, Buda V, Bengtsson M, Ryrholm N, Svensson M, Priesner E (1994) J Chem Ecol 20:91
146. Ando T, Ohno R, Kokuryu T, Funayoshi A, Nomura M (1995) J Chem Ecol 21:1181
147. Ma PWK, Roelofs WL (2002) Zool Science 19:501
148. Tóth M, Buser H (1992) J Chromatogr 598:303
149. Millar JG (2002) Compre Anal Chem 37:669
150. Borg-Karlson A, Mozuraitis R (1996) Z Naturforsch 51c:599
151. Mozûraitis R, Bûda V, Jonušaite V, Borg-Karlson A, Noreika R (2000) Entomol Exp Appl 94:15
152. Frérot B, Malosse C, Cain A (1997) J High Resol Chromatogr 20:340
153. Norin T (2001) Pure Appl Chem 73:607
154. Schnaider D (1969) Science 163:1031
155. Roelofs WL (1984) Electroanntenogram assays: rapid and convenient screening procedures for pheromones. In: Hummel HE, Millar TA (eds) Techniques in pheromone research. Springer, Berlin Heidelberg New York, pp 131–159
156. Arn H, Städler E, Rauscher S (1975) Z Naturforsch 30c:722
157. Svatős A, Kalinová B, Hoskovec M, Kindl J, Hovorka O, Hrdý I (1999) Tetrahedron Lett 40:7011
158. Gries G, McBrien HL, Gries R, Li J, Judd GJR, Slessor KN, Borden JH, Smith RF, Christie M, Troubridge JT, Wimalaratne PDC, Underhill EW (1993) J Chem Ecol 19:1789
159. Ferrao P, Gries G, Wimalaratne PDC, Maier CT, Gries R, Slessor KN, Li J (1998) J Chem Ecol 24:2059
160. Horiike M, Gu Y, Hirano C (1990) Org Mass Spectrom 25:329
161. Kuwahara Y, Yonekawa Y, Kamikihara T, Suzuki T (1986) Agric Biol Chem 50:2017
162. Oldham NJ, Svatos A (1999) Rapid Comm Mass Spectrom 13:331
163. Ando T, Ogura Y, Uchiyama M (1988) Agric Biol Chem 52:1415
164. Nishida T, Vang LV, Yamazawa H, Yoshida R, Naka H, Tuchida K, Ando T (2003) Biosci Biotechnol Biochem 67:822

165. Jones GR, Oldham NJ (1999) J Chromatogr A 843:199
166. Buser HR, Arn H, Guerin P, Rauscher S (1983) Anal Chem 55:818
167. Vincenti M, Guglielmetti G, Cassani G, Tonini C (1987) Anal Chem 59:694
168. Young DC, Vouros P, Holick MF (1990) J Chromtogr 522:295
169. McElfresh JS, Millar JG (1999) J Chem Ecol 25:687
170. Ando T, Kishi H, Akashio N, Qin X, Saito N, Abe H, Hashimoto S (1995) J Chem Ecol 21:299
171. Ando T, Ohsawa H, Ueno T, Kishi H, Okamura Y, Hashimoto S (1993) J Chem Ecol 19:787
172. Carlson DA, Bernier UR, Sutton BD (1998) J Chem Ecol 24:1845
173. Schulz S (2001) Lipids 36:637
174. Pomonis JG, Nelson DR, Fatland CL (1980) J Chem Ecol 6:965
175. Katritzky AR, Chen K, Maran U, Carlson DA (2000) Anal Chem 72:101
176. Junkes BS, Amboni RDMC, Heinzen VEF, Yunes RA (2002) Chromatogr 55:707
177. Ozawa RA, Ando T, Nagasawa H, Kataoka H, Suzuki A (1993) Biosci Biotech Biochem 57:2144
178. Wei W, Miyamoto T, Endo M, Murakawa T, Pu G, Ando T (2003) Insect Biochem Mol Biol 33:397
179. Qin X, Ando T, Yamamoto M, Yamashita M, Kusano K, Abe H (1997) J Chem Ecol 23:1403
180. Yamazawa H, Yamamoto M, Karasawa KI, Pu G, Ando T (2003) J Mass Spectrom 38: 328
181. Ohtani I, Kusumi T, Ishitsuka MO, Kakisawa H (1989) Tetrahedron Lett 30:3147
182. Ando T, Kurotsu Y, Kaiya M, Uchiyama M (1985) Agric Biol Chem 49:141
183. Ando T, Ohsawa H (1993) J Chem Ecol 19:119
184. Dorman DE, Jautelat M, Roberts JD (1971) J Org Chem 36:2757
185. Cardé RT, Minks AK (1995) Annu Rev Entomol 40:559
186. Ohtani K, Witjaksono, Fukumoto T, Mochizuki F, Yamamoto M, Ando T (2001) Entomol Exp Appl 100:203
187. Hiyori T, Kainoh Y, Ninomiya Y (1986) Appl Entomol Zool 21:153
188. Mochizuki F, Fukumoto T, Noguchi H, Sugie H, Morimoto T, Ohtani K (2002) Appl Entomol Zool 27:299

Topics in Current Chemistry (2004) 239: 97–132
DOI 10.1007/b95450

Insect Pheromone Biosynthesis

Russell Jurenka (✉)

Department of Entomology, Iowa State University, 3222 Science II, 50011-3222 Ames IA, USA
rjurenka@iastate.edu

Abstract Pheromones are utilized by many insects in a complex chemical communication system. This review will look at the biosynthesis of sex and aggregation pheromones in the model insects, moths, flies, cockroaches, and beetles. The biosynthetic pathways involve altered pathways of normal metabolism of fatty acids and isoprenoids. Endocrine regulation of the biosynthetic pathways will also be reviewed for the model insects. A neuropeptide named pheromone biosynthesis activating neuropeptide regulates sex pheromone biosynthesis in moths. Juvenile hormone regulates pheromone production in the beetles and cockroaches, while 20-hydroxyecdysone regulates pheromone production in the flies.

Keywords Pheromone · Biosynthetic pathways · Hormonal regulation · PBAN · Juvenile hormone · Ecdysone

List of Abbreviations

Abbreviations for pheromone molecules follows the following format: example (*Z*)-7-dodecen-1-yl acetate is shortened to Z7–12:OAc where Z denotes the double bond configuration, 7 the double bond position, 12 the number of carbons in the chain; OAc indicates the functional group as an acetate ester.

Ald	Aldehyde
CoA	Coenzyme A
D	Deuterium
epo	Epoxide
H	Hydrocarbon
HMG	3-Hydroxy-3-methyl-glutaryl
JH	Juvenile hormone
Me	Methyl group
OAc	Acetate ester
OH	Alcohol
one	Ketone
PBAN	Pheromone biosynthesis activating neuropeptide
PGN	PBAN-encoding gene neuropeptides
SEG	Subesophageal ganglion
yne	Triple bond

1 Introduction

Pheromones are utilized by a variety of organisms for chemical communication. Amongst the insects, several different types of pheromones are recognized based on behaviors produced by the compounds. The most common behavioral traits mediated by pheromones include attraction of the conspecific sex for mating (sex pheromones), aggregating both sexes to a specific site for feeding and/or mating (aggregation pheromones), and marking sites or forming trails (marking pheromones). In addition, social insects utilize a variety of pheromones to regulate colony and caste behaviors (see elsewhere in this book). The first sex pheromone identified was (*E,Z*)-10,12-hexadecadien-1-ol, bombykol, from the silkworm moth, *Bombyx mori* [1]. Since then pheromone components from well over three thousand insect species have been determined with the majority concentrating on sex pheromones from economically important pests [2]. This review will highlight our current knowledge about how these pheromone components are biosynthesized, concentrating on sex and aggregation pheromones from moths, beetles, flies, and cockroaches. Due to my area of specialty more attention will be paid to moth pheromone production. Current ideas on the hormonal regulation of pheromone production will also be discussed.

The types of structures identified as pheromones will not be discussed in this review but can be found in the other chapters of this book. However, the type of structures utilized will give clues as to how these compounds are biosynthesized. In addition the volatility of the compound will indicate how it will be utilized as a pheromone [3]. Most sex pheromones of moths are straight chain compounds with 12, 14, and 16 carbons and a functional group consisting of alcohols, aldehydes, and acetate esters [2]. Double bond location is critical with specific blends consisting of one or more double bonds of (*Z*) and (*E*) isomers. Another major group of moth sex pheromones include hydrocarbons and their epoxides. Structures are discussed in detail elsewhere in this book. Many of the pheromones from flies are also hydrocarbons of 21 carbons and longer, and due to decreased volatility function as short range or contact pheromones. A number of coleopteran pheromones are isoprenoid in origin [4]. In addition to isoprenoids, the pheromones from coleopterans also consist of components derived from fatty acids and amino acids. Many species of cockroaches utilize both volatile and cuticular contact pheromones [5]. However, few chemical structures have been identified and the biosynthesis of only the hydrocarbon-derived contact pheromone of the German cockroach has been determined [6].

A question that was posed early on in determining biosynthetic pathways of the pheromones was the origin of the precursors. There was some indication that plant derived compounds could be ingested and modified by the insect into a pheromone. We now know that in some cases this occurs [7], but for the most part pheromones are biosynthesized de novo by the insect [8]. For most of the pheromones studied to date it is apparent that biosynthetic pathways of 'normal' metabolism have been altered to produce specific pheromone components. Several enzymes in these biosynthetic pathways have been modified to produce species specific pheromone components.

2
Site of Pheromone Biosynthesis

The site of pheromone production is varied amongst the insects just as there are variable structures in the different orders. Several reviews are available detailing the ultrastructure of these glands [9–11]. Evidence that pheromone biosynthesis occurs in these cells and tissues requires that the isolated tissue be shown to incorporate labeled precursors into pheromone components. In the more studied model insects this criteria has been met.

In moths the pheromone gland is usually located as modified epidermal cells between the 8th and 9th abdominal segments [11]. It has been well established that these cells are the site of synthesis for oxygenated pheromone components in a variety of moths [12–14]. Although a large number of moths utilize a gland located between the 8th and 9th segments, exceptions occur. In *Theresimima ampelophaga* (Zygaenidae) the gland is located on the dorsal part of abdominal segments 3 to 5 [15], although it has yet to be demonstrated that the pheromone is biosynthesized de novo in this gland. The site of synthesis of hydrocarbon sex pheromones involves another group of cells called oenocytes that are usually associated with epidermal cells throughout the abdomen. Oenocytes are the cells that produce hydrocarbons in insects [16]. Once hydrocarbons are produced they are transported by lipophorin throughout the body and hydrocarbon sex pheromones are picked up by pheromone gland cells for release [17].

The site of pheromone biosynthesis in the Coleoptera is varied but is usually located in the abdomen [8]. In some cases a defined gland is present; an example of which is the production of aggregation pheromone in the male nitidulid beetle, *Carpophilus freemani* [18]. In this insect the gland, which is probably formed from modified oenocyte cells, is connected through ducts to the tracheal system where the pheromone is released. The site of synthesis of monoterpenoid pheromones in the bark beetles is the midgut [19]. Here the midgut cells are involved in both digestion of food and biosynthesis of pheromone. The pheromone is secreted into the gut lumen where it is released along with the frass.

The site of pheromone production in flies and cockroaches that utilize hydrocarbons is similar to that of the moths. Oenocyte cells produce the hydrocarbon pheromone which is transported by lipophorin in the hemolymph to epidermal cells throughout the body for release from the cuticular surface in general [20, 21].

3 Biosynthetic Pathways

3.1 Pheromone Biosynthesis in Moths

About 20 years ago the biosynthetic pathway for the major sex pheromone component of the cabbage looper moth was elucidated (Fig. 1) [22]. Since the majority of the moth sex pheromones have a straight-chain carbon backbone with an even number of carbons, it was surmised that these compounds are biosynthesized through modified fatty acid biosynthetic pathways. Several key enzymes involved in production of the Type I oxygenated compounds are fatty acid synthesis enzymes to produce the carbon chain. Modification of the chain includes introduction of a double bond by a desaturase and shorter chain lengths are produced by specific β-oxidation enzymes. The oxygenated functional groups are modified by oxidases, reductases, and acetyltransferases. Each of these enzymes are key in producing the specific pheromone blend of compounds used by each species.

3.1.1 Fatty Acid Biosynthesis

The key enzymes involved in the biosynthetic pathways of the Type I compounds are the fatty acid synthesis enzymes acetyl-CoA carboxylase and fatty acid synthetase. These enzymes are similar to those that produce the normal fatty acids used by all organisms. The resulting products are palmitic (16 car-

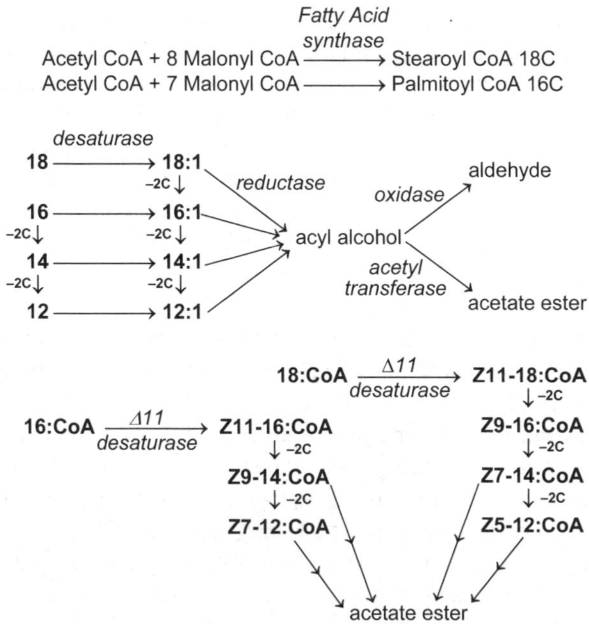

Fig. 1 General biosynthetic pathways for the production of alcohol, aldehyde, and acetate ester pheromone components in female moths. *Top*: production of saturated fatty acids. *Middle*: production of monounsaturated fatty acids and limited chain shortening produces intermediate compounds that can be reduced to an alcohol. Aldehyde and acetate ester pheromones are produced by an oxidase and acetyl-transferase, respectively. *Bottom*: biosynthetic pathway for the production of the acetate ester pheromone components in the cabbage looper moth, *Trichoplusia ni*. The CoA derivatives are reduced and acetylated to form the acetate esters. Additional pheromone components include 12:OAc and 11–12:OAc

bons) and stearic acids (18 carbons), with stearic acid the main product [23–26]. These even numbered carbon chains are modified by either a desaturase or chain shortening enzymes in either order.

3.1.2
Chain Shortening Enzymes

Specific chain length fatty acids could be produced in two ways. One is through the action of a thioester hydrolase that interacts with fatty acid synthetase to produce fatty acids shorter in length. Aphids produce myristic acid (14 carbons) and a specific thioester hydrolase releases the fatty acid from fatty acid synthetase after 6 additions of malonyl-CoA. If the hydrolase is not present then the fatty acid synthetase produces stearic acid [27]. A specific thioester hydrolase was ruled out in the biosynthesis of moth sex pheromones because labeling studies showed that longer chain length fatty acids were incorporated into shorter chain length pheromone components [22, 28].

The chain shortening pathway has not been characterized in detail at the enzymatic level in insects. It presumably is similar to the characterized pathway as it occurs in vertebrates. These enzymes are a partial β-oxidation pathway located in peroxisomes [29]. The key enzymes involved are an acyl-CoA oxidase (a multifunctional protein containing enoyl-CoA hydratase and 3-hydroxyacyl-CoA dehydrogenase activities) and a 3-oxoacyl-CoA thiolase [30]. These enzymes act in concert to chain shorten acyl-CoAs by removing an acetyl group. A considerable amount of evidence in a number of moths has accumulated to indicate that limited chain shortening occurs in a variety of pheromone biosynthetic pathways.

Evidence for the limited chain shortening enzymes has utilized stable isotope labeled precursors and analysis by mass spectrometry. The use of stable isotopes allowed researchers to unequivocally demonstrate that the labeled atoms did not migrate from the precursor and that the label is present in the pheromone product. Positioning the label on the terminal methyl carbon makes it difficult for any type of rearrangement to occur. By applying deuterium-labeled longer chain fatty acids like (16,16,16)-$^2H_3$-hexadecanoic acid to pheromone glands and monitoring for incorporation into pheromone it can be determined that the deuteriums end up in a pheromone component with fewer carbons (14 or 12) [12, 13, 31].

In addition to the labeling studies an in vitro enzyme assay was utilized to demonstrate substrate preferences in a study using cabbage looper moths, *T. ni* [25]. A mutant strain of cabbage loopers was isolated that produced a greatly increased amount of Z9–14:OAc [32], whereas normal cabbage loopers have the major pheromone component as Z7–12:OAc [33]. Increased amounts of Z9–14:OAc indicates that perhaps chain shortening was affected in the mutant cabbage loopers (Fig. 1). To determine substrate specificity of chain shortening enzymes for both normal and mutant cabbage loopers an in vitro enzyme assay was developed utilizing deuterium labeled acyl-CoA derivatives and NAD as a cofactor [25]. The products were monitored by GC/MS and in all cases the deuteriums were located on the methyl side of the double bond and could not be lost due to chain shortening. The results indicate that pheromone glands from normal cabbage loopers preferred to chain shorten Z11–16:CoA to Z7–12:CoA. In addition, other substrates were chain shortened to a lesser extent. Chain lengths less than 12 carbons were not observed if monoene fatty-acyl CoAs were used as substrates. This indicates that limited chain shortening stopped when the chain length became 12 carbons. If Z9–14:CoA was utilized the only product formed was Z7–12:CoA. It appeared that at most only two rounds of chain shortening occurred. This is what happens in the normal biosynthesis of the pheromone components. Chain shortening of Z11–18:CoA would need three rounds of chain shortening to produce Z5–12:CoA. To achieve this Z11–18:CoA would be chain-shortened to Z9–16:CoA or Z7–14:CoA and these products would then be chain shortened again by 2 or 1 round(s), respectively. Pheromone glands from the mutant cabbage looper apparently have the ability to chain shorten by only one round. Therefore Z11–16:CoA was chain shortened to Z9–14:CoA. This is the reason why the mutant had higher levels of Z9–14:OAc. Z7–12:OAc was also produced but at lower levels because Z9–14:CoA was the starting substrate for one round of chain shortening.

Interestingly chain shortening has been implicated in the alteration of pheromone ratios in several other species [13]. An example is the turnip moth, *Agrotis segetum*, where two populations with differing pheromone ratios were found in Sweden and Zimbabwe. The Swedish population has a ratio of Z9–14:OAc/Z7–12:OAc/Z5–10:OAc of 29/59/12 whereas the Zimbabwean population has a ratio of 2/20/78. By conducting labeling studies it was determined that chain shortening enzymes could be affected to produce the alteration in pheromone ratios [34]. Apparently the Swedish population has a reduced ability to chain shorten, although a change in reductase activity can not be ruled out. These studies indicate that alteration in chain shortening enzymes can have a major effect on pheromone blends. In fact, the combination of desaturases and chain shortening enzymes can produce many of the possible intermediates that can be converted to identified pheromones (Fig. 2).

3.1.3 Desaturases

A variety of desaturases have been described that are involved in the biosynthesis of female moth sex pheromones. The desaturases identified so far include enzymes that act on saturated and monounsaturated substrates. The desaturases that utilize saturated substrates are Δ5 [35], Δ9 [36], Δ10 [37], Δ11 [22], and Δ14 [38] The action of these desaturases along with chain shortening can account for the majority of double bond positions in the various chain-length monounsaturated pheromones so far identified. Figure 2 illustrates the large number of monounsaturated compounds that can be generated through a combination of desaturation and chain shortening. Addition of various functional groups, acetate esters, alcohols, and aldehydes increases the potential number of pheromone components. Notice that some of the intermediate compounds could be produced in two different ways. Therefore, although the desaturation and chain-shortening steps occur in a wide variety of moths, the order in which they occur and the type of desaturase must still be determined experimentally.

Some pheromone components are dienes and these can be produced by either the action of two desaturases or one desaturase and isomerization around the double bond. Some dienes with a 6,9-double bond configuration are produced using linoleic acid. Desaturases that utilize monounsaturated acyl-CoA substrates include Δ5 [39], Δ9 [36, 40], Δ11 [41], Δ12 [42], and Δ13 [43]. These can act sequentially to produce the diene [41, 42] or conjugated dienes could be produced by the action of one desaturase followed by isomerization [44–47].

A unique enyne pheromone component, Z13-hexadecen-11-ynyl acetate, found in *Thaumetopoea pityocampa* was found to be produced by the sequential action of a Δ13 desaturase and a Δ11 acetylenase to form yne11–16:acid [43]. The dienyl intermediate Z13,yne11–16:acid was formed by the action of

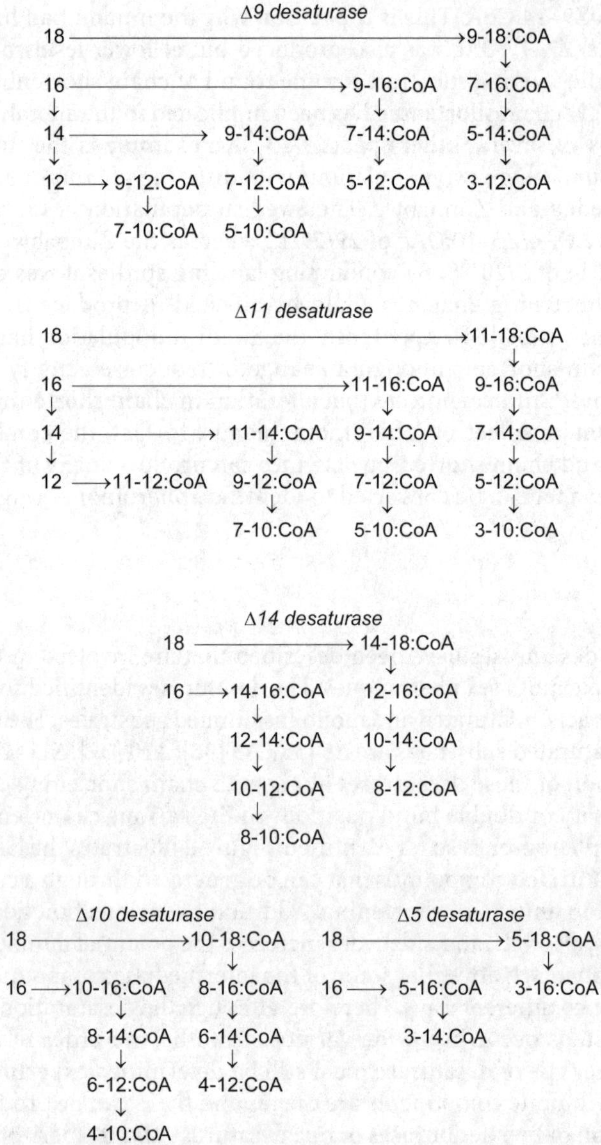

Fig. 2 Action of desaturases and limited chain shortening can produce a variety of monounsaturated acyl-CoA precursors that can be modified to form unsaturated pheromone compounds. The *arrow pointing down* indicates limited chain shortening by two carbons. Modification of all 16-, 14-, 12-, and 10-carbon acyl-CoA derivatives on the carbonyl carbon can account for the majority of monounsaturated acetate esters, aldehydes, and alcohols identified as sex pheromones

a Δ13 desaturase [43]. A closely related species, *Thaumetopoea processionea*, uses the diene Z11,Z13–16:OAc produced by the sequential action of a Δ11 and Δ13 desaturase [48]. The two species differ in that *T. processionea* lacks the Δ11 acetylenase and that a specific reductase is present in *T. pityocampa* [48].

Biosynthesis of triene pheromone components with a triene double bond system that is n-3 (3,6,9-) are probably produced from linolenic acid [49]. Moths in the families Geometridae, Arctiidae, and Noctuidae apparently utilize linoleic and linolenic acid as precursors for their pheromones that must be obtained in the diet, since moths can not synthesize these fatty acids [50]. Most of the Type II pheromones are produced by chain elongation and decarboxylation to form hydrocarbons [51]. Oxygen is added to one of the double bonds in the polyunsaturated hydrocarbon to produce an epoxide [49].

A variety of desaturases have been identified at the gene level with the first gene encoding a Δ11 desaturase being identified in the cabbage looper, *T. ni* [52]. Due to the homologous nature of the sequences a number of desaturase genes have been cloned [53, 54]. To demonstrate functionality the cloned genes are expressed in a strain of yeast lacking an endogenous desaturase. This strain of yeast will not grow in media lacking added unsaturated fatty acids but will grow if a functional desaturase is inserted into the yeast genome. After growth, the fatty acids are analyzed to determine double bond positions and thus the desaturase can be characterized regarding double bond insertion and chain-length specificity.

Several desaturase encoding cDNAs from pheromone glands have been identified and expressed into yeast cells. Interestingly additional desaturases have been identified in the glands than are required for pheromone production. A desaturase that is found in most glands is the Δ9 desaturase which is comparable to the metabolic desaturase found in the fat body [53]. The desaturases that have been cloned and expressed so far in addition to the Δ9 desaturase include a *Z*10-desaturase from *Planotortrix octo* [55], a Δ14 desaturase from *Ostrinia furnicalis* [56], and several Δ11-desaturases. The Δ11 desaturases from *T. ni* [52] and *H. zea* [57] produce primarily *Z*11–16:acid. A single Δ11-desaturase was characterized from *A. velutinana* that produces both *Z*11- and *E*11–14:acid [58]. This desaturase is unique in that it produces both isomers and uses 14:acid as a substrate. The characterization of this desaturase also helps answer the question of whether or not two desaturases are involved in producing the Z and E isomers. A study on the Δ11 desaturase of *Spodoptera littoralis*, using stable isotopes on carbons 11 and 12 suggested that a single enzyme can produce both Z and E isomers [59]. However, desaturases will be found that are Z and E specific as was the case with another unique Δ11-desaturase from *Epiphyas postvittana*. It was shown that this desaturase not only produced E11–14:acid and E11–16:acid from saturated precursors, but also E9,E11–14:acid from a E9–14:acid precursor (produced by chain-shortening E11–16:acid) [60].

Identification of desaturases in *Ostrinia nubilalis* and *Ostrinia furnicalis* produced the surprise finding that Δ11- and Δ14-desaturases are found in both moths [56]. *O. nubilalis* uses the pheromone components Z11- and E11–14:OAc, which are produced in a pathway using a Δ11-desaturase. *O. furnicalis* uses the pheromone components Z12- and E12–14:OAc, which are formed by a Δ14-desaturase producing Z14- and E14–16:acid followed by chain shortening. The genes encoding both desaturases were functionally expressed in yeast cells where it was determined that the Δ11-desaturase produced both *Z*11- and *E*11–14:acid and the Δ14-desaturase produced both *Z*14- and *E*14–16:acids. Although both genes are thought to be transcribed it is unknown whether the enzyme is translated and if so is the enzyme functional. However, this finding has implications in describing the speciation of these two moths [61].

3.1.4 Functional Group Modifications

Once a specific chain length pheromone intermediate that has the appropriate double bonds is produced, the carbonyl carbon is modified to form a functional group. The majority of oxygenated pheromone components are acetate esters (or other esters), alcohols, and aldehydes. Production of these components requires the reduction of a fatty-acyl precursor to an alcohol which is thought to be a two step reaction requiring a fatty acid reductase and an aldehyde reductase [62]. Thus alcohol formation goes through an aldehyde intermediate. Therefore, aldehydes could be produced by direct reduction of fatty acids. Another route for aldehyde formation is oxidation of alcohols. A cuticular oxidase has been characterized from pheromone glands of *H. zea* and *Manduca sexta* that produce aldehydes as pheromones [63, 64]. In those insects that utilize both an alcohol and an aldehyde as part of their pheromone, it is unclear how the production of both components occurs.

A fatty-acyl reductase was recently identified from the pheromone glands of *B. mori* [65]. This enzyme was characterized by first isolating a cDNA clone encoding the enzyme and then expressing the gene in yeast cells. It was shown that the transformed yeast cells supplemented with specific fatty acids produced the corresponding alcohol. The enzyme could reduce several fatty acids but E10,Z12–16:acid was preferred, indicating substrate specificity for the direct precursor of the sex pheromone bombykol. An aldehyde intermediate could not be detected suggesting direct conversion of the acid or CoA-derivative to the alcohol. The identification of this reductase [65] will help identify additional moth reductases and will help explain how species specific sex pheromones are produced, especially in those insects that utilize both alcohols and aldehydes as sex pheromone components.

Production of acetate ester pheromone components utilizes an enzyme called acetyl-CoA:fatty alcohol acetyltransferase that converts a fatty alcohol to an acetate ester. Therefore, alcohols could be utilized as substrates for both aldehyde and acetate ester formation. In some tortricids an in vitro enzyme assay was utilized to demonstrate specificity of the acetyltransferase for the *Z* isomer of 11–14:OH [66]. This specificity contributes to the final ratio of pheromone components. These results indicate that the family Tortricidae has members that have an acetyltransferase that is specific for the *Z* isomer of monounsaturated fatty alcohols. In contrast, several studies have shown no substrate preference for the acetyltransferase in other moths [66–69].

3.1.5 Production of Specific Pheromone Blends

It is most unusual for female moths to utilize just one compound as the pheromone. Rather a blend of compounds produced in precise ratios make up the species-specific pheromone. The production of this precise blend of chemical components is regulated in the biosynthetic pathway. The inherent specificities found within key enzymes in the pathway and combinations of enzymes is what is responsible for producing species specific ratios [13, 31].

An excellent example illustrating this point is the cabbage looper moth that utilizes a biosynthetic pathway as shown in Fig. 1. The key enzymes are the Δ11 desaturase which prefers 16:CoA with the major product being Z11–16:CoA and the minor product Z11–18:CoA [52]. To produce shorter chain length intermediates the products of the Δ11 desaturase are chain shortened through limited ß-oxidation. Although a variety of substrates can be chain shortened, Z11–16:CoA was preferred producing Z7–12:CoA [25]. The pheromone is an acetate ester produced by reduction of Z7–12:CoA to an alcohol followed by the action of an acetyl-transferase. The minor components are produced in lower amounts due to the substrate specificities of the Δ11 desaturase and chain-shortening enzymes. Another example is *A. velutinana* that utilizes a blend of seven acetate esters produced in a biosynthetic pathway similar to the one just described. The key enzymes are the Δ11 desaturase that produces both *Z* and *E* isomers of 11–14:CoA in about a 6/1 ratio [58]. However, to make the final ratio of *Z*11- to *E*11–14:OAc of 92/8 a selective increase in the *Z* isomer occurs within the biosynthetic pathway. The key enzyme here is the acetyl-transferase because it has specificity for the *Z* isomer [66]. Therefore, selective acetylation of *Z*11–14:OH and production of >60% *Z*11–14:CoA indicates that these enzymes have the inherent specificity to produce the 92:8 ratio of the major pheromone components *Z*11- and *E*11–14:OAc. The last example to illustrate how pheromone blends are produced is the European corn borer, *O. nubilalis*. Two strains are known in which one produces a ratio of *Z*/*E* of about 97/3 (*Z* strain) and the other produces an opposite ratio of *Z*/*E* of about 1/99 (*E* strain). The Δ11 desaturase from both strains produced a product with about 30/70 *Z*/*E* [56, 70]. Labeled acids applied to glands in vivo were selectively incorporated into the correct pheromone ratio indicating that the reductase shows specificity [69]. Therefore, the final pheromone ratios produced by females of the European corn borer are made through the action of a Δ11 desaturase that can produce both *Z* and *E* isomers. The final acetate ester ratio is strain dependent and is produced through the specificity found in the reductase system.

The above three examples illustrate how a species-specific pheromone blend is produced by the concerted action of desaturases, chain shortening enzymes, a reductase, and an acetyltransferase. The specificity inherent in certain enzymes in the pathway produces the final blend of pheromone components.

3.1.6
Hydrocarbon Pheromones

Several families of moths utilize hydrocarbons or epoxides of hydrocarbons as their sex pheromone. Oenocyte cells produce hydrocarbons that are transported through the hemolymph by lipophorin [71]. In a study using arctiid moths it was shown that sex pheromone hydrocarbons are transported on the same lipophorin particle as the hydrocarbons destined for the cuticular surface [17]. Therefore, specific uptake of the sex pheromone hydrocarbon occurred in pheromone glands [17]. Similar findings have been found with other moths [72–74]. The mechanism behind this specific uptake of one hydrocarbon from a potential pool of other hydrocarbons is unknown.

Most moth sex pheromones that are straight chain hydrocarbons also usually have an odd number of carbons. Most of these are polyunsaturated with double bonds in the 3,6,9- or 6,9-positions, indicating that they are derived from linolenic or linoleic acid, respectively [49, 51]. Linolenic and linoleic acid cannot be biosynthesized by moths so they must be obtained from the diet [75]. A few even chain-length hydrocarbon sex pheromones have been identified that also have 3,6,9- or 6,9-double bond configurations [49], indicating they too are derived from linolenic or linoleic acids; however, it is not known how these even chain hydrocarbons are formed.

A major class of sex pheromones that are derived from hydrocarbons are the monoepoxides of the polyene hydrocarbons [49]. These compounds usually have double bonds and the epoxy group in the 3,6,9-positions or 6,9-positions, again indicating they are biosynthesized from linolenic or linoleic acids, respectively. Although the production of hydrocarbon occurs in oenocytes the epoxidation step takes place in the pheromone gland. This has been demonstrated in several studies utilizing deuterium labeled precursors. In a study on the Japanese giant looper, *Ascotis selenaria cretacea*, that uses 6,9,epo3–19:H as a sex pheromone component, deuterium labeled hydrocarbon precursor, D3–3,6,9–19:H, was topically applied to pheromone glands and found to be converted to the epoxide, indicating that epoxidation takes place in pheromone glands [76]. By using a variety of polyene precursors it was also determined that the monooxygenase regiospecifically attacked the n-3 double bond regardless of chain length or degree of unsaturation. This indicates that the epoxidation enzyme is regiospecific in this insect [76].

A study using the gypsy moth, *Lymantria dispar*, illustrates the overall pathways involved in production of epoxide pheromone components (Fig. 3) [77]. This insect uses disparlure, Me2,epo7–18:H, as a pheromone component. In-

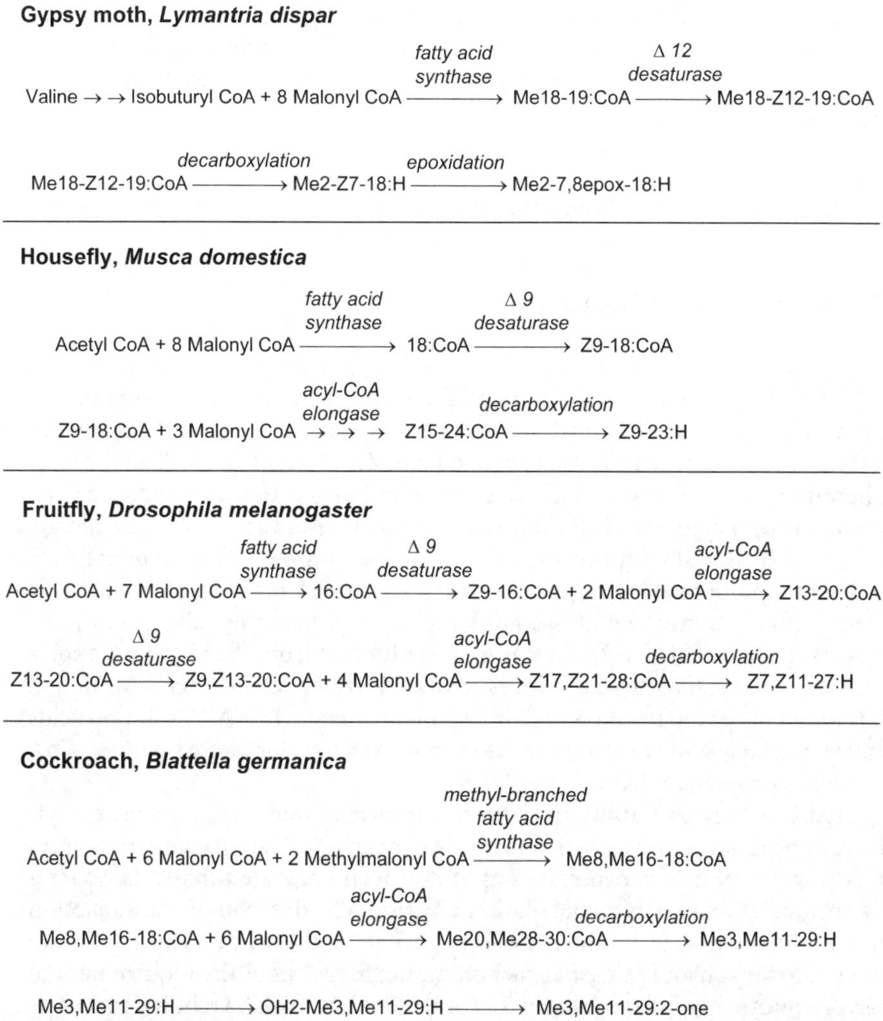

Fig. 3 Proposed biosynthetic pathways for the production of the sex pheromone components in the indicated insects. Common mechanisms include fatty acid synthesis, desaturation, chain elongation, and decarboxylation

cubation of isolated abdominal epidermal tissue with deuterium labeled valine resulted in incorporation into Me2,Z7–18:H. This indicates that the oenocyte cells associated with the epidermal tissues biosynthesize Me2,Z7–18:H using the carbons of valine to initiate the chain. The double bond is probably introduced by a Δ12 desaturase as determined by using specific deuterium labeled intermediates. Hemolymph transport of Me2,Z7–18:H is indicated by the finding of the alkene in the hemolymph [72]. Demonstration that Me2,Z7–18:H is converted to the epoxide in the pheromone gland was shown by using the deuterium labeled hydrocarbon and incubation with isolated pheromone glands. Disparlure is a stereoisomer that has the 7*R*,8*S* or (+) configuration and chiral chromatography indicated that only the (+)-isomer was produced by the pheromone glands [77].

3.2 Hydrocarbon Pheromone Production in Flies and Cockroaches

3.2.1 Pheromone Biosynthesis in Diptera

One of the sex pheromone components of the housefly, *Musca domestica*, is Z9–21:H that is found on the cuticular surface of the fly. This compound is formed by the elongation of Z9–18:CoA using malonyl-CoA and NADPH to Z15–24:CoA which is decarboxylated to form Z9–21:Hc (Fig. 3) [78–80]. Other pheromone components include an epoxide and ketone that are produced from Z9–21:Hc by a cytochrome P450 [81, 82] and methyl-branched alkanes that are produced by the substitution of methylmalonyl-CoA in place of malonyl-CoA at specific points during chain elongation [83, 84]. A novel microsomal fatty acid synthase is involved in production of methyl-branched alkanes in most insects [85–87]. This fatty acid synthase is different from the ubiquitous soluble fatty acid synthase that produces saturated straight chain fatty acids in that it is found in the microsomes and prefers methylmalonyl-CoA. The amino acids valine and isoleucine can provide the carbon skeletons for methylmalonyl-CoA as well as propionate [83].

Hydrocarbon formation involves the removal of one carbon from an acyl-CoA to produce a one carbon shorter hydrocarbon. The mechanism behind this transformation is controversial. It has been suggested that it is either a decarbonylation or a decarboxylation reaction. The decarbonylation reaction involves reduction to an aldehyde intermediate and then decarbonylation to the hydrocarbon and releasing carbon monoxide without the requirement of oxygen or other cofactors [88, 89]. In contrast, other work has shown that acyl-CoA is reduced to an aldehyde intermediate and then decarboxylated to the hydrocarbon, releasing carbon dioxide [90]. This reaction requires oxygen and NADPH and is apparently catalyzed by a cytochrome P450 [91]. Whether or not a decarbonylation reaction or a decarboxylation reaction produces hydrocarbons in insects awaits further research on the specific enzymes involved.

Drosophila melanogaster is another dipteran where pheromone biosynthesis has been studied [92]. Adult sexually mature female *D. melanogaster* utilizes primarily Z7,Z11–27:H as a contact sex pheromone. The biosynthesis of this compound follows the biosynthesis of other hydrocarbon-derived pheromones (Fig. 3). It is biosynthesized in oenocytes [93], transported through the hemolymph by lipophorin [94], and deposited on the cuticle surface. Biosynthesis in the oenocytes follows a similar pathway [95] as that described for the house fly except that a Δ9 desaturase produces Z9–16:CoA. A Δ9 desaturase has been identified in *D. melanogaster* that prefers 16:CoA as a substrate [96]. The Z9–16:CoA is elongated to Z13–20:CoA where it is acted on by another Δ9 desaturase to produce Z9,Z13–20:CoA which is further elongated to 28 carbons followed by decarboxylation to produce Z7,Z11–27:H. The identification of the second Δ9 desaturase awaits the functional expression and characterization of the remaining desaturases found in *Drosophila* [92].

3.2.2 Pheromone Biosynthesis in Cockroaches

The only cockroach pheromone that has received extensive investigation into its biosynthesis is the contact sex pheromone of *Blattella germanica*. A volatile sex pheromone is utilized for long range attraction while the contact sex pheromone produces a typical male mating response [97]. The contact sex pheromone consists of Me3,Me11–29:2-one and its two derivatives, an alcohol, OH29-Me3, Me11–29:2-one and an aldehyde, Ald29-Me3,Me11–29:2-one. Biosynthetic studies have shown that the ketone is produced from Me3,Me11–29:H found on the cuticular surface of males and females (Fig. 3) [6]. Radiolabeled Me3, Me11–29:H was converted to the ketone through an alcohol intermediate when applied to the cuticular surface of females. Hydroxylation and oxidation at carbon 29 would produce OH29- and Ald29-Me3,Me11–29:2-one.

3.3 Pheromone Biosynthesis in Beetles

Coleoptera comprise the largest order of insects and accordingly pheromone structures and biochemical pathways are diverse [98, 99]. Beetle pheromone biosynthesis involves fatty acid, amino acid, or isoprenoid types of pathways. In some cases dietary host compounds can be converted to pheromones, but it is becoming apparent that most beetle pheromones are synthesized de novo.

3.3.1 Isoprenoid Pheromones from Bark Beetles

Bark beetles primarily utilize isoprenoid derived pheromones [100, 101] and have been the most studied regarding their biosynthesis [8, 98]. Earlier work indicated that the isoprenoid pheromones could be produced by the beetle altering host derived isoprenoids; however more recent work indicates that for the most part bark beetles are producing pheromones de novo. The production of isoprenoids follows a pathway outlined in Fig. 4 which is similar to the isoprenoid pathway as it occurs in cholesterol synthesis in mammals. Insects cannot synthesize cholesterol but can synthesize farnesyl pyrophosphate. Insects apparently do not have the ability to cyclize the longer chain isoprenoid compounds into steroids. The key enzymes in the early steps of the isoprenoid

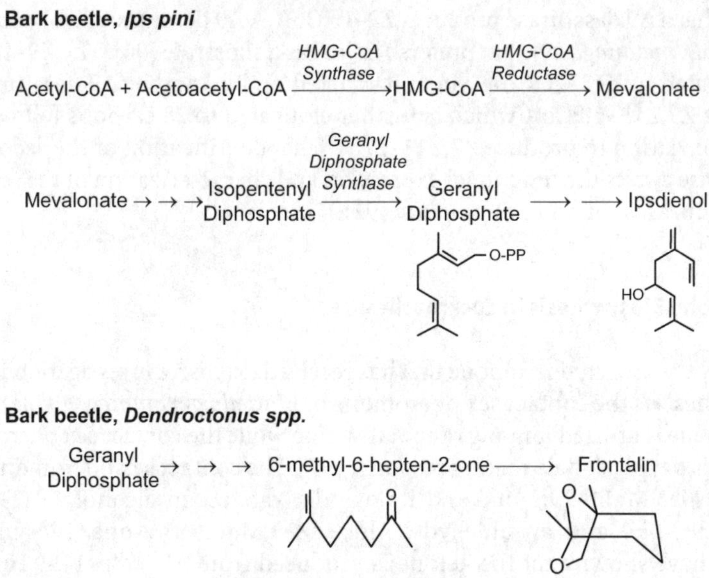

Fig. 4 Proposed biosynthetic pathways for the production of ipsdienol in *I. pini* and frontalin in *D. spp.* Both pathways utilize an isoprenoid biosynthetic pathway to produce geranyl diphosphate

pathway are HMG-CoA synthase and HMG-CoA reductase leading to formation of mevalonate. A key enzyme in the last part of the pathway is geranyl diphosphate synthase. Another pathway to formation of isoprenoid compounds involves the utilization of pyruvate and glyceraldehyde 3-phosphate to form 1-deoxy-xylulose 5-phosphate that is then converted to isopentenyl diphosphate [102]. These early non-mevalonate steps in the formation of isoprenoids has so far been found in bacteria and plants but not in insects.

Evidence for de novo synthesis of pheromone components was obtained by showing that labeled acetate and mevalonate were incorporated into ipsdienol by male *Ips pini* [103, 104]. Similarly, labeled acetate and other labeled intermediates were shown to be incorporated into frontalin in a number of *Dendroctonus* species [105]. Possible precursors to frontalin include 6-methyl-6-hepten-2-one, which was incorporated into frontalin by *D. ruffipennis* [106]. The precursor 6-methyl-6-hepten-2-one also was shown to be converted to brevicomin in the bark beetle, *Dendroctonus ponderosae* [107]. In addition, the expression patterns of HMG-CoA reductase and HMG-CoA synthase are tightly correlated with frontalin production in *Dendroctonus jeffreyi* [108, 109]. A geranyl diphosphate synthase cDNA from *I. pini* was also isolated, functionally expressed, and modeled [110]. These data indicate that the de novo isoprenoid biosynthetic pathway is present in bark beetles. A variety of other monoterpene alcohols such as myrcenol, pityol, and sulcitol are probably synthesized through similar pathways [111]

3.3.2
Fatty Acid-Derived and Other Pheromones

Beetles also utilize pheromones that are modified through fatty acid biosynthetic pathways. Examples include *Tenebrio molitor* that produces 4-methyl-1-nonanol from propionyl-, malonyl-, and methylmalonyl- precursors [112]. This is an example of carbon being shunted away from fatty-acyl elongation before long fatty acids are completed. An example of a beetle producing longer chain length methyl-branched pheromone components are the *Carpophilus* nitidulid beetles [113]. The branched hydrocarbons have 10–12 carbons with conjugated double bonds and methyl and ethyl branches and are of polyketide origin [114]. They are biosynthesized using propionate and butyrate to make the methyl and ethyl branches [114, 115]. The desaturated nature of these hydrocarbons is not due to fatty acyl desaturases, but to the inactivity of enoyl-ACP reductase during biosynthesis so that the enoyl-ACP intermediate formed during elongation is not reduced [115]. The enzymes involved in polyketide pheromone biosynthesis is unknown, but it will be interesting to determine if these enzymes are similar to the polyketide synthases found in bacteria and plants [116] or are they similar to the normal fatty acid biosynthetic enzymes.

Macrolide aggregation pheromones produced by male cucujid beetles are derived from fatty acids. Feeding experiments with labeled oleic, linoleic, and palmitic acids indicate incorporation into the macrolide pheromone component [117]. The biosynthesis of another group of beetle pheromones, the lactones, involves fatty acid biosynthetic pathways. Japonilure and buibuilactone biosynthesized by the female scarab, *Anomala japonica*, involves $\Delta 9$ desaturation of 16 and 18 carbon fatty acids to produce Z9–16:CoA and Z9–18:CoA, hydroxylation at carbon 8 followed by two rounds of limited chain shortening and cyclization to the lactone [118]. The hydroxylation step appears to be stereospecific [118].

Although not of fatty acid origin another group of scarab beetles utilizes amino acid derivatives as pheromones [119]. The large black chafer, *Holotrichia parallela*, uses L-isoleucine methyl ester [120] and the cranberry white grub, *Phyllophaga anxia*, uses both L-isoleucine and L-valine methyl esters [121]. More recently L-isoleucine methyl ester, *N*-formyl L-isoleucine methyl ester, and *N*-acetyl L-isoleucine methyl ester were identified in the scarab beetle *Phyllophaga elenans* [122]. These pheromone components are obviously derived from the amino acids isoleucine and valine.

3.4
Pheromone Biosynthesis in Other Insects

Moths, beetles, flies, and cockroaches have received the most attention regarding pheromone biosynthesis because their members contain prominent pest species and in addition are typically easy to rear in the laboratory. However several other insects have been investigated regarding pheromone biosynthesis, most notably the bees and butterflies.

Workers and queens of the common honey bee, *Apis mellifera*, produce various fatty acids in their mandibular glands. The queen mandibular pheromone consists primarily of OH9-E2–10:acid and 9-keto-E2–10:acids while the workers primarily have OH10-E2–10:acid and the corresponding diacid. Labeling studies using deuterium labeled precursors indicated that 18:acid served as the best substrate to produce the shorter chain acids indicating chain shortening reactions were involved in the biosynthetic pathway [123, 124]. The difference between queens and workers is that the hydroxylation of the 18:acid occurs preferentially on the ω carbon in workers and on the ω-1 carbon in queens. Chain shortening of the hydroxylated 18:acid to 8 or 10 carbons and subsequent hydroxy group oxidation produces the caste specific mandibular components [123–125].

Some male arctiid moths produce their courtship pheromone from dietary pyrrolizidine alkaloids acquired during feeding by the larvae [126]. Conversion of monocrotaline to hydroxydanaidal by males is accomplished by aromatization, ester hydrolysis and oxidation of an alcohol to the aldehyde [7]. In the case of *Utetheisa ornatirx* the stereo-configuration at C7 of the dietary alkaloid is the same as the pheromone released (*R*). In contrast, another arctiid, *Creatonotos transiens*, can convert a dietary precursor alkaloid with the (*S*) configuration at C7 (heliotrine) to (*R*)-hydroxydanaidal. The biosynthesis occurs by first oxidation-reduction at C7 to convert the stereochemistry and then proceeds through aromatization, hydrolysis, and oxidation [7].

4 Endocrine Regulation of Pheromone Production

In 1965 Barth proposed that the neuroendocrine control of pheromone production would occur in insects that were long-lived as adults and exhibited multiple reproductive cycles [127]. Cockroaches, beetles, and flies exhibit this type of life cycle and we now know that juvenile hormone (JH) and ecdysone are important regulators of pheromone production in these insects. Moths, on the other hand, typically are shorter lived as adults and exhibit continuous egg development. However, moths regulate pheromone production in order to produce the periodicity of pheromone release. Pheromone production in moths is primarily regulated by the neuropeptide pheromone biosynthesis activating neuropeptide (PBAN). Schal et al. [128] have proposed a revision of Barth's hypothesis taking into account the more current knowledge on the actual mechanisms of pheromone regulation in these insects.

4.1 PBAN Regulation in Moths

Most female moths release sex pheromones in a typical calling behavior in which the pheromone gland is extruded to release pheromone during a particular time of the photoperiod. In most cases pheromone biosynthesis coincides with calling behavior and the synchronization of these events is achieved by neuro-endocrine mechanisms present in the female that in turn are influenced by various environmental and physiological events such as temperature, photoperiod, host plants, mating, hormones, neurohormones, and neuromodulators. We now know that the main neuro-endocrine mechanism that regulates pheromone production in moths is pheromone biosynthesis activating neuropeptide (PBAN) [14, 129].

In the moth, *Helicoverpa zea,* PBAN was found in the subesophageal ganglion [130]. By dissection of brain-subesophageal ganglion complexes and purification through HPLC a peptide was sequenced [131]. Active fractions were identified based on induction of pheromone biosynthesis in head-ligated female moths during the photophase. The first PBAN identified had 33 amino acids with a C-terminal amidation and the core sequence FXPRLamide is required for activity, which places PBAN in a family of peptides with the C-terminal FXPRLamide motif (Table 1). The first member of this family to be identified was called leucopyrokinin based on its ability to stimulate hindgut contraction in the cockroach, *Leucophaea maderae* [132]. Subsequently it was found that peptides belonging to this family had other physiological roles. The peptide that induces melanization in lepidopteran larvae belongs to the PBAN/pyrokinin family [133]. The induction of embryonic diapause in *B. mori* is also mediated by PBAN-like peptides [134]. In addition it was determined that the white shrimp, *Penaeus vannamei*, has two peptides that can induce myotropic activity [135]. These results demonstrate the ubiquity and multifunctional nature of this family of peptides.

It was determined that the minimal peptide sequence required to stimulate pheromone biosynthesis was the C-terminal 5 amino acids, FXPRLamide, and that the carboxy terminus needs to be amidated [148, 149]. This sequence was also established as the minimal sequence required for myotropic activity in cockroaches [150] and induction of embryonic diapause in *B. mori* [151]. Cross-reactivity of peptides containing the FXPRLamide motif was also established for myotropic, diapause induction, and pheromone biosynthesis [152–154]. Therefore, the common C-terminal FXPRLamide defines this family of peptides. A partial listing of peptides identified to date is shown in Table 1.

Although structure activity studies established FXPRLamide as the minimal sequence for activity, considerably higher concentrations of the shortened peptides were required to stimulate pheromone biosynthesis to the same level as PBAN [148, 155]. Studies using pheromonotropic analogs also indicated that the FXPRLamide motif was required for activity [156, 157]. Nachman et. al. [156] demonstrated significant pheromonotropic activity with a conformationally constrained cyclic pyrokinin/PBAN analog, cyclo-[NTSFTPRL] octapeptide that retains a β-turn. NMR studies performed on the full PBAN sequence showed a type I β-turn conformation that encompassed the C-terminal region [158]. A lack of interaction between the C-terminal turn and the rest of the peptide molecule indicated that the C-terminal turn indeed represented the important conformation recognized by the PBAN receptor. Taking into account

Table 1 A partial listing of the PBAN family of peptides and the species where identified. The FXPRLamide motif is shown in bold. X=S, T, G, or V. Peptides are grouped together based on the primary function for which they were first identified

Function and Species	Peptide sequence		
PBAN			
Helicoverpa zea	LSDDMPATPADQEMYRQDPEQIDSRTKY**FSPRL**amide[a]		[131]
Bombyx mori	LSEDMPATPADQEMYQPDPEEMESRTRY**FSPRL**amide[a]		[136]
Pheromonotropic peptides			
Bombyx mori	α	II**FTPKL**amide[b]	[137]
Helicoverpa zea	PGN-7	VI**FTPKL**amide[b]	[138]
Bombyx mori	β	SVAKPQTHESLE**FIPRL**amide[b]	[137]
Helicoverpa zea	PGN-18	SLAYDDKSFENVE**FTPRL**amide[b]	[138]
Pseudaletia separata	β	KLSYDDKVFENVE**FTPRL**amide[b]	[139]
Bombyx mori	γ	TMS**FSPRL**amide[b]	[137]
Helicoverpa zea	PGN-8	TMN**FSPRL**amide[b]	[138]
Diapause Hormone			
Bombyx mori		TDMKDESDRGAHSERGALC**FGPRL**amide[a]	[134]
Helicoverpa zea	PGN-24	NDVKDGAASGAHSDRLGLW**FGPRL**amide[b]	[138]
Pyrokinins			
Leucophaea maderae		pETS**FTPRL**amide[a]	[132]
Locusta migratoria	(I)	pEDSGDGWPQQP**FVPRL**amide[a]	[140]
	(II)	pESVPT**FTPRL**amide[a]	[141]
Myotropins			
Periplaneta americana	PK1	HTAG**FIPRL**amide[a]	[142]
	PK2	SPP**FAPRL**amide[a]	[142]
	PK3	LVP**FRPRL**amide[a]	[143]
	PK4	DHLPHDVY**SPRL**amide[a]	[143]
	PK5	GGGGSGETSGMW**FGPRL**amide[a]	[143]
	PK6	SESEVPGMW**FGPRL**amide[a]	[144]
Drosophila melanogaster	CAP2b-3	TGPSASSGLW**FGPRL**amide[b]	[145]
	PK-2	SVPF**KPRL**amide[b]	[145]
	ETH-1	DDSSPGFFLKITKNV**PRL**amide[b]	[146]
	hug γ	pELQSNGIPAYRVR**TPRL**amide[b]	[147]
Penaeus vannamei		DFA**FSPRL**amide[a]	[135]
(Crustacea)		ADFA**FNPRL**amide[a]	[135]

[a] Identified from the amino acid sequence of a purified peptide.
[b] Deduced from the cloned gene sequence.
PGN=PBAN-encoding gene neuropeptide.

the cross-reactivity of these peptides in the different physiological functions, the peptide family share the binding requirements identified by this rigid, well-defined backbone structure. However, the requirement of the full sequence for full activity suggests that other parts of the PBAN sequence may be involved in the interaction with the receptor and that selective binding capabilities must occur [159].

Some of these questions can now be addressed in a more direct manner with the identification of a PBAN-receptor from pheromone glands of *H. zea* [160]. The receptor was identified based on sequence similarities with several G-protein coupled receptors (GPCR) identified from *D. melanogaster* [161, 162]. The receptor cDNA from pheromone glands was expressed in insect cells and functional activity tested using a calcium influx assay [160]. It was determined that the full length PBAN plus some shorter peptides had the highest affinity for the receptor. Leucopyrokinin was almost as effective as PBAN, which confirms that the FXPRLamide motif is essential for binding to the receptor. The ability to test peptides directly on the PBAN-receptor will help determine structural motifs required for binding activity.

PBAN binding to a receptor results in signal transduction events to stimulate the pheromone biosynthetic pathway (Fig. 5). Receptor activation results in the influx of extracellular calcium and has been demonstrated in a number of moths [163–168]. The increase in cytosolic calcium can directly stimulate pheromone biosynthesis in some moths [165–168] or it will stimulate the production of cAMP [169, 170]. So far cAMP has only been implicated in signal

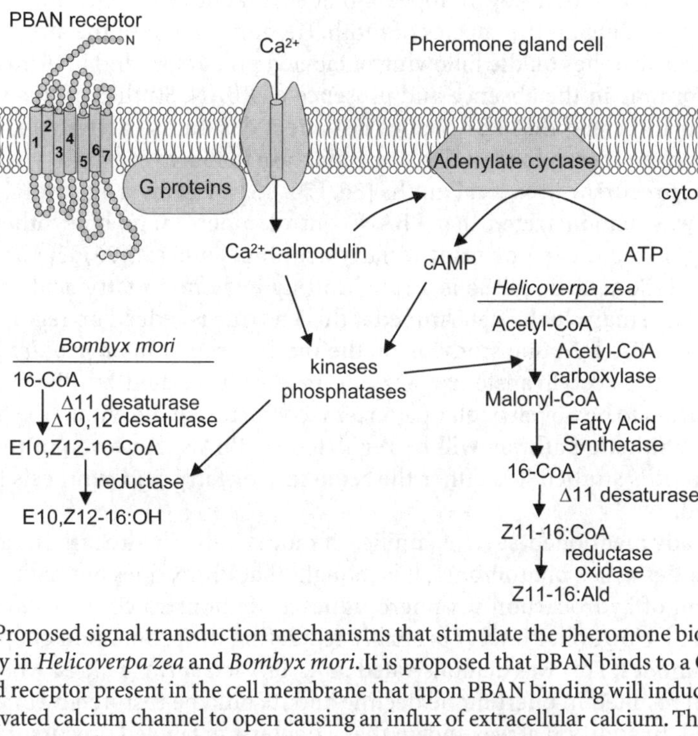

Fig. 5 Proposed signal transduction mechanisms that stimulate the pheromone biosynthetic pathway in *Helicoverpa zea* and *Bombyx mori*. It is proposed that PBAN binds to a G protein-coupled receptor present in the cell membrane that upon PBAN binding will induce a receptor-activated calcium channel to open causing an influx of extracellular calcium. This calcium binds to calmodulin and in the case of *B. mori* will directly stimulate a phosphatase that will dephosphorylate and activate a reductase in the biosynthetic pathway. In *H. zea* the calcium-calmodulin will activate adenylate cyclase to produce cAMP that will then act through kinases and/or phosphatases to stimulate acetyl-CoA carboxylase in the biosynthetic pathway

transduction in the heliothine moths [171]. The calcium channel is thought to be a receptor-activated calcium channel since voltage-gated calcium channel blockers did not inhibit PBAN stimulation [163]. Increase in calcium as a result of the pharmacological action of ionophores duplicated the action of PBAN on cAMP synthesis [172, 173].

Calcium is probably binding to calmodulin to form a complex that activates downstream events. A calmodulin was identified from pheromone glands of *B. mori* [174]. The calcium/calmodulin complex will then directly or indirectly activate a phosphoprotein phosphatase that will in turn activate an acyl-CoA reductase in the biosynthetic pathway [168]. Specific inhibitors of calcineurin (phosphoprotein phosphatase 2B) reduced the amount of pheromone when isolated glands were challenged with a pheromonotropic peptide [166]. Two genes encoding calcineurin heterosubunits were identified from the pheromone gland of *B. mori* and were found to be homologous to the catalytic subunit and regulatory subunits of other animal calcineurins [175]. The calcineurin complex will apparently dephosphorylate an acyl-CoA reductase which catalyzes the formation of bombykol in *B. mori*.

Several studies were undertaken to determine how PBAN affects the pheromone biosynthetic pathway. It appears that several different enzymes can be affected depending on the species of moth. Demonstration of the affected enzyme primarily relies on the following of labeled precursors and intermediates into pheromone in the absence and presence of PBAN. Studies of this nature have so far indicated that PBAN does not affect desaturase activity [43, 176]. PBAN has been shown to stimulate the reductase that converts an acyl-CoA to an alcohol precursor in several moths [36, 176–181]. However, in several other moths it was demonstrated that PBAN controls pheromone biosynthesis by regulating a step during or prior to fatty acid biosynthesis [26, 42, 163, 182–184]. Acetyl-CoA carboxylase is a rate limiting enzyme in fatty acid biosynthesis and it remains to be determined if this enzyme is indeed up regulated in response to PBAN. In one study using the moth *Sesamia nonagrioides* it was shown that the acetyltransferase enzyme may be regulated by PBAN [185]. There appears to be no particular pattern as to which enzyme within the pheromone biosynthetic pathway will be regulated by PBAN. However, in the majority of moths studied it is either the reductase or fatty acid synthesis that is stimulated.

As already mentioned, several families of moths utilize hydrocarbons and/or their epoxides as sex pheromones. It is thought that PBAN does not regulate the production of hydrocarbon sex pheromones as demonstrated in *Scoliopteryx libatrix* [73]. However, PBAN is probably regulating the production of epoxide sex pheromones. This was demonstrated in *Ascotis selenaria cretacea* where decapitation resulted in pheromone decline and it could be restored by injecting PBAN [76]. In addition, it was shown that a deuterium labeled precursor of the pheromone applied to the pheromone gland was converted to the epoxide only in PBAN injected females [76]. Decapitation also decreases the epoxide pheromone titer in the gypsy moth, *Lymantria dispar*, and injection of PBAN can restore pheromone production [186]. However, decapitation did not decrease the levels of the hydrocarbon precursor in the gypsy moth (Jurenka, unpublished). These findings indicate that PBAN may regulate the epoxidation step in those moths that utilize epoxide pheromones but not the production of the alkene precursor.

PBAN is produced by three groups of neurons present in the subesophageal ganglion (SEG). The gene encoding PBAN has been identified and this information has been used to conduct in situ hybridization studies that show the SEG neurons as containing the mRNA encoding for PBAN. PBAN-like immunoreactivity has also been localized to these same neurons in the SEG [187–189]. The three groups of neurons project to the corpus cardiacum, which is the main brain neurohemal organ for releasing neuropeptides into circulation [187, 190]. In addition, two pairs of maxillary neurons within the SEG send projections into the paired ventral nerve cord and travel its entire length to terminate in the terminal abdominal ganglion. Arborizations arising from these paired projections were found in each segmental ganglion [190]. Thoracic and abdominal segmental ganglia also have neurons that contain PBAN-like activity and these neurons can release peptides into the hemolymph [188–190]. Although in situ hybridization studies indicate that the segmental ganglia do not express the same gene as does the SEG [191–194].

The gene encoding PBAN was first characterized from *H. zea* and *B. mori* [134, 137, 138, 195]. The cDNA was found to encode the 33 amino acid PBAN plus four additional peptides with a common C-terminal FXPRL sequence motif, including that of the diapause hormone of *B. mori* (Fig. 6). Three additional peptides with the common C-termini and sequence homology to those of *H. zea* and *B. mori* have been deduced from cDNA isolated from pheromone glands of several other moths [194, 196–200]. Studies conducted to find the post-translational processed peptides indicated that PBAN was found to a greater extent in the mandibular and maxillary clusters than in the labial cluster of neurons

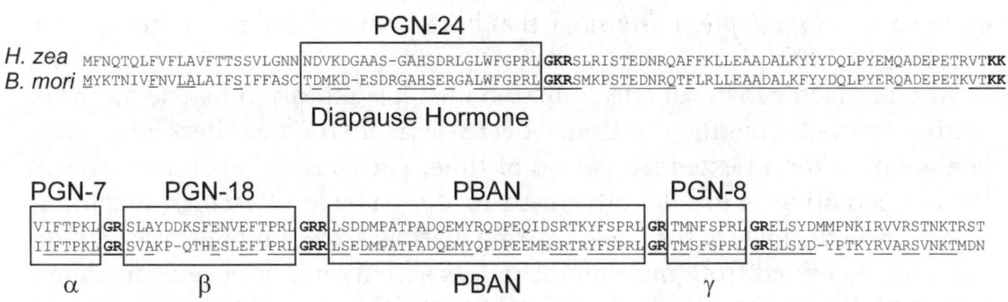

Fig. 6 Sequence alignment of the deduced amino acid sequence from the identified cDNA encoding PBAN and related peptides from *Helicoverpa zea* and *Bombyx mori*. The putatively expressed peptides are shown in *boxes*. The conserved amino acids are *underlined* in the *B. mori* sequence. Putative proteolytic posttranslational processing sites are shown in *bold* with glycine contributing the C-terminal amide. Sequences of PBAN-like peptides are also shown in Table 1. GenBank accession numbers: *H. zea* – P11159 and *B. mori* – BAA05971

in the SEG [201]. The other neuropeptides were found in all clusters. In addition, MALDI-MS data indicated that some larger peptide fragments were found suggesting alternative processing of the precursor protein [201].

Several factors can mediate the action of PBAN. The production of pheromone can be mediated by the action of other hormones. Juvenile hormones play an important role in reproductive development of many moth species. Although JH probably does not regulate pheromone biosynthesis directly it has been shown to be involved in the release of PBAN in some migratory moths [202–205]. In addition, JH has been shown to prime the pheromone glands in pharate adults of the non-migratory moth *H. armigera* [206]. JH II, in an in vitro assay, primed pheromone glands of pharate adults to respond to PBAN and induced earlier pheromone production by intact newly emerged females [206]. This induction could be mediated by JH up-regulation of a putative PBAN-receptor in pharate adults [207].

The role of the nervous system in pheromone biosynthesis in moths is not clearly understood. Christensen and co-workers [208–211] proposed that the neurotransmitter octopamine may be involved as an intermediate messenger during the stimulation of sex pheromone production in *H. virescens*. These workers suggested that octopamine was involved in the regulation of pheromone production and that PBAN's role lies in the stimulation of octopamine release at nerve endings. However, contradicting results concerning octopamine-stimulated pheromone production were reported in the same species as well as other moth species [163, 172, 212–214].

Studies, using octopamine and related biogenic amines, suggested that octopamine may play an important role as neuromodulator of the PBAN-induced pheromonotropic response in *H. armigera* [215–217]. Octopamine and several octopaminergic analogs inhibited pheromone production in studies using both in vitro and in vivo bioassays [215–218]. Active inhibition of pheromone biosynthesis during the photophase could be a role for octopamine acting as an endogenous inhibitory factor [216]. The role of the ventral nerve cord in pheromone production still requires clarification. It has been suggested, in some moth species (*Spodoptera littoralis*), that both humoral and neural regulation occurs [219, 220].

Another factor that can affect pheromone biosynthesis in female moths is mating. In moths, mating results in a decrease in pheromone titers for at least one night, or for an extended period of time. The cause of pheromonostasis following mating is usually attributed to the passage of factors, including sperm, from the male to the female during mating. A pheromonostatic peptide has been identified from male moths but its activity has not been confirmed [221, 222]. However, recently a truncated form of the pheromonostatic peptide was shown to inhibit pheromone production in *H. armigera* [223]. In addition the sex peptide and a duplex-derived peptide from *D. melanogaster* have been shown to inhibit pheromone biosynthesis in the moth *H. armigera* [224, 225]. These findings indicate that factors transferred from the male to the female during mating can influence pheromone production in post-mated females.

4.2
Juvenile Hormone Regulation in Beetles

It appears that, in beetles, pheromone production is regulated by JH III, despite the variations in biosynthetic pathways. JH apparently regulates pheromone production in beetles that utilize both fatty acid and isoprenoid biosynthetic pathways [8, 98]. Environmental and physiological factors will in turn regulate production of JH. The endocrine regulation of pheromone production in the beetles has been best studied with regard to the bark beetles.

Bark beetles primarily produce aggregation pheromones in response to feeding on a new host tree [101]. However, unfed beetles could be induced to produce pheromone by treatment with JH III or a JH analog [104, 226, 227]. Feeding on a new host tree will elevate JH titers by stimulating activity of the corpus allatum, the endocrine gland that produces JH [104]. The resulting increased JH hemolymph titer will trigger pheromone biosynthesis that occurs in the midgut cells [19]. The action of JH at target tissues is to stimulate the production of enzymes involved in the biosynthetic pathways.

The elevation of enzyme levels following the increase in JH levels after feeding or by treatment has been observed for key enzymes in the isoprenoid biosynthetic pathway. In *Ips paraconfusus*, *I. pini*, and *D. jeffreyi* JH-treated male beetles had increased HMG-CoA reductase mRNA levels compared to controls [98, 109, 228]. The response was dose- and time-dependent and resulted in increased pheromone production. These findings indicate that JH controls the up-regulation of HMG-CoA reductase, a key enzyme in the isoprenoid biosynthetic pathway. mRNA levels of other enzymes in the pathway are also affected by JH treatment including HMG-CoA synthase [108] and geranyl diphosphate synthase [98, 110]. Enzymes further along in the pathway can also be elevated due to JH treatment. For example, male-specific myrcene synthase activity was increased in *I. pini* beetles treated with JH [229]. Genomic approaches to understanding the regulation of enzymes in the biosynthesis of bark beetle pheromones will yield insights into the regulatory function of JH [98].

4.3
Juvenile Hormone Regulation in Cockroaches

Most female cockroaches exhibit reproductive cycles of egg development, embryogenesis and egg laying. Juvenile hormone is the primary hormone responsible for egg development and thus there is a correlation between egg development and hemolymph JH titers or production by the corpora allata. In addition the sex pheromone gland undergoes corresponding cyclic maturational changes in the gland in relation to the ovarian cycle [128]. Several cockroaches utilize volatile sex pheromones; however, little is known about the biosynthesis of these compounds and thus little is known about the hormonal regulation of pheromone biosynthesis. In contrast more information is available

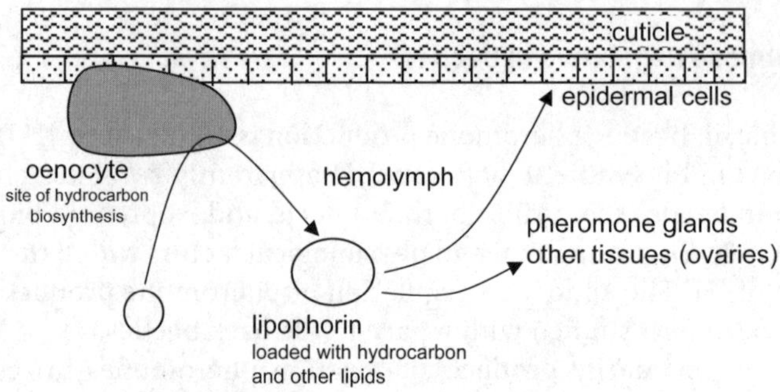

Fig. 7 Transport of hydrocarbons and other lipids by lipophorin from site of synthesis (oenocytes) to various tissues. In the case of pheromone glands specific hydrocarbons are unloaded to be used directly as a pheromone or modified with the addition of oxygen and then released as a pheromone

on the hormonal regulation of the biosynthesis of the contact sex pheromone of the German cockroach, *B. germanica* [128].

Production of the contact sex pheromone, Me3,Me11–29:2-one, produced by female *B. germanica* is regulated by JH. A correlation between the amount and biosynthesis (as measured by incorporation of radiolabeled propionate) of pheromone on the cuticular surface and active vitellogenesis indicated regulation by JH [230, 231]. Reduction of JH titer by allatectomy reduced pheromone amounts whereas treatment with JH III or JH analogs stimulated pheromone production [6, 230]. The step that is thought to be under JH regulation is the hydroxylation of Me3,Me11–29:H because this conversion occurs in only females [6]. Contact pheromone production in *B. germanica* is also regulated through the production of its precursor, Me3,Me11–29:H, which is thought to be in part regulated by food intake [230].

The biosynthesis of Me3,Me11–29:H takes place in oenocyte cells, released into the hemolymph and transported by lipophorin to peripheral tissues (Fig. 7) [71, 231, 232]. Direct evidence for oenocytes biosynthesizing hydrocarbon has come recently with the dissociation of oenocytes from epidermal cells and in vitro incubation with labeled propionate [233]. Differential uptake of some hydrocarbons in different tissues has also been documented although the exact mechanism behind the differential placement of hydrocarbons is unknown [20, 128, 230, 232, 234]. Although the biosynthesis of hydrocarbons may not be under direct endocrine regulation supply of precursor hydrocarbon that is converted to the sex pheromone is a requirement.

4.4 20-Hydroxyecdysone Regulation in Diptera

The Diptera regulate ovarian development in a different manner than do cockroaches and beetles. Whereas in most insects JH directly stimulates the fat body to produce vitellogenin and the ovaries to undergo oogenesis, in the Diptera JH primarily plays a role in making the ovaries and fat body competent for further development. The hormone that stimulates egg development directly is 20-hydroxyecdysone which is produced by the follicle cells of the ovary. In the housefly, *M. domestica*, sex pheromone production is correlated with egg development. It has been well established that 20-hydroxyecdysone is the main hormone for stimulation of the 23 carbon sex pheromones of the housefly [235].

The evidence for 20-hydroxyecdysone stimulating sex pheromone production in the housefly comes from both direct and indirect studies. A correlation was found between ovarian development and sex pheromone production in female flies [236, 237]. Surgical removal of ovaries immediately after adult emergence resulted in no sex pheromone production, whereas allatectomized (which removes the source of JH production) females produced the same amount of pheromone as did normal females [238, 239]. Additionally, ovariectomized females that received ovary implants produced sex pheromone [238]. These data demonstrate that 20-hydroxyecdysone and not JH regulates pheromone production in the housefly [111].

Additional evidence came from the finding that sex pheromone production could be stimulated in male houseflies that do not normally produce detectable sex pheromone components. Male houseflies were found to have longer chain alkenes, Z9–27:H, but did not have Z9–23:H. Implantation of ovaries into male houseflies resulted in a change in hydrocarbon biosynthesis such that the longer chain alkenes were not made but rather they produced the shorter chain length Z9–23:H [240]. Likewise, injection of 20-hydroxyecdysone into males induced sex pheromone production in a dose-dependent manner. These studies demonstrated that males possess the biosynthetic capability to produce sex pheromone, but normally do not produce the 20-hydroxyecdysone necessary to induce sex pheromone production. Males became an excellent model in which to study the hormonal regulation of pheromone biosynthesis in the housefly.

Since it was established that 20-hydroxyecdysone regulates sex pheromone and that a switch was taking place from producing longer chain-length hydrocarbons to producing the shorter chain-length sex pheromones, a question became how this switch occurred. The shorter chain-length hydrocarbons could be produced by an alteration in the decarboxylation reactions such that Z9–24:CoA is preferred or that the fatty acyl-CoA elongation system is altered such that chain elongation stops at Z9–24:CoA. It was determined that the reduction of various chain length acyl-CoAs to alkenes was not different in females of all ages examined or males treated with 20-hydroxyecdysone [91, 241]. However a correlation was observed with females actively producing Z9-23:H and the chain elongation system. This was shown by comparing the activity obtained from microsomes prepared from sexually mature females to immature females and males. The microsomes from sexually mature females did not elongate Z9–18:CoA past Z9–24:CoA, but microsomes from immature females and males elongated Z9–18:CoA and Z9–24:CoA to Z9–28:CoA [241, 242]. These results indicate that 20-hydroxyecdysone is regulating the fatty acyl-CoA elongases. Further research will determine the exact mechanism behind this regulation, but it could be that a new elongation system is induced by 20-hydroxyecdysone.

As with the other insects studied that utilize hydrocarbon sex pheromones, once Z9–23:H is produced by oenocyte cells it is released into the hemolymph. Lipophorin is the transport protein that will move the hydrocarbon to cuticular tissue [21]. It was found that about 24 h were required once Z9–23:H was induced to actual deposition on the cuticular surface [237]. As is the case with other insects selective partitioning of the sex pheromone was observed with relatively larger proportions of Z9–23:H being found on the cuticular surface than in other tissues [21].

5 Conclusions

Pheromone biosynthesis has been primarily studied in several model insects that include moths, flies, and cockroaches. A considerable amount of information is now known about how these insects biosynthesize and regulate pheromone production. As additional pheromones are identified [243] work will be carried out on their biosynthesis. Current and future work will also concentrate on the molecular mechanisms behind the enzymes involved in biosynthetic pathways and also on the signal transduction pathways involved in the endocrine regulation. Additional information will provide valuable insights into these pathways especially with regard to receptor identification [160]. Another area of research that will provide valuable information is the specific uptake of hydrocarbons by pheromone glands and other tissues and how these pheromone components can be transported to the cuticular surface [128]. Work on pheromone and odorant binding proteins may help solve this problem [244]. The future should also bring research on pheromone production that will produce practical applications to insect pest population management.

References

1. Butenandt A, Beckman R, Stamm D, Hecker E (1959) Z Naturforsch 14B:283
2. Arn H (2003) http://wwwnysaescornelledu/pheronet/ or http://www-pherolistsluse/
3. Schulz S (2001) In: Barth FG, Schmid A (eds) Ecology of sensing. Springer, Berlin Heidelberg New York, p 237
4. Francke W, Bartels J, Meyer H, Schroder F, Kohnle U, Baade RE, Vite JP (1995) J Chem Ecol 21:1043
5. Gemeno C, Schal C (2004) In: Cardé RT, Millar J (eds) Advances in insect chemical ecology. Cambridge University Press

6. Chase J, Touhara K, Prestwich GD, Schal C, Blomquist GJ (1992) Proc Natl Acad Sci USA 89:6050
7. Schulz S, Francke W, Boppre M, Eisner T, Meinwald J (1993) Proc Natl Acad Sci USA 90:6834
8. Tillman JA, Seybold SJ, Jurenka RA, Blomquist GJ (1999) Insect Biochem Mol Biol 29:481
9. Billen J, Morgan ED (1998) In: Vander Meer RK, Breed MD, Winston ML, Espelie KE (eds) Pheromone communication in social insects. Westview Press, Boulder, CO, p 3
10. Quennedy A (1998) In: Harrison FW, Locke M (eds) Microscopic anatomy of invertebrates, vol 11A. Wiley-Liss, New York, p 177
11. Ma PWK, Ramaswamy SB (2003) In: Blomquist GJ, Vogt RG (eds) Insect pheromone biochemistry and molecular biology. Academic Press
12. Bjostad LB, Wolf WA, Roelofs WL (1987) In: Prestwich GD, Blomquist GJ (eds) Pheromone biochemistry. Academic Press, Orlando, FL, p 77
13. Jurenka RA (2003) In: Blomquist G, Vogt R (eds) Pheromone biochemistry and molecular biology. Academic Press, New York
14. Rafaeli A, Jurenka RA (2003) In: Blomquist G, Vogt R (eds) Pheromone biochemistry and molecular biology. Academic Press, New York
15. Hallberg E, Subchev M (1997) Int J Insect Morphol Embryol 25:381
16. Wigglesworth VB (1970) Tissue Cell 2:155
17. Schal C, Sevala V, Cardé RT (1998) Naturwissenschaften 85:339
18. Nardi JB, Dowd PF, Bartelt RJ (1996) Tissue Cell 28:43
19. Hall GM, Tittiger C, Andrews GL, Mastick GS, Kuenzli M, Luo X, Seybold SJ, Blomquist GJ (2002) Naturwissenschaften 89:79
20. Gu X, Quilici D, Juarez P, Blomquist GJ, Schal C (1995) J Insect Physiol 41:257
21. Schal C, Sevala V, Capurro M, Snyder T, Blomquist G, Bagnères A (2001) J Insect Sci 1. Available online: insectscience.org/1.12
22. Bjostad LB, Roelofs WL (1983) Science 220:1387
23. Bjostad LB, Roelofs WL (1984) J Chem Ecol 10:681
24. Jurenka RA, Jacquin E, Roelofs WL (1991) Arch Insect Biochem Physiol 17:81
25. Jurenka RA, Haynes KF, Adlof RO, Bengtsson M, Roelofs WL (1994) Insect Biochem Mol Biol 24:373
26. Tang JD, Charlton RE, Jurenka RA, Wolf WA, Phelan PL, Sreng L, Roelofs WL (1989) Proc Natl Acad Sci USA 86:1806
27. Ryan R, de Renobales M, Dillwith J, Heisler C, Blomquist G (1982) Arch Biochem Biophys 213:26
28. Bjostad LB, Roelofs WL (1981) J Biol Chem 256:7936
29. Hashimoto T (1996) Ann N Y Acad Sci 804:86
30. Bosch H, Schutgens RBH, Wanders RJA, Tager JM (1992) Annu Rev Biochem 61:157
31. Jurenka RA, Roelofs WL (1993) In: Stanley-Samuelson DW, Nelson DR (eds) Insect lipids: chemistry, biochemistry and biology. University of Nebraska Press, Lincoln, p 353
32. Haynes KF, Hunt RE (1990) J Chem Ecol 16:1249
33. Berger RS (1966) Ann Entomol Soc Am 59:767
34. Wu WQ, Zhu JW, Millar J, Löfstedt C (1998) Insect Biochem Mol Biol 28:895
35. Foster SP, Roelofs WL (1996) Arch Insect Biochem Physiol 33:135
36. Martinez T, Fabriàs G, Camps F (1990) J Biol Chem 265:1381
37. Foster SP, Roelofs WL (1988) Arch Insect Biochem Physiol 8:1
38. Zhao C, Löfstedt C, Wang X (1990) Arch Insect Biochem Physiol 15:57
39. Ono A, Imai T, Inomata S-I, Watanabe A, Ando T (2002) Insect Biochem Mol Biol 32:701
40. Abad JL, Camps F, Fabriás G (2001) Insect Biochem Mol Biol 31:799–803

41. Foster SP, Roelofs WL (1990) Experientia 46:269
42. Jurenka RA (1997) Cell Mol Life Sci 53:501
43. Arsequell G, Fabriàs G, Camps F (1990) Arch Insect Biochem Physiol 14:47
44. Ando T, Hase T, Arima R, Uchiyama M (1988) Agric Biol Chem 52:473–478
45. Löfstedt C, Bengtsson M (1988) J Chem Ecol 14:903
46. Fang N, Teal PEA, Doolittle RE, Tumlinson JH (1995) Insect Biochem Mol Biol 25:39
47. Rodriguez S, Clapes P, Camps F, Fabrias G (2002) J Org Chem 67:2228
48. Villorbina G, Rodriguez S, Camps F, Fabrias G (2003) Insect Biochem Mol Biol 33:155
49. Millar JG (2000) Annu Rev Entomol 45:575
50. Renobales M, Cripps C, Stanley-Samuelson DW, Jurenka RA, Blomquist GJ (1987) Trends Biochem Science 12:364
51. Rule GS, Roelofs WL (1989) Arch Insect Biochem Physiol 12:89
52. Knipple DC, Rosenfield C, Miller SJ, Liu W, Tang J, Ma PWK, Roelofs WL (1998) Proc Natl Acad Sci USA 95:15287
53. Knipple DC, Rosenfield C-L, Nielsen R, You KM, Jeong SE (2002) Genetics 162:1737
54. Knipple DC, Roelofs WL (2003) In: Blomquist GJ, Vogt RG (eds) Insect pheromone biochemistry and molecular biology. Academic Press, New York, p 81
55. Hao G, Liu W, O'Connor M, Roelofs WL (2002) Insect Biochem Mol Biol 32:961
56. Roelofs WL, Liu W, Hao G, Jiao H, Rooney AP, Linn CE Jr (2002) Proc Natl Acad Sci USA 99:13621
57. Rosenfield C, You KM, Marsella-Herrick P, Roelofs WL, Knipple DC (2001) Insect Biochem Mol Biol 31:949
58. Liu W, Jiao H, O'Connor M, Roelofs WL (2002) Insect Biochem Mol Biol 32:1489
59. Pinilla A, Camps F, Fabriàs G (1999) Biochemistry 38:15272
60. Liu W, Jiao H, Murray NC, O'Connor M, Roelofs WL (2002) Proc Natl Acad Sci USA 99:620
61. Roelofs WL, Rooney AP (2003) Proc Natl Acad Sci USA 100:9179
62. Morse D, Meighen EA (1987) In: Prestwich GD, Blomquist GJ (eds) Pheromone biochemistry. Academic Press, Orlando, FL, p 121
63. Teal PEA, Tumlinson JH (1988) J Chem Ecol 14:2131
64. Fang N, Teal PEA, Tumlinson JH (1995) Arch Insect Biochem Physiol 29:243
65. Moto K, Yoshiga T, Yamamoto M, Takahashi S, Okano K, Ando T, Nakata T, Matsumoto S (2003) Proc Natl Acad Sci USA 100:9156
66. Jurenka RA, Roelofs WL (1989) Insect Biochem 19:639
67. Bestmann HJ, Herrig M, Attygalle AB (1987) Experientia 43:1033
68. Teal PEA, Tumlinson JH (1987) Arch Insect Biochem Physiol 4:261
69. Zhu JW, Zhao CH, Lu F, Bengtsson M, Lofstedt C (1996) Insect Biochem Mol Biol 26:171
70. Wolf WA, Roelofs WL (1987) J Chem Ecol 13:1019
71. Schal C, Sevala VL, Young HP, Bachmann JAS (1998) Am Zool 38:382
72. Jurenka RA, Subchev M (2000) Arch Insect Biochem Physiol 43:108
73. Subchev M, Jurenka RA (2001) Arch Insect Biochem Physiol 47:35
74. Wei W, Miyamoto T, Endo M, Murakawa T, Pu G-Q, Ando T (2003) Insect Biochem Mol Biol 33:397
75. Stanley-Samuelson DW, Jurenka RA, Cripps C, Blomquist GJ, deRenobales M (1988) Arch Insect Biochem Physiol 9:1
76. Miyamoto T, Yamamoto M, Ono A, Ohtani K, Ando T (1999) Insect Biochem Mol Biol 29:63
77. Jurenka RA, Subchev M, Abad J-L, Choi M-Y, Fabriàs G (2003) Proc Natl Acad Sci USA 100:809
78. Dillwith JW, Blomquist GJ, Nelson DR (1981) Insect Biochem 11:247

79. Blomquist GJ, Dillwith JW, Adams TS (1987) In: Blomquist GJ, Prestwich GD (eds) Pheromone biochemistry. Academic Press, New York, p 217
80. Vaz AH, Blomquist GJ, Wakayama EJ, Reitz RC (1987) Insect Biochem 18:177
81. Ahmad S, Kirkland KE, Blomquist GJ (1987) Arch Insect Biochem Physiol 6:121
82. Guo L, Latli B, Prestwich GD, Blomquist GJ (1991) J Chem Ecol 17:1769
83. Dillwith JW, Nelson JH, Pomonis JG, Nelson DR, Blomquist GJ (1982) J Biol Chem 257:11305
84. Chase J, Jurenka RA, Schal C, Halarnkar PP, Blomquist GJ (1990) Insect Biochem 20: 149
85. Juarez P, Chase J, Blomquist GJ (1992) Arch Biochem Biophys 293:333
86. Gu P, Welch WH, Blomquist GJ (1993) Insect Biochem Mol Biol 23:263
87. Gu P, Welch WH, Guo L, Schegg KM, Blomquist GJ (1997) Comp Biochem Physiol 118B:447
88. Cheesbrough TM, Kolattukudy PE (1984) Proc Natl Acad Sci USA 81:6613
89. Yoder JA, Denlinger DL, Dennis MW, Kolattukudy PE (1992) Insect Biochem Mol Biol 22:237
90. Reed JR, Vanderwel D, Choi S, Pomonis JG, Reitz RC, Blomquist GJ (1994) Proc Natl Acad Sci USA 91:10000
91. Reed JR, Quilici DR, Blomquist GJ, Reitz RC (1995) Biochemistry 34:16
92. Jallon J-M, Wicker-Thomas C (2003) In: Blomquist GJ, Vogt RG (eds) Insect sex pheromone biochemistry and molecular biology. Academic Press
93. Ferveur JF, Savarit F, O'Kane CJ, Sureau G, Greenspan RJ, Jallon JM (1997) Science 276:1555
94. Pho DB, Pennanećh M, Jallon JM (1996) Arch Insect Biochem Physiol 31:289
95. Pennanećh M, Bricard L, Kunesch G, Jallon J-M (1997) J Insect Physiol 43:1111
96. Wicker-Thomas C, Henriet C, Dallerac R (1997) Insect Biochem Mol Biol 27:963
97. Liang D, Schal C (1993) Tissue Cell 25:763
98. Seybold SJ, Tittiger C (2003) Annu Rev Entomol 48:425
99. Seybold SJ, Vanderwel D (2003) In: Blomquist GJ, Vogt RG (eds) Insect pheromone biochemistry and molecular biochemistry. Academic Press, p 137
100. Schlyter F, Birgersson GS (1999) In: Hardie J, Minks AK (eds) Pheromones of non-lepidopteran insects associated with agricultural plants. CABI Publishing, p 113
101. Seybold SJ, Bohlmann J, Raffa KF (2000) Can Entomol 132:697
102. Rohmer M (1999) Nat Prod Rep 16:565
103. Seybold SJ, Quilici DR, Tillman JA, Vanderwel D, Wood DL, Blomquist GJ (1995) Proc Natl Acad Sci USA 92:8393
104. Tillman JA, Holbrook GL, Dallara PL, Schal C, Wood DL, Blomquist GJ, Seybold SJ (1998) Insect Biochem Mol Biol 28:705
105. Barkawi LS, Francke W, Blomquist GJ, Seybold SJ (2003) Insect Biochem Mol Biol 33:773
106. Perez AL, Gries R, Gries G, Oehlschlager AC (1996) Bioorg Med Chem 4:445
107. Vanderwel D, Oehlschlager AC (1992) J Am Chem Soc 114:5081
108. Tittiger C, O'Keeffe C, Bengoa CS, Barkawi LS, Seybold SJ, Blomquist GJ (2000) Insect Biochem Mol Biol 30:1203–1211
109. Tittiger C, Barkawi LS, Bengoa CS, Blomquist GJ, Seybold SJ (2003) Mol Cell Endocrinol 199:11
110. Gilg-Young A, Welch WH, Tittiger C, Blomquist GJ (unpublished)
111. Blomquist GJ, Jurenka R, Schal C, Tittiger C (2004) In: Gilbert LI, Iatrou K, Gill S (eds) Comprehensive insect physiology, biochemistry, pharmacology and molecular biology. Elsevier (in press)
112. Islam N, Bacala R, Moore A, Vanderwel D (1999) Insect Biochem Mol Biol 29:201

113. Bartelt RJ, Weisleder D, Dowd PF, Plattner RD (1992) J Chem Ecol 18:379–402
114. Bartelt RJ, Weisleder D (1996) Bioorg Med Chem 4:429
115. Petroski RJ, Bartelt RJ, Weisleder D (1994) Insect Biochem Mol Biol 24:69
116. Staunton J, Weissman K (2001) Nat Prod Rep 18:380
117. Vanderwel D, Pierce HD, Oehlschlager AC, Borden JH, Pierce AM (1990) Insect Biochem 20:567
118. Leal WS (1996) Proc Natl Acad Sci USA 93:12112
119. Leal WS (1999) In: Minks AK (ed) Pheromones of nonlepidopteran insects in agriculture. CAB International, New York, p 51–68
120. Leal WS, Matsuyama S, Kuwahara Y, Wakamura S, Hasegawa M (1992) Naturwissenschaften 79:184–185
121. Zhang A, Robbins PS, Leal WS, Linn CE Jr, Villani MG, Roelofs WL (1997) J Chem Ecol 23:231
122. Leal WS, Oehlschlager AC, Zarbin PH, Hidalgo E, Shannon PJ, Murata Y, Gonzalez L, Andrade R, Ono M (2003) J Chem Ecol 29:15
123. Plettner E, Sutherland GRJ, Slessor KN, Winston ML (1995) J Chem Ecol 21:1017
124. Plettner E, Slessor KN, Winston ML, Oliver JE (1996) Science 271:1851
125. Plettner E, Slessor KN, Winston ML (1998) Insect Biochem Mol Biol 28:31
126. Boppré M (1990) J Chem Ecol 16:165
127. Barth RH (1965) Science 149:882
128. Schal C, Fan Y, Blomquist GJ (2003) In: Blomquist GJ, Vogt RG (eds) Insect pheromone biochemistry and molecular biology. Academic Press, New York, p 283
129. Rafaeli A (2002) Int Rev Cytol 213:49
130. Raina AK, Jaffe H, Klun JA, Ridgway RL, Hayes DK (1987) J Insect Physiol 33:809
131. Raina AK, Jaffe H, Kempe TG, Keim P, Blacher RW, Fales HM, Riley CT, Klun JA, Ridgway RL, Hayes DK (1989) Science 244:796
132. Holman GM, Cook BJ, Nachman RJ (1986) Comp Biochem Physiol 85C:219
133. Matsumoto S, Kitamura A, Nagasawa H, Kataoka H, Orikasa C, Mitsui T, Suzuki A (1990) J Insect Physiol 36:427
134. Imai K, Konno T, Nakazawa Y, Komiya T, Isobe M, Koga K, Goto T, Yaginuma T, Sakakibara K, Hasegawa K, Yamashita O (1991) Proc Jpn Acad 67(B):98
135. Torfs P, Nieto J, Cerstiaens A, Boon D, Baggerman G, Poulos C, Waelkens E, Derua R, Calderon J, De Loof A, Schoofs L (2001) Eur J Biochem 268:149
136. Kitamura A, Nagasawa H, Kataoka H, Inoue T, Matsumoto S, Ando T, Suzuki A (1989) Biochem Biophys Res Commun 163:520
137. Kawano T, Kataoka H, Nagasawa H, Isogai A, Suzuki A (1992) Biochem Biophys Res Commun 189:221
138. Ma PWK, Knipple DC, Roelofs WL (1994) Proc Natl Acad Sci USA 91:6506
139. Matsumoto S, Fónagy A, Kurihara M, Uchiumi K, Nagamine T, Chijimatsu M, Mitsui T (1992) Biochem Biophys Res Commun 182:534
140. Schoofs L, Holman GM, Hayes TK, Nachman RJ, DeLoof A (1991) Gen Comp Endocrinol 81:97
141. Schoofs L, Holman GM, Nachman R, Proost P, Vandamme J, Deloof A (1993) Comp Biochem Physiol C 106:103
142. Predel R, Kellner R, Kaufmann R, Penzlin H, Gäde G (1997) Peptides 18:61
143. Predel R, Kellner R, Nachman RJ, Holman GM, Rapus J, Gäde G (1999) Insect Biochem Mol Biol 29:139
144. Predel R, Eckert M (2000) J Comp Neurol 419:352
145. Choi M-Y, Rafaeli A, Jurenka R (2001) Cell Tissue Res 306:459
146. Park Y, Zitnan D, Gill SS, Adams ME (1999) FEBS Lett 463:133

147. Meng X, Wahlstrom G, Immonen T, Kolmer M, Tirronen M, Predel R, Kalkkinen N, Heino T, Sariola H, Roos C (2002) Mech Dev 117:5
148. Raina A, Kempe T (1990) Insect Biochem 20:849
149. Raina AK, Kempe TG (1992) Insect Biochem Mol Biol 22:221
150. Nachman RJ, Holman GM, Cook BJ (1986) Biochem Biophys Res Commun 137:936
151. Suwan S, Isobe M, Yamashita O, Minakata H, Imai K (1994) Insect Biochem Mol Biol 24:1001
152. Fónagy A, Schoofs L, Matsumoto S, DeLoof A, Mitsui T (1992) J Insect Physiol 38:651
153. Fónagy A, Matsumoto S, Schoofs L, DeLoof A, Mitsui T (1992) Biosci Biotechnol Biochem 56:1692
154. Kuniyoshi H, Nagasawa H, Ando T, Suzuki A, Nachman RJ, Holman GM (1992) Biosci Biotech Biochem 56:167
155. Rafaeli A (1994) Arch Insect Biochem Physiol 25:287
156. Nachman RJ, Roberts VA, Dyson HJ, Holman GM, Tainer JA (1991) Proc Natl Acad Sci USA 88:4518
157. Altstein M, Ben-Aziz O, Daniel S, Schefler I, Zeltser I, Gilon C (1999) J Biol Chem 274:17573
158. Clark BA, Prestwich GD (1996) Int J Pept Prot Res 47:361
159. Nachman RJ, Holman GM, Haddon WF (1993) Arch Insect Biochem Physiol 22:181
160. Choi M-Y, Fuerst E-J, Rafaeli A, Jurenka R (2003) Proc Natl Acad Sci USA 100:9721
161. Hewes RS, Taghert PH (2001) Genome Res 11:1126
162. Park Y, Kim Y-J, Adams ME (2002) Proc Natl Acad Sci USA 99:11423
163. Jurenka RA, Jacquin E, Roelofs WL (1991) Proc Natl Acad Sci USA 88:8621
164. Jurenka RA, Fabriàs G, DeVoe L, Roelofs WL (1994) Comp Biochem Physiol 108C:153
165. Fónagy A, Matsumoto S, Uchhiumi K, Mitsui T (1992) J Pest Sci 17:115
166. Fónagy A, Yokoyama N, Ozawa R, Okano K, Tatsuki S, Maeda S, Matsumoto S (1999) Comp Biochem Physiol B 124:51
167. Ma PWK, Roelofs WL (1995) Insect Biochem Mol Biol 25:467
168. Matsumoto S, Ozawa RA, Nagamine T, Kim G, Uchiumi K, Shono T, Mitsui T (1995) Biosci Biotech Biochem 59:560
169. Rafaeli A, Soroker V (1989) Mol Cel Endocrinol 65:43
170. Rafaeli A, Soroker V (1994) In: Borkovec A, Loeb MJ (eds) Insect neurochemistry and neurophysiology, 1993. CRC Press, Boca Raton, FL, p 223
171. Jurenka RA (1996) Arch Insect Biochem Physiol 33:245
172. Rafaeli A, Gileadi C (1996) Insect Biochem Mol Biol 26:797
173. Soroker V, Rafaeli A (1995) Insect Biochem Mol Biol 25:1
174. Iwanaga M, Dohmae N, Fonagy A, Takio K, Kawasaki H, Maeda S, Matsumoto S (1998) Comp Biochem Physiol B 120:761
175. Yoshiga T, Yokoyama N, Imai N, Ohnishi A, Moto K, Matsumoto S (2002) Insect Biochem Mol Biol 32:477
176. Gosalbo L, Fabriàs G, Camps F (1994) Arch Insect Biochem Physiol 26:279
177. Arima R, Takahara K, Kadoshima T, Numazaki F, Ando T, Uchiyama M, Nagasawa H, Kitamura A, Suzuki A (1991) Appl Entomol Zool 26:137
178. Ozawa RA, Ando T, Nagasawa H, Kataoka H, Suzuki A (1993) Biosci Biotech Biochem 57:2144
179. Fabriàs G, Marco M-P, Camps F (1994) Arch Insect Biochem Physiol 27:77
180. Fang N, Teal PEA, Tumlinson JH (1995) Arch Insect Biochem Physiol 29:35
181. Tumlinson JH, Fang N, Teal PEA (1997) In: Cardé RT, Minks AK (eds) Insect pheromone research: new directions. Chapman & Hall, New York, p 54
182. Jacquin E, Jurenka RA, Ljungberg H, Nagnan P, Löfstedt C, Descoins C, Roelofs WL (1994) Insect Biochem Mol Biol 24:203

183. Rafaeli A, Soroker V, Kamensky B, Raina AK (1990) J Insect Physiol 36:641
184. Zhu J, Millar J, Löfstedt C (1995) Arch Insect Biochem Physiol 30:41
185. Mas E, Lloria J, Quero C, Camps F, Fabriàs G (2000) Insect Biochem Mol Biol 30:455
186. Thyagaraja BS, Raina AK (1994) J Insect Physiol 40:969
187. Kingan TG, Blackburn MB, Raina AK (1992) Cell Tissue Res 270:229
188. Ma PWK, Roelofs WL (1995) J Insect Physiol 41:339
189. Ma PWK, Roelofs WL, Jurenka RA (1996) J Insect Physiol 42:257
190. Davis NT, Homberg U, Teal PEA, Altstein M, Agricola H-J, Hildebrand JG (1996) Microsc Res Tech 35:201
191. Sato Y, Ikeda M, Yamashita O (1994) Gen Comp Endocrin 96:27
192. Sato Y, Shiomi K, Saito H, Imai K, Yamashita O (1998) J Insect Physiol 44:333
193. Ma PWK, Knipple DC, Roelofs WL (1998) Insect Biochem Mol Biol 28:373
194. Choi MY, Tanaka M, Kataoka H, Boo KS, Tatsuki S (1998) Insect Biochem Mol Biol 28:759
195. Davis MB, Vakharia VN, Henry J, Kempe TG, Raina AK (1992) Proc Natl Acad Sci USA 89:142
196. Jacquin-Joly E, Burnet M, Francois M, Ammar D, Meillour P, Descoins C (1998) Insect Biochem Mol Biol 28:251
197. Duportets L, Gadenne C, Couillaud F (1999) Peptides 20:899
198. Iglesias F, Marco P, François M-C, Camps F, Fabriàs G, Jacquin-Joly E (2002) Insect Biochem Mol Biol 32:901
199. Zhang T, Zhang L, Xu W, Shen J (2001) GenBank Direct Submission AAL05596
200. Lee JM, Choi MY, Han KS, Boo KS (2001) GenBank Direct submission
201. Ma PWK, Garden RW, Niermann JT, O'Connor M, Sweedler JV, Roelofs WL (2000) J Insect Physiol 46:221
202. Cusson M, McNeil JN (1989) Science 243:210
203. Cusson M, Yu CG, Carruthers K, Wyatt GR, Tobe SS, McNeil JN (1994) J Insect Physiol 40:129
204. Gadenne C, Renou M, Sreng L (1993) Experientia 49:721
205. Picimbon J-F, Becard J-M, Sreng L, Clement J-L, Gadenne C (1995) J Insect Physiol 41:377
206. Fan Y, Rafaeli A, Gileadi C, Applebaum SW (1999) Insect Biochem Mol Biol 29:635
207. Rafaeli A, Zakharova T, Lapsker Z, Jurenka RA (2003) Insect Biochem Mol Biol 33:371
208. Christensen TA, Itagaki H, Teal PEA, Jasensky RD, Tumlinson JH, Hildebrand JG (1991) Proc Natl Acad Sci USA 88:4971
209. Christensen TA, Lehman HK, Teal PEA, Itagaki H, Tumlinson JH, Hildebrand JG (1992) Insect Biochem Mol Biol 22:841
210. Christensen TA, Lashbrook JM, Hildebrand JG (1994) Physiol Entomol 19:265
211. Christensen TA, Hildebrand JG (1995) Invert Neurosci 1:97
212. Delisle J, Picimbon J-F, Simard J (1999) Arch Insect Biochem Physiol 42:253
213. Park YI, Ramaswamy SB (1998) Ann Entomol Soc Am 91:329
214. Ramaswamy SB, Jurenka RA, Linn CE, Roelofs WL (1995) J Insect Physiol 41:501
215. Rafaeli A, Gileadi C (1995) Insect Biochem Mol Biol 25:827
216. Rafaeli A, Gileadi C, Cao M (1997) J Insect Physiol 43:261
217. Rafaeli A, Gileadi C, Hirashima A (1999) Pestic Biochem Physiol 65:194
218. Hirashima A, Eiraku T, Watanabe Y, Kuwano E, Taniguchi E, Eto M (2001) Pest Manag Sci 57:713
219. Marco M-P, Fabriàs G, Lázaro G, Camps F (1996) Arch Insect Biochem Physiol 31:157
220. Marco M-P, Fabriàs G, Camps F (1996) In: Cardé RT, Minks AK (eds) Proceedings of the 1st International Symposium on Insect Pheromones. Chapman & Hall, New York, p 46

221. Raina AK, Bird TG, Giebultowicz JM (1990) In: Borkovec AB, Masler EP (eds) Insect neurochemistry and neurophysiology. Humana Press Inc., Clifton Springs, NJ, p 299
222. Kingan TG, Bodnar WM, Hunt DF (1995) Proc Nat Acad Sci USA 92:5082
223. Eliyahu D, Nagalakshmi V, Applebaum SW, Kubli E, Choffat Y, Rafaeli A (2003) J Insect Physiol 49:569
224. Fan Y, Rafaeli A, Gileadi C, Kubli E, Applebaum SW (1999) J Insect Physiol 45:127
225. Fan Y, Rafaeli A, Moshitzky P, Kubli E, Choffat Y, Applebaum S (2000) Insect Biochem Mol Biol 30:805
226. Borden JH, Nair KK, Slater CE (1969) Science 166:1626
227. Ivarsson P, Birgerson G (1995) J Insect Physiol 41:843
228. Tittiger C, Blomquist GJ, Ivarsson P, Borgeson CE, Seybold SJ (1999) Cell Mol Life Sci 55:121
229. Martin D, Bohlmann J, Gershenzon J, Francke W, Seybold SJ (2003) Naturwissenschaften 90:173
230. Schal C, Gu X, Burns EL, Blomquist GJ (1994) Arch Insect Biochem Physiol 25:375
231. Sevala V, Shu S, Ramaswamy SB, Schal C (1999) J Insect Physiol 45:431
232. Fan Y, Chase J, Sevala VL, Schal C (2002) J Exp Biol 205:781
233. Fan Y, Zurek L, Dykstra MJ, Schal C (2003) Naturwissenschaften 90:121
234. Young HP, Larabee JK, Gibbs AG, Schal C (2000) J Chem Ecol 26:1245
235. Blomquist GJ (2003) In: Blomquist GJ, Vogt RG (eds) Insect pheromone biochemistry and molecular biology. Academic Press, New York, p 231
236. Dillwith JW, Adams TS, Blomquist GJ (1983) J Insect Physiol 29:377
237. Mpuru S, Blomquist GJ, Schal C, Kuenzli M, Dusticier G, Roux M, Bagneres A-G (2001) Insect Biochem Mol Biol 31:139
238. Adams TS, Dillwith JW, Blomquist GJ (1984) J Insect Physiol 30:287
239. Blomquist GJ, Adams TS, Halarnkar PP, Gu P, Mackay ME, Brown LA (1992) J Insect Physiol 38:309
240. Blomquist GJ, Adams TS, Dillwith JW (1984) J Insect Physiol 30:295
241. Tillman-Wall JA, Vanderwel D, Kuenzli ME, Reitz RC, Blomquist GJ (1992) Arch Biochem Biophys 299:92
242. Blomquist GJ, Tillman JA, Reed JR, Gu P, Vanderwel D, Choi S, Reitz RC (1995) Insect Biochem Mol Biol 25:751
243. Minks AK (1999) Pheromones of nonlepidopteran insects in agriculture. CAB International, New York
244. Steinbrecht RA (2003) The 3rd International Symposium on Insect Pheromones, Bäckaskog Castle, Sweden

Topics in Current Chemistry (2004) 239: 133–177
DOI 10.1007/b95452

Hymenopteran Semiochemicals

Christopher I. Keeling[1] (✉) · Erika Plettner[2] · Keith N. Slessor[2]

[1] Michael Smith Laboratories, University of British Columbia, 2185 East Mall, Vancouver, BC V6T IZ4, Canada
ckeeling@mac.com

[2] Department of Chemistry, Simon Fraser University, 8888 University Dr., Burnaby, BC V5A 1S6, Canada

Abstract Hymenoptera is a very large and diverse insect order that includes the majority of both the social and the parasitic insects. With such diversity comes a variety and complexity of semiochemicals that reflect the varied biology of members of this order. This chapter reviews the chemical identification of pheromones and semiochemicals in the order Hymenoptera since 1990. For this review, the species in Hymenoptera have been classified as solitary, parasitic, or social. The chemical diversity of semiochemicals in Hymenoptera and future trends in pheromone identification are also discussed.

Keywords Hymenoptera · Symphyta · Apocrita · Pheromone · Semiochemical

List of Abbreviations

CLG	Cephalic labial gland
CW	Cuticular wash
DG	Dufour's gland
HE	Head extract
HG	Hindgut (rectal bladder)
JH	Juvenile hormone
MG	Mandibular gland
NG	Nasonov gland
PG	Poison gland
PPyG	Postpygidial gland
PvG	Pavan gland
PyG	Pygidial gland
Q-	Queen
SA	Sting apparatus
VG	Venom gland
VQ-	Virgin queen
W-	Worker

1 Introduction

Hymenoptera is one of the most diverse insect orders. With an estimated 300,000 species [1], it includes ants, bees, wasps, and sawflies. They occupy many ecological niches as herbivores, predators, or parasitoids. Although most species in Hymenoptera are solitary, they also include most of the described species of social insects. With such biological diversity comes rich semiochemical diversity. Pheromones have been identified in Hymenoptera that change the behavior (releaser pheromones) or physiology (primer pheromones) of the recipient. Other semiochemicals have been identified that mediate communication between other species. Although pheromones and other semiochemicals have been identified in only a small fraction of the described species, they have interesting chemistry and biology.

Members of Hymenoptera have significant economic importance. Most well known is the crucial role bees play in pollination of agricultural crops, with honey bees alone contributing an estimated pollination value of US$15 billion per year in the United States [2]. In addition, some parasitoids are important allies in the biological control of other insect pests, whereas other species are significant agricultural, forestry, structural, or household pests in their own right. As hymenopteran semiochemicals are better understood, they can be utilized for integrated pest management as well as the management of economically important beneficial insects.

Because of the complexity of insect extracts and the likelihood of multiple, synergistic components, a bioassay-guided approach to the isolation and identification of pheromone components is the most productive method for identifying new pheromones. The search for a new pheromone begins with the observation of a behavior or physiology that implicates the presence of a chemical cue. The chemical cue must then be isolated from other cues (i.e., visual, auditory, and tactile) by preparing an extract, wash, or headspace sample of the releasing organism and then its biological activity confirmed with a bioassay. The active chemical component(s) must then be found by isolating activity to specific glands or through chemical separations while still maintaining the activity of a complete extract in a bioassay. The importance of a robust and effective bioassay to guide isolation cannot be overemphasized. A powerful technique that can sometimes be used is gas chromatography-electroantennographic detection (GC-EAD), allowing simultaneous separation and electrophysiological recording of an insect's response to a complex mixture. Chemical identification usually involves gas chromatography-mass spectral analysis with comparison to synthetic samples and then bioassay of the synthetic compounds. The many bioassay and chemical methods used to isolate and identify pheromone components, including GC-EAD, are well described by Millar and Haynes [3].

This chapter reviews the literature of semiochemical (mostly pheromone) identification in Hymenoptera published since 1990. For this review, we separate the order Hymenoptera into the following three, somewhat overlapping, classes to reflect their differences in biology and semiochemistry: solitary, parasitic, and social (Table 1). Although there is considerable literature on the semiochemical activity of specific glandular extracts and the chemical composition of specific glands, only those chemicals with demonstrated pheromonal (or semiochemical) activity will be specifically discussed here. The earlier literature of pheromones in social hymenoptera has previously been reviewed [4–6]. There have been more recent reviews of pheromones in social hymenoptera [7–10], parasitic wasps [11, 12], sawflies and seed wasps [13, 14], and mating pheromones across Hymenoptera [15].

Table 1 Classification in the order Hymenoptera [1, 11, 16]

SYMPHYTA (sawflies and horntails) (Sol.)	**APOCRITA (ACULEATA)** (Ants, bees, and wasps)	**APOCRITA (PARASITICA)**
CEPHOIDEA	APOIDEA (APIFORMES)	CERAPHRONOIDEA
Cephidae	(Para., Sol., Soc.)	(Para.)
MEGALODONTOIDAE	Andrenidae	Ceraphronidae
Megalodontidae	Anthophoridae	Megaspilidae
Pamphiliidae	Apidae	CHALCIDOIDEA
ORUSSOIDEA	Colletidae	(Para.)
Orussidae	Ctenoplectridae	Agaonidae
SIRICOIDEA	Fideliidae	Aphelinidae
Anaxyelidae	Halictidae	Chalcididae
Siricidae	Megachilidae	Elasmidae
Xiphydriidae	Melittidae	Encyrtidae
TENTHREDINOIDEA	Oxaeidae	Eucharitidae
Argidae	Stenotritidae	Eulophidae
Blasticotomidae	APOIDEA (SPHECIFORMES)	Eupelmidae
Cimbicidae	(Para., Sol., Soc.)	Eurytomidae
Diprionidae	Ampulicidae	Leucospidae
Pergidae	Astatidae	Mymaridae
Tenthredinidae	Crabronidae	Ormyridae
XYELOIDEA	Heterogynaidae	Perilampidae
Xyelidae	Mellinidae	Pteromalidae
	Nyssonidae	Rotoitidae
	Pemphredonidae	Signiphoridae
	Philanthidae	Tanaostigmatidae
	Sphecidae	Tetracampidae
	CHRYSIDOIDEA	Torymidae
	(Para., Sol., Soc.)	Trichogrammatidae
	Bethylidae	CYNIPOIDEA (Para.)
	Chrysididae	Austrocynipidae
	Dryinidae	Charipidae
	Embolemidae	Cynipidae
	Plumariidae	Eucoilidae
	Sclerogibbidae	Figitidae
	Scolebythidae	Ibaliidae
	VESPOIDEA (Sol., Soc.)	Liopteridae
	Bradynobaenidae	EVANIOIDEA (Para.)
	Formicidae	Evaniidae
	Masaridae	Gasteruptiidae
	Mutillidae	ICHNEUMONOIDEA
	Pompilidae	(Para.)
	Rhopalosomatidae	Braconidae
	Sapygidae	Ichneumonidae
	Scoliidae	MEGALYROIDEA
	Sierolomorphidae	(Para.)
	Tiphiidae	Megalyridae
	Vespidae (Soc.)	

Table 1 (continued)

SYMPHYTA (sawflies and horntails) (Sol.)	APOCRITA (ACULEATA) (Ants, bees, and wasps)	APOCRITA (PARASITICA)
		MYMAROMMATOIDEA (Para.)
		Mymarommatidae
		PLATYGASTROIDEA (Para.)
		Platygastridae
		Scelionidae
		PROCTOTRUPOIDEA (Para., Soc.)
		Austroniidae
		Diapriidae
		Heloridae
		Monomachidae
		Pelecinidae
		Peradeniidae
		Proctotrupidae
		Renyxidae
		Roproniidae
		Vanhorniidae
		STEPHANOIDEA (Para.)
		Stephanidae
		TRIGONALYOIDEA (Para.)
		Trigonalyidae

Extinct families excluded.
Abbreviations: Sol., solitary; Para., parasitic; Soc., social.

2 Solitary Hymenoptera

2.1 Introduction

All of the suborder Symphyta and many species in the superfamily Aculeata in the suborder Apocrita are solitary insects. Although not requiring the complex semiochemistry of parasitic or social insects, solitary insects employ pheromones for mating, territorial marking, and host marking. Unfortunately, very few of these have been chemically identified. The pheromones of sawflies and seed wasps were extensively reviewed in 1999 [14]. The semiochemicals recently identified in solitary hymenoptera, discussed below, are summarized in Table 2 and Fig. 1.

Table 2 Recently identified pheromones of solitary hymenoptera. Numbers in bold refer to chemical structures shown in Fig. 1. See Table 5 for abbreviations

Taxon	Source	Function	Chemical Name(s)	Ref.
SYMPHYTA				
Diprionidae				
Diprion jingyuanensis	CW	Virgin female-produced male attractant	Propanoate of (2*S*,3*R*,7*R*)-3,7-dimethyltridecan-2-ol **3**	[36]
Diprion nipponica	CW	Virgin female-produced male	Propanoate of (2*S*,3*R*,9*S*)-3,9-dimethylundecan-2-ol **2**	[18]
Diprion pini	CW	Virgin female-produced male attractant	Acetate & propanoate of (2*S*,3*R*,7*R*)-3,7-dimethyltridecan-2-ol **3**	[37, 38]
Macrodiprion nemoralis	CW	Virgin female-produced male attractant	Acetate of (2*S*,3*R*,7*R*,9*S*)-3,7,9-trimethyltridecan-2-ol **4**	[25]
Microdiprion pallipes	CW	Virgin female-produced male attractant	Propanoate of 3,7,11-trimethyltridecan-2-ol **5**	[26]
Neodiprion dailingensis	?	Previously identified compounds from other *Neodiprion* spp. attractive to males	Acetate & propanoate of (2*S*,3*S*,7*S*)-3,7-dimethylpentadecan-2-ol **1**	[39]
Neodiprion sertifer	?	Virgin female-produced male attractant	Acetate & propanoate of (2*S*,3*S*,7*S*)-3,7-dimethylpentadecan-2-ol **1**. Depending on the population, the (2*S*,3*R*,7*R*)-isomer of the acetate can be benign, inhibitory, or necessary.	[27, 40, 41]
Cephidae				
Cephus cinctus	CW	Produced & attractive to both sexes equally	9-Acetyloxynonanal **6**	[31, 32]
Janus integer	CW	Produced only in females and antennally active only in males	(9*Z*)-Octadec-9-en-4-olide **7**	[33]
APOIDEA				
Colletidae				
Colletes cunicularius	HE	Attraction of males to virgin females	(3*S*)-(+)-Linalool **8**	[35]

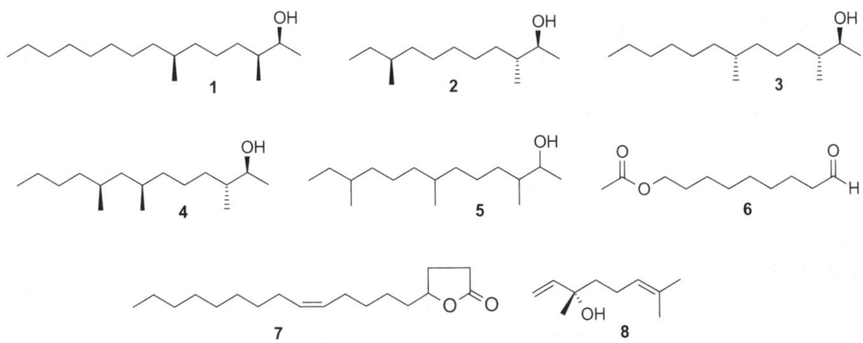

Fig. 1 Structures of recently identified pheromones of solitary hymenoptera

2.2 Symphyta

2.2.1 Diprionidae

2.2.1.1 Introduction

With the exception of the conifer sawflies (Symphyta:Diprionidae) the pheromone chemistry of the solitary hymenoptera is sparse. Conifer sawflies consist of more than 100 species and some are considered severe pests of pine trees. The early pheromone studies suggested that pine sawflies utilized one or more stereoisomers of female-produced 3,7-dimethylpentadecan-2-ol **1** as a sex pheromone, either as the free alcohol or the acetate or propanoate esters (Fig. 1) [13]. IUPAC nomenclature does not facilitate the identification of these alcohols and their esters, in that alcohols are named with the longest continuous carbon chain bearing the alcohol, whereas in the derivative ester the alkyl group numbering begins at the ester oxygen. The propanoate ester of 3,7-dimethylpentadecan-2-ol thus becomes 1,2,6-trimethyltetradecyl propanoate. For the purposes of this review we will refer to these as acetates or propanoates of the corresponding alcohol.

The importance of the three stereogenic centers became evident as two of the stereoisomers [(2*S*,3*S*,7*S*) and (2*S*,3*R*,7*R*)] were recognized early as sex pheromones and that other enantiomers and diastereoisomers were often found to be inhibitory to the attractive response. Recently, the sawfly pheromone field has undergone a major advance with the recognition that several sawfly species synthesize and utilize sex pheromones of different structural types than the 3,7-dimethylpentadecan-2-ols. Shorter and longer chain lengths (undecan-2-ols **2** and tridecan-2-ols **3**) and an additional methyl group in position 9 or 11 (**4** and **5**) characterize these new pheromone discoveries (Fig. 1). With an additional stereogenic methyl substituent, the complexity of the possible chiral isomers increases to 16 stereoisomers for a given skeleton. This makes positive identification even more difficult and the synthesis of chirally-pure materials for comparison and bioassay very challenging. A considerable amount of effort has been focused on synthetic methodology to effectively produce the stereoisomers in high chiral purity.

2.2.1.2
Synthesis

Various synthetic methodologies have been employed recently to establish the stereochemistry of the multiple asymmetric centers. The enantioselective synthesis of (2*R*,3*R*,7*S*)-3,7-dimethylpentadecan-2-ol from commercial (–)-isopulegol via a crystalline intermediate has been reported [17]. Commercially available methyl (2*R*)- and (2*S*)-3-hydroxy-2-methylpropanoate were utilized to fashion the chiral centers for the propanoate of (2*S*,3*R*,9*S*)-3,9-dimethylundecan-2-ol **2**, the sex pheromone for *Diprion nipponica* [18]. Asymmetric 1,3-dipolar cycloaddition with a chiral camphorsultam led to a 9:1 diastereomeric mixture of *trans*-3,4-disubstituted tetrahydrothiophene amides to be utilized as the *threo*-3-methylalkanol unit of acetate of (2*S*,3*R*,7*R*,9*S*)-3,7,9-trimethyltridecan-2-ol **4**, the sex pheromone of *Macrodiprion nemoralis* [19]. As in previous studies, many synthetic procedures utilize stereoselective lipase kinetic resolutions to isolate and purify intermediates [20]. Coupling of racemic [21] and chiral *cis*-3,4-dimethyl-γ-butyrolactones with racemic as well as each of the four stereoisomers of 1-lithio-2,4-dimethyloctane gave the 16 stereoisomers of 3,7,9-trimethyltridecan-2-ol [22]. The chiral methyl groups in the octane portion were established by alkylation of amide enolates of pseudoephedrine enantiomers. In an analogous fashion, the 16 stereoisomers of 3,7,11-trimethyltridecan-2-ols were generated from the butyrolactone and 1-lithio-2,6-dimethyloctanes. The latter were obtained by lipase-catalyzed kinetic separation of the chiral alcohols derived from optically pure citronellal enantiomers [23]. Syntheses of (2*S*,3*S*,7*S*,11*R*)- and (2*S*,3*R*,7*R*,11*R*)- diastereoisomers of 3,7,11-trimethyltridecan-2-ols and their esters were achieved through coupling of a (2*R*)-2-methylpropanoic acid derivative with a lipase-resolved 2,6-dimethylheptane-1,7-diol [24].

2.2.1.3
Pheromone Identification

Since the last review [14], the female-produced sex pheromones of *Diprion nipponica*, *Macrodiprion nemoralis*, and *Microdiprion pallipes* have been investigated and clarification of the active components has been made. GCMS of the GC-EAD-active material in virgin female extracts of *D. nipponica* have suggested that the propanoate ester of (2*S*,3*R*,9*S*)-3,9-dimethylundecan-2-ol **2** is the sex pheromone. Field tests with synthetic material confirmed its activity, but also indicated that the propanoates of (2*S*,3*R*,9*R*)-3,9-dimethylundecan-2-ol, (2*S*,3*R*,8*S*)-3,8-dimethylundecan-2-ol, and (2*S*,3*R*,7*S*)-3,7-dimethylundecan-2-ol could attract male sawflies as pheromone mimics [18]. The structure of the female *Macrodiprion nemoralis* sex pheromone has been identified as the acetate of (2*S*,3*R*,7*R*,9*S*)-3,7,9-trimethyltridecan-2-ol **4** by chiral GC comparison and field testing of synthetic material [25]. In *Microdiprion pallipes*, at least one stereoisomer of the propanoate of 3,7,11-trimethyltridecan-2-ol **5** attracts males [26]. The pheromones identified during the last decade are tabulated in Table 2.

Most diprionid species appear to rely principally on one substance for sexual attraction [14] and very often the female appears to produce a single major compound. Whether biosynthesis is species specific has been difficult to establish given the sub-nanogram quantities of potential trace components. Correspondingly, the specificity of the male antennal response has been equally difficult to assess by electroantennography until very recently with access to adequate libraries of synthetic test materials of very high stereochemical purity. As noted below, some evidence of diastereomeric involvement has surfaced in the extensive field trapping studies of *Neodiprion sertifer*. Further investigations will likely reveal that sawfly sex pheromone communication is more complex than a reliance on a single compound for each species.

2.2.1.4 Field and Population Studies

The field response of the European pine sawfly, *Neodiprion sertifer*, to the acetate of (2*S*,3*S*,7*S*)-3,7-dimethylpentadecan-2-ol **1** and the effect of the (2*S*,3*R*,7*R*)-isomer was studied from Japan in the east to Canada in the west. The addition of the (2*S*,3*R*,7*R*)-isomer was benign in Japan, inhibitory in Europe with the inclusion of as little as 1%, and its presence essential in Siberia. These findings will allow for the development of a more efficient *N. sertifer* monitoring scheme without having to determine the stereochemistry produced in each population [27].

In Kentucky, the acetate of (2*S*,3*S*,7*S*)-3,7-dimethylpentadecan-2-ol **1** was used to assess the population density and flight activity of the European pine sawfly, *N. sertifer*, amongst the loblolly pine sawfly, *N. taedae linearis*, and the redheaded pine sawfly, *N. lecontei*. [28]. The results indicated that, although present, *N. sertifer* was a minor participant in sawfly populations.

Release rates of acetate esters of (2*S*,3*S*,7*S*)-3,7-dimethylpentadecan-2-ol**1** and (2*S*,3*R*,7*R*)-3,7-dimethyltridecan-2-ol **3** from polyethylene and cotton dispensers have been measured at different temperatures and loadings. Adjustment of the initial pheromone load on polyethylene for expected temperatures should permit the formulation of a constant release rate during the entire flight period. Successful field trapping trials for *N. sertifer* were carried out in Sweden, the Czech Republic, Italy, and Greece [29].

Haplodiploidy in *N. sertifer* leads to male production from unfertilized eggs, which seriously complicates the issue of mating disruption with the acetate of

3,7-dimethylpentadecan-2-ol. Recently, a three-year mating disruption study in isolated pine stands in Northern Italy has resulted in a significant reduction of the sawfly population in treated stands. Nearly half of the egg clusters resulted in only male cocoons and the number of males caught within the treated areas was significantly lower than in control areas [30] providing encouraging results for potential mating disruption in sawflies.

2.2.2
Cephidae

Volatile collection from male and female wheat stem sawflies, *Cephus cinctus*, revealed 13 GC-EAD materials active to both male and female antennae. Amongst the C_9–C_{16} aldehydes and C_8–C_{10} acids, were phenylacetic acid, 13-acetyloxytridecanal and 9-acetyloxynonanal **6**, which elicited the strongest GC-EAD response [31]. The latter produced "consistent although fairly subtle" attraction to both males and females in field traps at the 3–1000 μg range [32]. The acetyloxyaldehydes present in volatile collections were absent in cuticular extracts, and their presence and resulting activity appear to be formed from the air oxidation of (9*Z*)-alk-9-en-1,ω-diyl diacetates, with males being the predominant producers [31]. Placing several males together in a collection tube also triggered wing-fanning and release of phenylacetic acid.

Volatiles and cuticular extracts from both sexes of the currant stem girdler, *Janus integer*, were analyzed by GC-EAD using antenna of both sexes. A female specific compound, (9*Z*)-octadec-9-en-4-olide **7**, was identified as active only on male antennas [33]. Separation by chiral GC has shown that only one enantiomer is produced in females. The synthesis of both enantiomers has recently been described [34] and the field testing results are forthcoming.

2.2.3
Other Symphyta

Studies done before the scope of this review have chemically identified pheromones in a few species of Tenthredinidae and Pamphiliidae [14]. There have been no studies to show whether pheromones are used in the remaining families in Symphyta.

2.3
Apocrita

In recent years, there is only one example of a pheromone in solitary Apocrita being chemically identified. Chiral GC and chiral GC-EAD provided identification of (3*S*)-(+)-linalool **8** >99.9% e.e. as a mandibular gland mate attractant in both males and females of *Colletes cunicularius*. Male contact with a scented source could be initiated with ~5 ng per lure (3*S*)-(+)-linalool, which may act as both a sex attractant and a food attractant [35].

3
Parasitic Hymenoptera

3.1
Introduction

Parasitic hymenoptera are a large and very diverse group of species, accounting for 80% of all parasitic insect species [11]. Almost 75% of all Apocrita Hymenoptera are parasitoids at the larval stage [1] and have representatives in all super-families (Table 1). These insects have significant ecological and economic importance in regulating host populations, many of which are, or otherwise would be, pest species to humans. Parasitic hymenoptera use pheromones for sexual attraction, spacing (to avoid excessive competition for a limited resource), and alarm much like the solitary hymenoptera. In addition, their interactions with their host require a greater repertoire of semiochemicals.

Unlike parasitoids of other insect orders that have host-seeking larvae, most parasitic hymenoptera lay their eggs on, in, or very close to a host individual [11]. This requires the adult female to find a suitable host, often with the aid of chemical cues from host frass, pheromones, plant volatiles emitted upon host feeding or egg-deposition, silk, honeydew and other secretions. She may then chemically mark the host following oviposition to reduce superparasitism by herself or intra- and inter-specific insects [11].

Tetratrophic interactions between a host plant, a phytophagous pest (primary host), a hymenopteran parasitoid or symbiont (secondary host) and a hymenopteran hyperparasitoid (which parasitizes the secondary host) are of considerable importance, because hyperparasitism can significantly reduce populations of economically beneficial parasitoids [11]. Hyperparasitoids use host-marking (=spacing) pheromones, sex pheromones [12], and host-detection cues [42], but they also show additional chemically mediated interactions with the other partners. These include detection of the primary host's secretions by the hyperparasitoid [43], detection of plant volatiles by the hyperparasitoid [44], and detection of the hyperparasitoid's secretions by the primary host [45] or by the secondary host. The latter causes the secondary host to avoid locations where the hyperparasitoid is foraging [46].

Parasitic hymenoptera hold promise in integrated pest management schemes, because they parasitize many economically important insect pests in a species- and stage-selective manner. The pheromones and kairomones of the parasitic hymenoptera have been studied for a long time, and there are many examples where there is evidence of chemical mediation of parasitoid behavior. This review emphasizes work done since the last major reviews [11, 12, 42] and, where it is available, on the primary bioassay-guided chemical identification of the semiochemical (Fig. 2 and Tables 3 and 4).

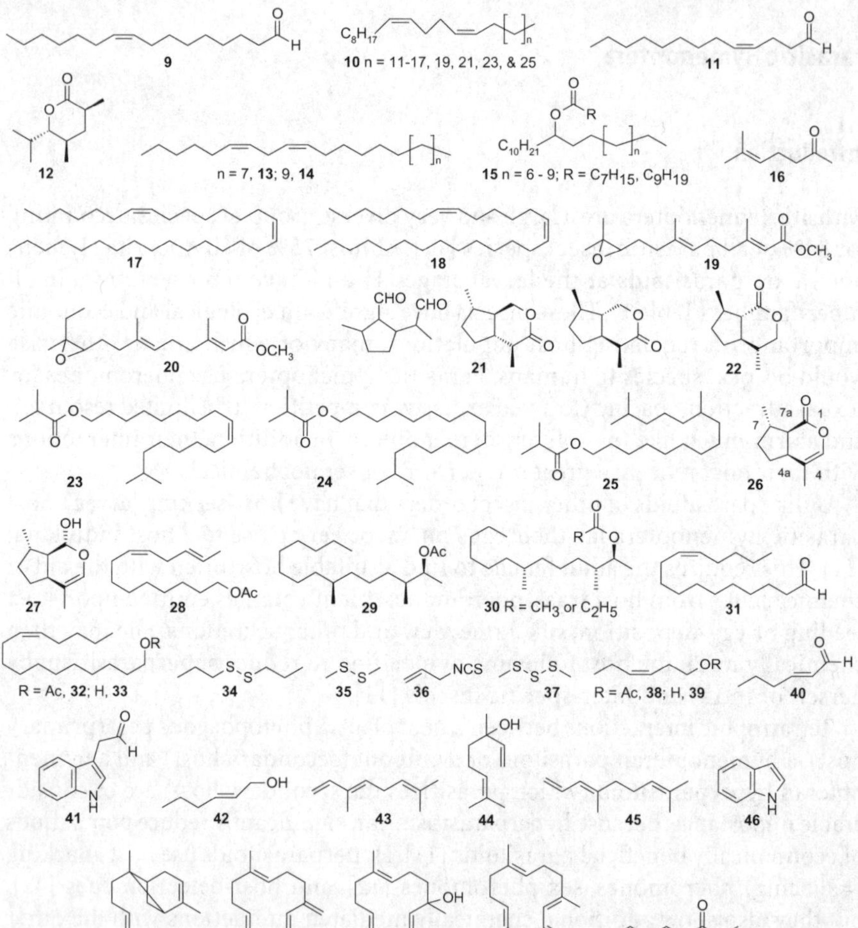

Fig. 2 Structures of recently identified semiochemicals of parasitic hymenoptera

3.2
Sex Pheromones

Parasitic hymenoptera utilize sex pheromones at two levels: long-range (>2 cm) for attraction and short-range (contact) for species recognition and courtship. The long-range sex attractants have practical implications for monitoring parasitoid populations in the field and, in turn, in assessing the impact of parasitoids on their hosts. Courtship involves a well-defined sequence of behaviors, which may include antennation of the substrate, trail-following, wing-fanning, and mounting. The short-range pheromones, spread by antennal contact during courtship, are thought to be involved in species recognition and in the induction of female receptivity.

Several long-range attractant sex pheromones have been identified (Table 3) or evidence for such a pheromone has been obtained in the past 13 years. The parasitoid *Ascogaster reticulatus* produces (9*Z*)-hexadec-9-enal **9** in a tibial gland [47]. The pheromone is spread by the females much like a trail on the substrate, and the males follow the mark to the source by close antennal contact with the substrate [48]. EAG studies have revealed that the males respond much more strongly to (9*Z*)-hexadec-9-enal than the females and that the response to the

Table 3 Recently identified sex pheromones of parasitic hymenoptera. Numbers in bold refer to chemical structures shown in Fig. 2

Taxon	Host	Detection Mode	Chemical Name(s)	Ref.
CHALCIDOIDEA				
Eurytomidae				
Eurytoma amygdali	Almond	Olfaction	(6*Z*,9*Z*)-Tricosa-6,9-diene **13** & (6*Z*,9*Z*)-pentacosa-6,9-diene **14**	[54]
CYNIPOIDEA				
Charipidae				
Alloxysta victrix	*Aphidius uzbekistanikus*	Olfaction	6-Methylhept-5-en-2-one **16**	[60]
ICHNEUMONOIDEA				
Braconidae				
Ascogaster reticulates	*Adoxophyes* sp.	Short-range olfaction & substrate contact	(9*Z*)-Hexadec-9-enal **9**	[47–49]
Cardiochiles nigriceps	*Heliothis virescens*	Olfaction	(7*Z*,15*Z*)-Hentriaconta-7,15-diene **17** & (7*Z*,13*Z*)-heptacosa-7,13-diene **18**	[66]
Macrocentrus grandii	*Ostrinia nubilalis*	Olfaction	(4*Z*)-Tridec-4-enal **11** & (3*S*,5*R*,6*S*)-3,5-dimethyl-6-(methylethyl)-3,4,5,6-tetrahydropyran-2-one **12**	[50, 51, 53]
Ichneumonidae				
Eriborus terebrans	*Ostrinia nubilalis*	Olfaction	Polar comp. & hydrocarbons	[58]

(*Z*)-isomer is much stronger than the response to the (*E*)-isomer (which is not pheromonally active) [49]. The parasitoid *Macrocentrus grandii* produces a series of homologous long-chain (9*Z*,13*Z*)-alka-9,13-dienes **10**, which oxidize to a series of homologous aldehydes. One of these aldehydes, (4*Z*)-tridec-4-enal **11**, is active as a sex attractant in wind tunnel bioassays [50]. It is not clear whether the oxidation is spontaneous or enzyme-mediated. A second synergistic component of the sex attractant of *M. grandii* was identified as the lactone (3*S*,5*R*,6*S*)-3,5-dimethyl-6-(methylethyl)-3,4,5,6-tetrahydropyran-2-one **12** [51]. The relative stereochemistry was determined by NMR from the material isolated from female *M. grandii*. The absolute stereochemistry was then determined by synthesis of both enantiomers via the *meso* precursor (2*S*,4*R*)-2,4-dimethylpenta-1,5-diol [52]. The (3*R*,5*S*,6*R*)-enantiomer was neither active by itself, nor inhibitory or synergistic with the natural (3*S*,5*R*,6*S*)-enantiomer or the aldehyde **11** [53].

Male almond seed wasps, *Eurytoma amygdali* (Eurytomidae), display wing raising and attraction to virgin female extracts as well as SPME fibers exposed to the volatiles of virgin females [54]. Two attractive alkadienes, (6*Z*,9*Z*)-tricosa-6,9-diene **13** and (6*Z*,9*Z*)-pentacosa-6,9-diene **14**, were subsequently identified using silver nitrate-impregnated silica gel chromatography.

Some progress has been made in the identification of the sex pheromone of the aphid parasitoids *Aphidius rhopalosiphi* and *Praon volucre*. In *A. rhopalosiphi*, methanol and dichloromethane extracts of females were found to be attractive to males and to cause wing-fanning as well as mounting behavior [55]. In *P. volucre*, hexane extracts of females caused both wing-fanning and mounting of males. In field trials with these species, live virgin females only caught conspecific males, suggesting that the attractants in these two species are different [55]. More recently, *P. volucre* was studied in more detail by GC-EAD and by a coupled GC-behavioral assay. These studies clearly revealed three behaviorally active minor peaks in a complex GC trace of air-entrapped volatiles from the insects. The most strongly behaviorally active peak also gave a distinct EAD signal. This latter peak was estimated to be produced by the females in the order of pg per hour [56]. This result underscores one of the difficulties with the identification of parasitic hymenopteran pheromones: the insects produce very complex, multi-purpose signals, which may consist of only trace amounts of very powerful attractants.

Indication of a sex attractant has also been obtained for the noctuid pupal parasitoid *Diapetimorpha introita* (Ichneumonidae). Antennae from *D. introita* males gave EAG responses to diethyl ether extracts of female head, thorax or abdomen. Antennae from females did not respond to this chemical signal and male extracts elicited no activity. This suggested the presence of an extractable female-produced pheromone, to which the males respond. While live females were able to attract males, extracts were not active. This may be due to very low levels of biologically active material in the extracts [57].

Synergy between polar and nonpolar components is indicated for sex attraction in the larval parasitoid *Eriborus terebrans* (Ichneumonidae) [58]. Males show several behaviors when exposed to a polar component with chemical properties of a functionalized carboxylic acid from females. These behaviors increased when a hydrocarbon fraction, which is inactive alone, was added.

Close-range signals are important for the initiation of male courtship behavior. Because this signal is involved in sex recognition, comparison of male and female extracts can be used to tentatively identify potential candidate compounds. In *Diglyphus isaea* Walker, evidence was obtained that (a) hexane-soluble female-specific compound(s) is (are) involved in close-range sex recognition [59]. Comparison of male and female extracts pinpointed a series of unusual esters of octanoic and decanoic acids with saturated C_{21}–C_{25} 11-alcohols **15** [59]. The chirality at C-11 has not been determined and the racemic ester standards used for identification were not tested for biological activity.

In *Alloxysta victrix*, 6-methylhept-5-en-2-one **16**, which is produced by both males and females, was identified as potentially attractive to the males and slightly repellent to the females in Y-tube olfactometer assays [60]. In this study, the activity was also dependent on prior exposure of the insects to the compound. Naïve insects responded more strongly than previously exposed ones. This underscores a second difficulty in the bioassay-guided identification of parasitoid hymenopteran pheromones: the responses are very dependent on the context and on prior exposure. Learning has been demonstrated in several species of parasitic hymenoptera [61–65].

The next stage after initial attraction is recognition and initiation of courtship. Pheromones have also been shown to be involved in this stage [12]. In *Cardiochiles nigriceps* (Braconidae), only the females contain a series of C_{25}–C_{35} (*Z*,*Z*)-alkadienes with double bond positions of 5 or 6 for the first double bond and 13, 14, or 15 for the second [66]. The alkadienes did not elicit long-range attraction of males, but they elicited wing-fanning, antennation and mounting at close range. The most abundant female alkadiene, (7*Z*,15*Z*)-hentriaconta-7,15-diene **17**, elicited the male behaviors. The third most abundant diene, (7*Z*,13*Z*)-heptacosa-7,13-diene **18** elicited male antennation when synergized by male monoenes (see below) [66].

Contact sex pheromones appear to be transferred during courtship. Male hymenopteran parasitoids have pheromonal glands on the distal antennal segments. These glands have been identified in multiple species by ultrastructural studies [67–69]. When males stroke female's antennae during courtship using their own antennae, chemical cues (which increase female receptivity) are transferred. Preliminary evidence for this transfer of chemicals comes from experiments in which increasing portions of antennal segments have been removed or in which the glandular ducts were occluded with glue [67]. Again, either learning or priming appears to be important in this context. Previously exposed females require much less stroking than naïve females [67]. One contact sex pheromone has been identified in *C. nigriceps* by comparison of male and female cuticular hydrocarbons and bioassay. The male-derived series of monoenes synergized with the female-produced (7*Z*,13*Z*)-heptacosa-7,13-diene (see above) in eliciting antennation [66]. The particular monoene responsible for the synergy was not identified, although (12*Z*)-heptacos-12-ene and (13*Z*)-heptacos-13-ene stand out as the most abundant monoenes in male cuticular hydrocarbon extracts.

3.3
Host Marking, Spacing (Epideictic), and Defense Compounds

Host marking pheromones are important in many species of parasitic hymenoptera, because they ensure that a female parasitoid focuses on non-parasitized hosts. This, in turn, ensures a more effective use of limited host resources. Marking pheromones can be internal (injected into the host at the time of oviposition) or external (applied to the host during inspection and/or oviposition). The internal markers can be detected by sensory hairs on the parasitoid ovipositor [11]. The internal markers often also delay the development of the host.

Venoms from the female are often used as internal markers. Venoms contain components that can cause paralysis, reduce growth rates, and other specific metabolic effects within the host. For example, the two major venom proteins from the parasitoid *Aphidius ervi* (Braconidae) have recently been identified as being responsible for detrimental effects on the reproductive tract of the host aphid, *Acythosiphon pisum* [70]. Recently, juvenile hormone (JH) III **19** and JH II **20** were identified as the compounds used by the hyperparasitoid *Dendrocerus carpenteri* (Megaspilidae) females as both a marking pheromone, which deters oviposition by other females, and as a primer allomone, which delays pupation and adult emergence of the host parasitoid *Aphidius uzbekistanicus* [71, 72].

The existence of volatile spacing cues has been demonstrated in a number of species. Recently, 6-methylhept-5-en-2-one **16** was shown to be slightly repellent in Y-tube olfactometer assays to *Alloxysta victrix* (see above). The same compound appears to be involved in defense (see below).

Defensive substances are often general irritants that can be used in a variety of contexts. For example, the alloxystine wasps (Cynipoidea), all hyperparasitoids of other hymenopteran parasitoids, produce a large number of compounds in their cephalic (mainly mandibular) glands. These compounds include *m*/*p*-xylol, 6-methylhept-5-en-2-one **16**, various iridoids **21** and *trans*-dihydronepetalactone **22** [46, 73].

3.4
Host Kairomones

Parasitic hymenoptera often eavesdrop on the pheromone communication of their host species. The type of host pheromone recognized depends on the host stage parasitized. Phoretic egg parasitoids are often attracted by the host sex pheromone, while species that parasitize later stages (larval, pupal) often do not respond to host sex pheromone components [11, 42]. Larval parasitoids often recognize volatiles from the damaged host plant and/or host larval frass volatiles. Parasitoids of forest beetles respond to the beetle aggregation pheromones [42].

The attraction of several egg parasitoids to their host sex pheromone has been studied in the past 13 years (Table 4). Both males and females of the phoretic egg parasitoid, *Telenomus euproctidis* (Scelionidae), are strongly attracted to the host's (Tussock moth, *Euproctis taiwana*) sex pheromone. Both components of the sex pheromone, (9*Z*)-16-methylheptadec-9-en-1-yl isobutyrate **23** and 16-methylheptadec-1-yl isobutyrate **24** were required for field trapping of the parasitoids [74]. Differences in local strains of the parasitoids were also observed, where parasitoids of *Euproctis pseudoconspersa* responded to racemic 10,14-dimethylpentadecyl isobutyrate **25**, but not to the *E. taiwana* component **23** [75]. Aphid parasitoids *Aphidius ervi*, *Aphidius rhopalosiphi*, and *Praon volucre* responded to (4a*S*,7*S*,7a*R*)-nepetalactone **26** and (1*R*,4a*S*,7*S*,7a*R*)-nepetalactol **27**, both components of the sex pheromone of many aphid species in the subfamily Aphidinae [76, 77]. The parasitoids *A. ervi* and *P. volucre* responded only to the (7*S*)-enantiomer of nepetalactone; the (7*R*)-enantiomer was neither attractive nor antagonistic [78]. The parasitoid *Trichogramma evanescens*, which parasitizes eggs of stored-product pests *Ephestia* spp. and *Plodia interpunctella*, is attracted to the main sex pheromone component of the host, (9*Z*,12*E*)-tetradeca-9,12-dienyl acetate **28**. The acetate **28** attracted and arrested female parasitoids that have had previous oviposition experience [64]. Two species of parasitoids of sawfly eggs responded to the sex pheromones of their hosts *Diprion pini* and *Neodiprion sertifer*. In a four arm airflow olfactometer, *Chrysonotomyia ruforum* females responded to the (2*S*,3*S*,7*S*)- but not the (2*S*,3*R*,7*R*)-configuration of the acetate of 3,7-dimethylpentadecan-2-ol **29**. *Dipriocampe diprioni* females responded to the acetate and propanoate of (2*S*,3*R*,7*R*)-3,7-dimethyltridecan-2-ol **30** [79]. *Trichogramma chilonis* (Trichogrammatidae), an egg parasitoid, was attracted to the sex pheromone of the diamondback moth, *Plutella xylostella* [80]. Strongest responses were seen to the complete blend ((11*Z*)-hexadec-11-enal **31**, (11*Z*)-hexadec-11-en-1-yl acetate **32** and the alcohol **33** 1:1:0.01), but the parasitoids were also attracted to the acetate alone and to a mixture of acetate and aldehyde. They were not attracted to the alcohol **33** component alone [80].

An example of a larval parasitoid that responds to the host sex pheromone is seen with *Cotesia plutellae* (Braconidae), also a parasitoid of the diamondback moth. These insects were attracted equally to the pheromone blend (**31, 32, 33**, see above), the acetate **32**, or aldehyde **31** components [80]. This larval parasitoid, however, was also strongly attracted to host frass volatiles, in particular, dipropyl disulfide **34**, dimethyl disulfide **35**, allyl isothiocyanate **36**, and dimethyl trisulfide **37**. In contrast, the egg parasitoid *Trichogramma chilonis* was only weakly attracted to **36**. In both, *T. chilonis* and *C. plutellae*, plant volatiles, in particular (3*Z*)-hex-3-en-1-yl acetate **38**, significantly enhanced attraction by the pheromone [80].

Several other examples of host plant recognition by hymenopteran parasitoids have been described recently. Six species of aphid parasitoids, *Aphidius ervi*, *Trioxys* sp., *Praon* sp., *Aphelinus flavus*, *Lysiphlebus fabarum*, and *Aphidius rophalosiphi* were most strongly attracted to their host aphid in combination with the damaged host plant [62]. For *A. rhopalosiphi*, three wheat volatiles were

Table 4 Recently identified parasitoid semiochemicals. Numbers in bold refer to chemical structures shown on Fig. 2

Taxon	Host [stage]	Kairomone [parasitoid response]	Chemical Name(s)	Ref.
CHALCIDOIDEA				
Aphelinidae				
Encarsia formosa	*Trialeurodes vaporariorum*	Damaged plant volatiles [attractant]	(3*Z*)-Hex-3-en-1-ol **39** & octan-3-one **51**	[84]
Eulophidae				
Chrysonotomyia ruforum	*Diprion pini* & *Neodiprion sertifer*	Host sex pheromone [female arrestant]	Acetate of (2*S*,3*S*,7*S*)-3,7-dimethyl-pentadecan-2-ol **29** & acetate & propanoate of (2*S*,3*R*,7*R*)-3,7-dimethyltridecan-2-ol **30**	[79]
Dipriocampe diprioni	*Diprion pini*	Host sex pheromone [female arrestant]	Acetate & propanoate of (2*S*,3*R*,7*R*)-3,7-dimethyltridecan-2-ol **30**	[79]
Trichogrammatidae				
Trichogramma chilonis	*Plutella xylostella*	Host sex pheromone blend [attraction]	(11*Z*)-Hexadec-11-enal **31**, (11*Z*)-hexadec-11-enyl acetate **32** & (11*Z*)-hexadec-11-en-1-ol **33**	[74, 75]
		Host frass volatiles	Allyl isothiocyanate **36**	[80]
		Plant volatiles [enhancement of pheromone attraction]	(3*Z*)-Hex-3-en-1-yl acetate **38**	[80]
Trichogramma evanescens	*Plodia interpunctella* & *Ephestia* spp.	Host sex pheromone component [attraction, arrestment]	(9*Z*,12*E*)-Tetradeca-9,12-dienyl acetate **28**	[64]

Table 4 (continued)

Taxon	Host [stage]	Kairomone [parasitoid response]	Chemical Name(s)	Ref.
ICHNEUMONOIDEA				
Braconidae				
Aphidius ervi	Aphid sp. (Aphidinae)	Host sex pheromone [attraction, increased parasitization]	(4a*S*,7*S*,7a*R*)-Nepetalactone **26** & (1*R*,4a*S*,7*S*,7a*R*)-nepetalactol **27**	[77, 78]
Aphidius rhopalosiphi	*Sitobion avenae*	Host sex pheromone [attraction, increased parasitization]	(4a*S*,7*S*,7a*R*)-Nepetalactone **26** & (1*R*,4a*S*,7*S*,7a*R*)-nepetalactol **27**	[76]
		Wheat volatiles [attraction]	(3*Z*)-Hex-3-en-1-yl acetate **38**, (3*Z*)-hex-3-en-1-ol **39** & (2*E*)-hex-2-enal **40**	[62]
		Aphid honeydew [attraction]	Indole 3-acetaldehyde **41**	[62]
Aphidius uzbekistanicus	*Aphidius* sp.	Host alarm signal [attraction]	(*E*)-β-Farnesene **48**	[88]
Cotesia marginiventris	Various lepidopteran sp. (*Spodoptera* spp., *Heliothis* spp.)	Plant volatiles [attraction]	(3*Z*)-Hex-3-enal, (2*E*)-hex-2-enal **40**, (3*Z*)-hex-3-en-1-ol **39**, (3*Z*)-hex-3-en-1-ylacetate **38,** linalool **44**, (3*E*)-4,8-dimethylnona-1,3,7-triene **45**, indole **46**, α-*trans*-bergamotene **47**, (*E*)-β-farnesene **48**, (*E*)-nerolidol **49** & (3*E*,7*E*)-4,8,12-trimethyltrideca-1,3,7,11-tetraene **50**	[61]

Table 4 (continued)

Taxon	Host [stage]	Kairomone [parasitoid response]	Chemical Name(s)	Ref.
Cotesia plutellae	*Plutella xylostella*	Host sex pheromone [attraction]	(11*Z*)-Hexadec-11-enal **31**, (11*Z*)-hexadec-11-en-1-yl acetate **32** & (11*Z*)-hexadec-11-en-1-ol **33**	[80]
		Host frass volatiles [attraction]	Dipropyl disulfide **34**, dimethyl disulfide **35**, allyl isothiocyanate **36** & dimethyl trisulfide **37**	[80]
		Plant volatiles [enhancement of attraction]	(3*Z*)-Hex-3-en-1-yl acetate **38**	[80]
Microplitis crocipes		Green leaf volatiles [attraction]	Hexanal, (2*E*)-hex-2-enal **40**, (2*E*)-hex-2-en-1-ol **42**, (3*Z*)-hex-3-en-1-ol **39**, acetates of **42** the propanoate & **39**, butyrate of **39**, & β-ocimene **43**	[63, 81, 82]
Praon volucre	Various aphid sp. (Aphidinae)	Host sex pheromone [attraction, increased parasitization]	(4a*S*,7*S*,7a*R*)-Nepetalactone **26** & (1*R*,4a*S*,7*S*,7a*R*)-nepetalactol **27**	[77, 78]
PLATYGASTROIDEA				
Scelionidae				
Telenomus euproctidis	*Euproctis taiwana*	Host sex pheromone [attraction]	(9*Z*)-16-Methylheptadec-9-en-1-yl isobutyrate **23** & 16-methylheptadec-1-yl isobutyrate **24**	[74, 75]
	Euproctis pseudoconspersa	Host sex pheromone [attraction]	Racemic 10,14-dimethylpentadec-1-yl isobutyrate **25**	[75]

identified as attractants: (3*Z*)-hex-3-en-1-yl acetate **38**, (3*Z*)-hex-3-en-1-ol **39**, and (2*E*)-hex-2-enal **40**. A decomposition product of aphid honeydew, indole 3-acetaldehyde **41**, was also identified as attractive [62]. Similarly, the larval parasitoid *Microplitis crocipes* (Braconidae) was attracted to green leaf volatiles: hexanal, (2*E*)-hex-2-enal **40**, (2*E*)-hex-2-en-1-ol **42** and its acetate, (3*Z*)-hex-3-en-1-ol **39** and its acetate, propanoate and butyrate [81]. *M. crocipes* was shown to exhibit EAG responses to 29 cotton plant volatiles, with β-ocimene **43** being the most effective of the monoterpenes tested [82]. The larval parasitoid of various lepidopteran pests (*Spodoptera* spp., *Heliothis* spp.), *Cotesia marginiventris*, was also attracted to the (3*Z*)-hex-3-enal, the alcohol **39**, and the acetate **38** [61]. In addition, analytical and preparative GC were used to isolate corn plant volatiles active in long-range attraction (as determined in a flight tunnel bioassay). Spectroscopic identification of active peaks revealed that linalool **44**, (3*E*)-4,8-dimethylnona-1,3,7-triene **45**, indole **46**, α-*trans*-bergamotene **47**, (*E*)-β-farnesene **48**, (*E*)-nerolidol **49**, and (3*E*,7*E*)-4,8,12-trimethyltrideca-1,3,7,11-tetraene **50** are attractive to *C. marginiventris* [61]. The larval parasitoid *Diglyphus isaea* attacks the American serpentine leaf miner (*Liriomyza trifolii*). Olfactometric studies revealed that strongest parasitoid attraction occurred to mined bean leaves. Leaf extracts and volatiles were collected, but no specific compounds were identified as being responsible for the attraction [83]. Recently, analysis of volatiles from whitefly-damaged bean plants has revealed increased levels of (3*Z*)-hex-3-en-1-ol **39**, (3*E*)-4,8-dimethylnona-1,3,7-triene **45**, and octan-3-one **51**. All three compounds were found to attract the whitefly parasitoid *Encarsia formosa* [84]. Analysis of sawfly-damaged and jasmonic acid-treated pine twigs revealed that more (*E*)-β-farnesene **48** was consistently released. Both sawfly-damaged and jasmonic acid-treated pine twigs were attractive to the sawfly egg parasitoid *Chrysonotomyia ruforum* [85].

Egg-deposition also can induce the host plant to emit volatiles that attract egg parasitoids. For example, egg deposition by the elm leaf beetle (*Xanthogaleruca luteola*) causes its host plant, the field elm (*Ulmus minor*), to release a blend of mostly terpenoids that attract the egg parasitoid *Oomyzus gallerucae* (Eulophidae) [86]. Although the specific compounds that initiate the volatile emission and that attract the egg parasitoid are unknown, the host plant response can be induced with jasmonic acid.

One recent study addressed the response of a parasitoid to the host's spacing pheromone. *Aphidius rhopalosiphi* did not respond to the spacing pheromone of one host, the cherry-oat aphid *Rhopalosiphum padi*, which consists of 6-methylhept-5-en-2-one **16**, 6-methylhept-5-en-2-ol, tridecan-2-one and methyl salicylate. These compounds did not attract or repel the parasitoid. Because the aphid spacing pheromone can potentially be used to cause aphids to disperse, the failure of the parasitoid to respond to the spacing pheromone makes simultaneous use of the spacing pheromone and the parasitoid possible in aphid management [87].

Alarm signals produced by stressed hosts also attract parasitoids. For example, stressed aphids (*Aphidius* sp.) were attractive to female parasitoids of two species: *A. uzbekistanicus* and *A. volucre*. The cornicle secretions, produced by the aphids under stress, contain (*E*)-β-farnesene **48**, which was found to be weakly active in a Y-tube olfactometer. However, live stressed aphids were more attractive than (*E*)-β-farnesene alone, suggesting the existence of other attractive compounds [88].

3.5 Multi-Level Interactions

Kairomones also play a role in hyperparasitism and multi-trophic interactions. For example, *Euneura augarus* (Pteromalidae), a hyperparasitoid of conifer lachnid parasitoids, utilizes volatile compounds released by the conifers for both attraction to a host site and for host acceptance [44]. Recent studies suggest that detection of the damaged host plant by a parasitoid is specific for damage induced by its herbivorous host. Thus, corn seedlings damaged and treated with *N*-(17-hydroxylinolenoyl)-L-glutamine (volicitin, a saliva compound from *Spodoptera exigua*) were significantly more attractive to the parasitoid *Cotesia marginiventris* than buffer-treated control plants. The blend of volatiles released was significantly different between the volicitin and buffer-treated plants and the total amount of volatiles, in particular indole **46**, (3*E*)-4,8-dimethylnona-1,3,7-triene **45**, and (*E*)-β-farnesene **48**, was much higher in the volicitin treated plants [89]. This type of tritrophic interaction was clearly demonstrated to be host-specific in the case of *Cardiochiles nigriceps*, a parasitoid of *Heliothis virescens*. Plants (tobacco, cotton, or corn) infested with *H. virescens* emitted a different blend of volatiles than plants infested with *H. zea*. The plants infested with *H. virescens* attracted significantly more *C. nigriceps* females than plants infested with *H. zea* [90].

Detection of the hyperparasitoid by the primary parasitoid has also been recently described. The parasitoid *Aphidius uzbekistanikus* detects *trans*-fused iridoids **21** produced by females of the hyperparasitoid *A. victrix* as part of their defensive cephalic gland secretion. The iridoids cause avoidance behavior in *A. uzbekistanikus* [46].

4 Social Hymenoptera

4.1 Introduction

Most social insects are found in the order Hymenoptera. Sociality in insects is defined by the presence of one or more of the following traits: (1) individuals of the same species cooperate in caring for the young; (2) there is a reproductive division of labor, with usually sterile individuals working on behalf of fecund individuals; and (3) there is an overlap of at least two generations in life stages capable of contributing to colony labor, so that offspring assist parents during some period of their life [4]. The number of these traits a species possess defines the level of sociality. Eusocial insects exhibit all three traits, whereas presocial insects exhibit some but not all of these traits. Most of the social hymenoptera are found in the Aculeata (Table 1). Some species of bees (Apoidea), wasps (Vespidae), and all ants (Formicidae) are eusocial. Presocial hymenoptera are represented in several families. Although only a small fraction of the hymenopteran species are social, the increased complexity of communication requires an abundance of semiochemicals that have intrigued (and frustrated) chemical ecologists for decades [4]. As a result, pheromones from social hymenoptera account for most of the chemical identifications in this order (for overview, see [7]).

Most social hymenoptera have a reproductively dominant female – the queen. Thus, chemical cues can originate from the queen, workers, brood, and males. These different members of the colony may coordinately regulate the behavior or physiology of other colony members, making for an interesting and complex communication system. The particular response to a pheromone can be classified into one of the following twelve functional categories [6] but will be grouped by source in the discussion below:

1. Alarm
2. Simple attraction
3. Recruitment, as to a new food source or nest site
4. Grooming
5. Trophallaxis (exchange of oral and anal liquid)
6. Exchange of solid food particles
7. Group effect: either facilitating or inhibiting a given activity
8. Recognition of both nestmates and members of particular castes, including discrimination of injured and dead individuals
9. Caste determination, either by inhibition or by stimulation
10. Control of competing reproductives
11. Territorial and home range signals and nest markers
12. Sexual communication, including species recognition, sex recognition, synchronization of sexual activity, and assessment during sexual competition

Although many chemicals have been identified, much research is still needed to identify the semiochemicals involved in the many semiochemical-mediated responses observed in social hymenoptera. The chemical communication in wasps [9] and ants [8] has recently been reviewed. The semiochemicals recently identified in social hymenoptera are summarized in Table 5 and Fig. 3, and some examples are discussed in detail below.

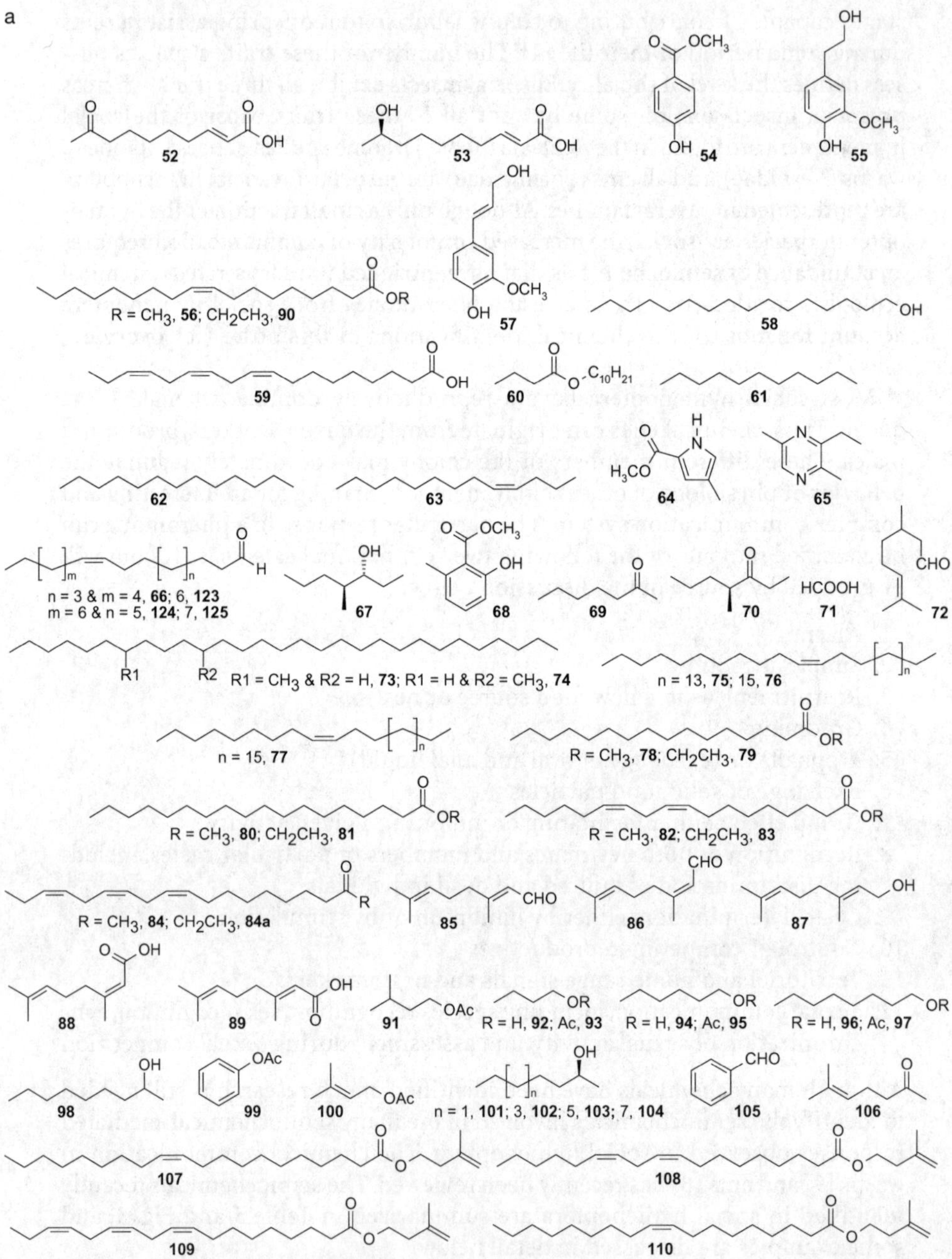

Fig. 3 Structures of recently identified semiochemicals in social hymenoptera

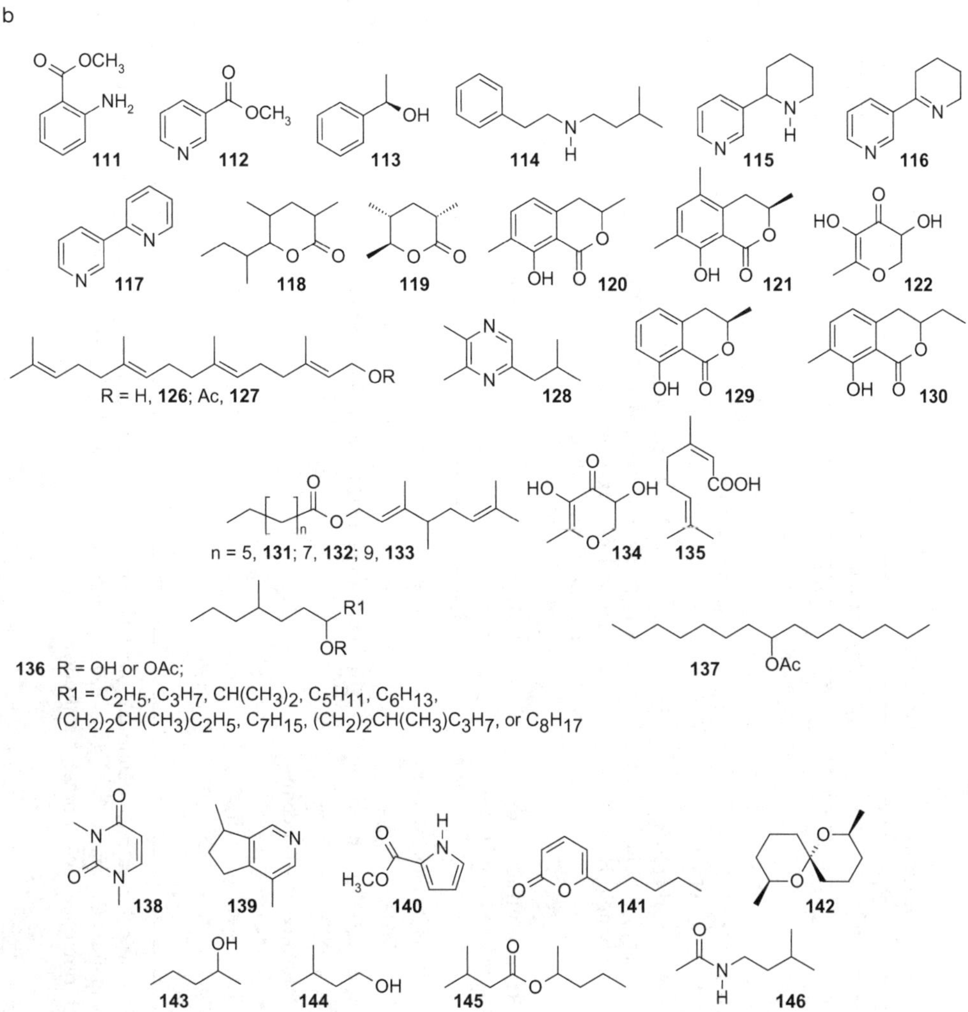

Fig. 3 (continued)

Table 5 Recently identified pheromones and semiochemicals of social hymenoptera. Numbers in bold refer to chemical structures shown on Fig. 3

Taxon	Source	Function	Chemical Name(s)	Ref.
APOIDEA				
Apidae				
Apis florea	W-MG	Repels workers from sucrose solution	Heptan-2-one **69**	[137]
Apis mellifera	Q-MG	Delays worker foraging ontogeny & reduces their hemolymph JH titre	(2*E*)-9-Oxodec-2-enoic acid **52**, (2*E*)-9-hydroxydec-2-enoic acid (71% (*R*)-(-)) **53**, methyl 4-hydroxybenzoate **54**, & 4-hydroxy-3-methoxyphenylethanol **55** (collectively called queen mandibular pheromone "QMP")	[138]
	Q-MG, CLG, DG, & ?	Retinue attraction of workers to mated queen	QMP, methyl oleate **56**, (2*E*)-3-(4-hydroxy-3-methoxy)-prop-2-en-1-ol **57**, hexadecan-1-ol **58**, & linolenic acid **59**	[96, 99]
	Q-MG	Inhibits JH biosynthesis in workers	(2*E*)-9-Oxodec-2-enoic acid **52**	[139]
	Q-MG	Suppresses queen rearing in queenless colonies	QMP or just (2*E*)-9-oxodec-2-enoic acid **52**	[140, 141]
	Q-MG	Influences worker comb-building behavior: decreases comb cell diameter, increases comb area & weight	QMP	[142]

Abbreviations: CLG, cephalic labial gland; CW, cuticular wash; DG, Dufour's gland; HE, head extract; HG, hindgut (rectal bladder); JH, juvenile hormone; MG, mandibular gland; NG, Nasonov gland; PG, poison gland; PvG, Pavan gland; PyG, pygidial gland; PPyG, postpygidial gland; Q-, queen; SA, sting apparatus; VG, venom gland; VQ-, virgin queen; W-, worker; ?, unknown

Table 5 (continued)

Taxon	Source	Function	Chemical Name(s)	Ref.
	Q-MG	Attracts drones on mating flights	QMP	[143, 144]
	Q-MG	Transiently & chronically regulates gene expression in worker brains	QMP	[145]
	Q-MG & W-NG	Long- & short-range attraction of swarms	QMP, (*E*)-citral **85**, (*Z*)-citral **86**, geraniol **87**, nerolic acid **88**, & geranic acid **89**	[146]
	Larva	Inhibits worker ovarian development	Ten component blend of "brood pheromone": methyl & ethyl palmitate **78** & **79**, stearate **80** & **81**, oleate **56** & **90**, linoleate **82** & **83**, & linolenate **84** & **84a**, specifically ethyl palmitate & methyl linolenate	[121, 122]
	Larva	Decreases worker response proboscis threshold to sucrose in a extension response & increases number of pollen foragers	Ten component blend of "brood pheromone"	[123]
	Larva	Delays onset of worker foraging & suppresses JH titre in worker hemolymph	Ten component blend of "brood pheromone"	[124]
	Larva	Trigger workers to cap larval cells	Methyl linoleate **82**, linolenate **84**, oleate **56**, & palmitate **78**	[147]

Table 5 (continued)

Taxon	Source	Function	Chemical Name(s)	Ref.
	Larva	Stimulate development of worker hypo-pharyngeal glands	Ethyl oleate **90** & methyl palmitate **78**	[148]
	Larva	Increases acceptance of queen cells	Methyl stearate **80** [149]	
	Larva	Increases deposition of royal jelly into queen cells	Methyl linoleate **82**	[149]
	Larva	Increases queen larval weight	Methyl palmitate **78**	[149]
	Q-pupa	Worker recognition of queen pupa	Methyl linoleate **82**, linolenate **84**, & oleate **56**	[150]
	W-SA	Previously identified alarm pheromone components that (a) change flight activity, (b) increase recruitment, or (c) help localize target	Isopentyl acetate **91** (a,b,c), butan-1-ol **92** (b), hexyl acetate **95** (b), butyl acetate **93** (b,c), hexan-1-ol **94** (b,c), octan-1-ol **96** (b), octyl acetate **97** (c), nonan-2-ol **98** (a,b), benzyl acetate **99** (a)	[151]
Apis mellifera (africanized)	W-SA	Worker recruitment at colony entrance	3-Methylbut-2-en-1-yl acetate **100** & isopentyl acetate **91**	[152]
Apis mellifera carnica & *A. m. scutellata*	W-SA	Stimulates walking, wing beating & avoidance by drones	Isopentyl acetate **91**	[153]
	W-NG	Attracts drones	Geraniol **87**	[153]

Table 5 (continued)

Taxon	Source	Function	Chemical Name(s)	Ref.
Scaptotrigona postica	VQ-HE	Attracts males to virgin queen	(*S*)-(+)-Heptan-2-ol **101**, (*S*)-(+)-nonan -2-ol **102**, (*S*)-(+)-undecan-2-ol **103**, & (*S*)-(+)-tridecan-2-ol **104**	[154, 155]
Trigona (*Tetragonisca*) *angustula*	MG	Defensive recruitment	Benzaldehyde **105**	[156]
	Head	Kairomone components from head of robber bee *Lestrimelitta limao* that induce defensive behavior	Citral **85** & **86**, & 6-methylhept-5-en-2-one **106**	[156]
Halictidae				
Lasioglossum (*Evylaeus*) *malachurum*	CW	Attraction of males to virgin females	3-Methylbut-3-enyl eicos-11-enoate **107**, linoleate **108**, linolenoate **109**, & oleate **110**	[157]
VESPOIDEA				
Formicidae				
Acromyrmex ubterraneus subterraneus	W-VG	Trail following	Methyl 4-methylpyrrole-2-carboxylate **64**	[158]
Aenictus species	W-PPyG	Trail following	Methyl 2-aminobenzoate **111** (releaser) & methyl pyridine-3-carboxylate **112** (primer)	[159]
Aphaenogaster albisetosus	W-PG	Trail following	(*S*)-4-Methylheptan-3-one **70** ((±) active, 56% e.e. (*S*) found in gland)	[160]

Table 5 (continued)

Taxon	Source	Function	Chemical Name(s)	Ref.
Aphaenogaster cockerelli	W-PG	Trail following	(*R*)-(+)-1-Phenylethanol **113**	[160]
	W-PG	Recruitment	(*S*)-4-Methylheptan-3-one **70**	[160]
Aphaenogaster rudis	W-PG	Trail following	*N*-Isopentyl-2-phenylethylamine **114**, 3-(2-piperidinyl) pyridine **115**, 3,4,5,6-tetrahydro-2,3′-bipyridine **116**, & 2,3′-bipyridyl **117**	[161]
Atta bisphaerica & *A. capiguara*	W-MG	Alarm	4-Methylheptan-3-one **70**	[162]
Atta sexdens sexdens	W-VG	Trail following	3-Ethyl-2,5-dimethylpyrazine **65** & methyl 4-methylpyrrole-2-carboxylate **64**	[163]
Atta texana	W-?	Antennal response of previously identified trail pheromone component	Methyl 4-methylpyrrole-2-carboxylate **64**	[164]
	W-?	Antennal response of previously identified alarm pheromone component	Heptan-2-one **69** & 4-methylheptan-3-one **70**	[164]
Camponotus atriceps	W-HG	Trail following	3,5-Dimethyl-6-(1′-methylpropyl)-tetrahydropyran-2-one **118**	[165]
Camponotus floridanus	W-HG	Trail following	Nerolic acid **88**	[165]
Camponotus herculeanus	W-HG	Trail following	2,4-Dimethyl-5-hexanolide **119** (without stereochemistry)	[166]

Table 5 (continued)

Taxon	Source	Function	Chemical Name(s)	Ref.
Camponotus rufipes	W-HG	Trail following	3,4-Dihydro-8-hydroxy-3,7-dimethyl-isocoumarin **120**	[167]
Camponotus silvicola	W-HG	Trail following	(3*R*)-(-)-3,4-Dihydro-8-hydroxy-3,5,7-trimethyl-isocoumarin **121**	[167]
Camponotus socius	W-HG	Trail following	(2*S*,4*R*,5*S*)-2,4-Dimethyl-5-hexanolide **119** & 2,3-dihydro-3,5-dihydroxy-6-methylpyran -4-one **122**	[168]
	W-PG	Recruitment	Formic acid **71**	[168]
	W-DG	Alarm	Undecane **61**	[168]
Dolichoderus thoracicus	W-PvG	Trail following	(9*Z*)-Hexadec-9-enal **66**, (9*Z*)-octadec9 -enal **123**, (11*Z*)-eicos-11-enal **124**, & (13*Z*)-docos-13-enal **125**	[169]
Ectatomma ruidum	W-DG	Recruitment	(2*E*,6*E*,10*E*)-Geranylgeraniol **126** & (2*E*,6*E*,10*E*)-geranylgeranyl acetate **127**	[170]
Eutetramorium mocquerysi	W-PG	Trail following	2,3-Dimethyl-5-(2-methylpropyl)pyrazine **128**	[171]
Formica fusca	W-HG	Trail following & antennal responses	(3*R*)-(-)-3,4-Dihydro-8-hydroxy-3,5,7-trimethylisocoumarin **121**	[172, 173]
Formica lugubris	Q-DG?	Attraction of males to alate queens	Undecane **61**, tridecane **62**, & (4*Z*)-tridec -4-ene **63**	[112]
Formica rufa	W-HG	Trail following & antennal responses	3,4-Dihydro-8-hydroxy-3-methylisocoumarin **129** (without stereochemistry)	[172, 173]

Table 5 (continued)

Taxon	Source	Function	Chemical Name(s)	Ref.
Formica sanguinea	W-HG	Trail following & antennal responses	3,4-Dihydro-8-hydroxy-3,7-dimethyliso-coumarin **120** & 3-ethyl-3,4-dihydro-8-hydroxy-7-methylisocoumarin **130**	[172, 173]
Gnamptogenys striatula	W-DG	Trail following	4-Methylgeranyl octanoate **131**, decanoate **132**, & dodecanoate **133**	[174]
Lasius fuliginosus	W-HG	Trail following & antennal responses	(3*R*)-(-)-3,4-Dihydro-8-hydroxy-3-methyliso-coumarin **129** & 2,3-dihydro-3,5-dihydroxy-6-methyl-4*H*-pyran-4-one **134**	[175]
Lasius niger	W-HG	Trail following & antennal responses	(3*R*)-(-)-3,4-Dihydro-8-hydroxy-3,5,7-trimethylisocoumarin **121**	[172, 173]
Leptogenys diminuta	W-PG	Trail following & antennal responses	(3*R*,4*S*)-4-Methylheptan-3-ol **67**	[176, 177]
	W-PyG	Antennal response of previously identified recruitment pheromone	*cis*-Isogeraniol **135**	[177]
Leptogenys peuqueti	W-PG	Trail following	Series of 14 related methyl-branched secondary alcohols & acetates **136** & **137**	[178]
Linepithema humile	W-PvG	Previously identified trail pheromone enhances consumption of sucrose solution	(9*Z*)-Hexadec-9-enal **66**	[179]

Table 5 (continued)

Taxon	Source	Function	Chemical Name(s)	Ref.
Mayriella overbecki	W-PG	Trail following	Methyl 6-methylsalicylate **68**	[180]
Megaponera foetens	W-PG	Trail following	*N,N*-Dimethyluracil **138**	[181]
	W-PyG	Alarm-exodus from nest	Actinidine **139**	[181]
Metapone madagascarica & *M. new species*	W-PG	Trail following	Methyl pyrrole-2-carboxylate **140**	[182]
Pachycondyla marginata	W-PyG	Trail following	Citronellal **72**	[183]
Pogonomyrmex barbatus, *P. maricopa*, *P. occidentalis*, & *P. rugosus*	W-PG	Recruitment	3-Ethyl-2,5-dimethylpyrazine **65**	[184]
Polyergus rufescens	Q-DG	Worker repellent produced by slave-making queen	Decyl butanoate **60**	[111]
Pristomyrmex pungens	W-PG	Trail following & antennal response	Deca-2,4-dien-5-olide **141**	[185]
Tetramorium impurum	W-PG	Trail following	Methyl 6-methylsalicylate **68**	[186]

Table 5 (continued)

Taxon	Source	Function	Chemical Name(s)	Ref.
Vespidae				
Pheidole pallidula	W-PG	Attracts & stimulates locomotion in workers	3-Ethyl-2,5-dimethylpyrazine **65**	[187]
Polistes dominulus	W-CW	Nestmate recognition cues	7-Methylnonacosane **73**, 11-methylnonacosane **74**, (11*Z*)-nonacos-11-ene **75**, (11*Z*)-hentriacont-11-ene **76**, & (9*Z*)-nonacos-9-ene **77**	[116]
Polybia occidentalis	W-VG	Alarm	(2*S*,6*R*,8*S*)-2,8-Dimethyl-1,7-dioxaspiro-[5.5]undecane **142**	[188]
Vespa mandarinia	W-VG	Alarm	Pentan-2-ol **143**, 3-methylbutan-1-ol **144**, & 1-methylbutyl 3-methylbutanoate **145**	[189]
Vespula maculifrons	W-SA	Alarm	*N*-3-Methylbutylacetamide **146**	[190]

4.2 Queen Pheromones

The queen is usually reproductively dominant within the colony and uses chemical cues as both primer and releaser pheromones to suppress the production or fecundity of other sexuals, inhibit reproduction by worker castes, modulate reproductive behaviors (e.g., inhibit swarming and orient swarms), attract males, regulate worker tasks and worker ontogeny, and produce host repellents in slave-making species. Considering the importance of queen semiochemicals in social hymenoptera, few queen pheromones have been chemically identified. The queens of most social hymenopteran colonies are attractive to workers, allowing them to be properly tended as well as to facilitate the dissemination of other pheromone cues. However, the retinue pheromone has been chemically identified in very few species. In the 1980s, queen pheromone components were identified in the fire ant, *Solenopsis invicta* [91, 92], and in the Pharaoh's ant, *Monomorium pharaonis* [93].

Although the first component of the honey bee (*Apis mellifera*) retinue pheromone, (2*E*)-9-oxodec-2-enoic acid **52**, was identified from queen mandibular glands more than 40 years ago [94, 95], this retinue pheromone has been better understood only recently. In 1988, a total of five synergistic components were identified from queen mandibular glands, adding (2*E*)-9-hydroxydec-2-enoic acid (71:29 (*R*)-(–):(*S*)-(+)) **53**, methyl 4-hydroxybenzoate **54**, and 4-hydroxy-3-methoxyphenylethanol **55**, dubbed queen mandibular pheromone (QMP) [96]. However, this was not the complete pheromone; additional retinue activity originates from the head [97]. However, identifying additional components was difficult because head extracts also contain highly attractive QMP. Fortunately, the retinue response to QMP is a heritable trait, allowing the selective breeding of workers that respond poorly to synthetic QMP while still responding strongly to their queen [98]. By using these workers in a retinue bioassay, four additional active components have recently been identified, methyl oleate **56**, (2*E*)-3-(4-hydroxy-3-methoxy)-prop-2-en-1-ol **57**, hexadecan-1-ol **58**, and linolenic acid **59** [99]. Thus, retinue attraction is attributed to at least nine synergistic components now called queen retinue pheromone (QRP). Although the potential these nine components have to elicit other responses in the colony has only begun to be studied [100], synthetic QMP has already been found to elicit many other responses (Table 5). Chemicals with multiple pheromonal roles are common in social insects [101].

In bumblebees, the queen is thought to modulate the onset of worker reproduction by regulating JH biosynthesis in workers [102]. The cuticle wash or whole body extract of *Bombus terrestris* bumblebee queens inhibits JH biosynthesis in workers. However, no specific gland extract, or synthetic compound identified from the mandibular gland, have proven effective [103]. Analogously, the queen of the ponerine ant *Pachycondyla apicalis* regulates the switch from trophic to reproductive worker-laid eggs [104].

Although many queen-produced pheromones are actively being studied in ants, particularly in the fire ant (*Solenopsis invicta*) [10, 105–110], chemical identifications remain elusive. However, there have been a few recent advances in ants. The queen of the slave-making species *Polyergus rufescens* produces decyl butanoate **60** that repels host workers when upsurping a colony [111]. Males of *Formica lugubris* are attracted to alate queens by undecane **61**, tridecane **62** and (4*Z*)-tridec-4-ene **63** [112].

4.3 Worker Pheromones

Within Hymenoptera, pheromones produced by workers in social colonies are the best studied across many genera, principally in ants [6], with those eliciting trail following most extensively studied. The distinct behavior and the relative ease of the bioassay have resulted in chemical identifications in many species [113, 114]. Those that have been recently identified are listed in Table 5. In addition, several alarm and recruitment signals have recently been identified. Many of the compounds recently identified in ants have previously been reported as trail or alarm pheromones in other ant species. For example, methyl 4-methylpyrrole-2-carboxylate **64**, 3-ethyl-2,5-dimethylpyrazine **65**, (9*Z*)-hexadec-9-enal **66**, 4-methylheptan-3-ol **67**, and methyl 6-methylsalicylate **68** have been identified as trail pheromone components, and heptan-2-one **69**, 4-methylheptan-3-one **70**, formic acid **71**, undecane **61**, 4-methylheptan-3-ol **67**, methyl 6-methylsalicylate **68**, and citronellal **72** have been identified as alarm pheromone components [6]. The use of the same chemicals across genera, with some used for very different functions, is an interesting phenomenon.

Workers are often cited as the source of nestmate or kin recognition cues (e.g., [115]). Many GC analysis studies have shown that the cuticular hydrocarbon profiles are different between nestmates and non-nestmates. However, only recently have specific hydrocarbons, 7-methylnonacosane **73**, 11-methylnonacosane **74**, (11*Z*)-nonacos-11-ene **75**, (11*Z*)-hentriacont-11-ene **76**, and (9*Z*)-nonacos-9-ene **77**, been shown to mediate nestmate recognition in the paper wasp, *Polistes dominulus* [116].

Many interesting worker-produced pheromones have recently been documented but have not been chemically identified. For example, in ants of the genus *Diacamma*, which lack a queen caste, wingless reproductive workers use a sex pheromone to attract mates [117]. Workers may also influence the physiology of other members of the colony, such as in the honey bee, where older foragers influence the ontogeny of age-related division of labor in younger workers [118]. Foraging bees, such as bumblebees [119], also appear to chemically mark flowers they have visited to temporarily prevent conspecifics from visiting a flower depleted of its nectar. Much research remains to be completed in order to fully understand worker pheromones in social hymenoptera.

4.4
Brood Pheromones

Although the immature members of most social hymenopteran colonies likely produce pheromones that modulate the behavior or physiology of their nestmates, only in the honey bee have these chemical cues begun to be identified. A blend of methyl and ethyl palmitate, stearate, oleate, linoleate, and linolenate **56, 78–84a,** and **90** first shown to attract the parasitic *Varroa* mite [120], also has been shown to have several pheromonal effects within the colony. These effects include inhibiting the ovarian development of workers [121, 122], increasing the number of pollen foragers while decreasing workers' response threshold to sucrose [123], and delaying the onset of worker foraging by suppressing the worker juvenile hormone (JH) hemolymph titer [124]. Some of these esters, particularly from brood destined to become queens, also change the behavior and physiology of the bees that care for the brood (Table 5). The recent identification of many of these esters in adult queens suggests that the queen and brood may use the same semiochemicals to coordinately modulate the behavior and physiology of colony members [99].

An example of an uncharacterized chemical cue from ant brood occurs in the obligatory slave-making ant, *Polyergus breviceps.* Pupae of this species are cared for by their enslaved host worker, while pupae of other species are consumed [125].

4.5
Male Pheromones

Although not studied extensively, males of social hymenoptera certainly produce pheromones. Male ants produce aggregation pheromones that attract both sexes to mating areas [6] as well as cause virgin alates to disperse from their colony [6]. However, these have not been chemically elucidated.

Bumblebee males mark objects on their flight paths with secretions from the labial glands [126, 127]. Although these marks can be attractive to both males and virgin queens, and are chemically well characterized in several species, the specific components responsible for attraction remain unknown.

4.6
Chemical Mimicry

Social hymenoptera must contend with other organisms chemically mimicking a semiochemical to gain advantage. Chemical mimicry may result from the organism biosynthesizing the compounds themselves or simply acquiring the chemical odor of the host colony once inside. The many interactions are well reviewed [4, 6, 128] and include mimicry by hymenoptera of the same or different species, other arthropods, and even orchids.

In the honey bee, two related examples have recently been examined. Egg-laying workers within a colony partially mimic the queen's attractive secretions

[129]. In the two sympatric races of honey bee in South Africa, queen-mimicking workers of *Apis mellifera capensis* invade queenright *A. m. scutellata* colonies and become a reproductive parasite. These pseudoqueens are not removed by workers or suppressed by the resident queen, and lay eggs that are acceptable to the colony [130]. Interestingly, these pseudoqueens may also fight to the death with the resident queen as multiple virgin queens do in a typical colony [131].

An example from ants is the acceptance of the caterpillar of the parasitic butterfly *Maculinea rebeli* by its host *Myrmica schencki* [132]. The caterpillar chemically mimics ant larvae and, in its final instar, drops from its food plant and waits for a foraging ant worker to bring it to the brood chambers of the host colony [133].

Flowers of some orchids mimic both the appearance and sex pheromone of virgin females of certain species of bees or wasps. This sexual deception results in pollination by male hymenoptera that would not normally visit flowers. Japanese honey bee drones (*Apis cerana japonica*) cluster on the oriental orchid (*Cymbidium pumilum*) while on their mating flights [134]. By comparing volatile profiles of orchids and the female hymenoptera they mimic, or by GC-EAD and GC-MS analysis of orchid volatiles, several compounds have been identified that may mediate this attraction for the solitary bee *Andrena nigroaenea* [135, 136] and the scoliid wasp *Campsoscolia ciliata* [135].

5
Chemical Diversity and Biochemistry

As can be seen in the tables and figures, the diversity of semiochemicals in Hymenoptera is outstanding. The large variation in polarity, volatility, functional groups, and stereochemistry mirrors the variety of situations that hymenoptera use semiochemicals. To date, no carbohydrate or proteinaceous semiochemical has been identified in Hymenoptera, although proteins have been implicated in the queen fire ant, *Solenopsis invicta* [109] and in the parasitoid *Aphidius ervi* (see above). There are many instances where identical compounds are used in different species for very different purposes. Many pheromones consist of multi-component blends. In some cases, strong synergy is seen between components, a situation that challenges our goal to fully understand a chemical cue.

The repertoire of chemicals that can be used for communication is limited by the biosynthetic ability of the insect. Compared to other insect orders, pheromone biosynthesis in Hymenoptera has received little study [191]. However, the biosynthetic origins of chemically diverse hymenopteran semiochemicals likely include aromatic, fatty acid, and terpenoid pathways as well as simple modifications of host-derived precursors. Notable recent studies include the biosynthesis of the fatty acid components (2*E*)-9-oxodec-2-enoic acid **52** and (2*E*)-9-hydroxydec-2-enoic acid of the honey bee queen mandibular pheromone from octadecanoic acid [192, 193], and the aliphatic alcohol and ester marking pheromones of bumblebee males from hexadecanoic acid [194]. Pheromone biosynthesis in Hymenoptera will likely prove to be a productive area of study in the future.

6 Looking to the Future

In preparing this review, it became clear just how much remains chemically unknown about the semiochemicals of hymenoptera. The use of semiochemicals has been postulated in most hymenoptera studied and the behavioral or physiological effect of extracts or washes have been examined in many species. In addition, there are many exhaustive reports of the chemical identification of the constituents of a particular gland or extract postulated to be involved in pheromone communication. However, for all these investigations, very few attempt to assign a biological activity to particular chemicals through biological testing. Since many pheromones are multi-component blends of synergistic components, with the possibility of originating from more than one glandular source, it is most worthwhile to use a bioassay-guided approach to isolate and identify the chemical(s) responsible. The synergy between chemists using modern, highly sensitive, analytical techniques and biologists conducting robust and sensitive bioassays is necessary to make progress in this area.

Although the traditional methods of semiochemical discovery and identification should not be neglected [195], the recent advances in molecular biology provide opportunities to study semiochemicals in different, and often more powerful, ways. For example, rather than examining general ovarian development in worker honey bees exposed to queen pheromone, the expression of specific genes involved in ovarian development might be monitored. Although a hymenopteran pheromone has not yet been chemically identified with such techniques, these techniques are being used to study the effects of previously identified pheromones. For example, cDNA microarrays have been used to identify genes involved in queen pheromone-mediated changes in the behavior and physiology of worker honey bees [145, 196, 197]. At the time of this writing, sequencing of the honey bee genome nears completion. In addition to being the most economically important beneficial insect, this eusocial hymenopterid is a perfect model organism to study the molecular biology, biochemistry, and genomics of semiochemicals in Hymenoptera, and the genome will lay the groundwork for such studies.

With adequate resources and effort, the tools are available to chemically identify many more semiochemicals in Hymenoptera. Much is still to be understood about the chemically-mediated communication in this large and diverse insect order. In addition, because many hymenoptera are significant beneficial or pest insects, and the use of semiochemicals in the management and monitoring of insects is becoming standard, the identification of additional semiochemicals in Hymenoptera is an economically worthwhile endeavor.

Acknowledgments We thank Adam L. Ludlow for assistance in collecting the references, drawing some of the structures, and editing. Support was given by the Natural Sciences and Engineering Research Council of Canada by grants to E.P. and K.N.S., and a postdoctoral fellowship to C.I.K.

References

1. Goulet H, Huber JT (eds) (1993) Hymenoptera of the world: an identification guide to families. Centre for Land and Biological Resources Research, Research Branch, Agriculture Canada, Ottawa
2. Morse RA, Calderone NW (2000) Bee Culture 128:1
3. Millar JG, Haynes KF (eds) (1998) Methods in chemical ecology. Chapman and Hall, New York
4. Wilson EO (1971) The insect societies. Harvard University Press, Cambridge, Mass
5. Free JB (1987) Pheromones of social bees. Chapman and Hall, London
6. Hölldobler B, Wilson EO (1990) The ants. Harvard University Press, Cambridge, Massachusetts
7. Billen J, Morgan ED (1998) Pheromone communication in social insects: sources and secretions. In: Vander Meer RK, Breed MD, Espelie KE, Winston ML (eds) Pheromone communication in social insects: ants, wasps, bees and termites. Westview Press, Boulder, Colorado, p 3
8. Vander Meer RK, Alonso LE (1998) Pheromone directed behavior in ants. In: Vander Meer RK, Breed MD, Espelie KE, Winston ML (eds) Pheromone communication in social insects: ants, wasps, bees and termites. Westview Press, Boulder, Colorado, p 159
9. Landolt PJ, Jeanne RL, Reed HC (1998) Chemical communication in social wasps. In: Vander Meer RK, Breed MD, Espelie KE, Winston ML (eds) Pheromone communication in social insects: ants, wasps, bees and termites. Westview Press, Boulder, Colorado, p 216
10. Vargo EL (1998) Primer pheromones in ants. In: Vander Meer RK, Breed MD, Espelie KE, Winston ML (eds) Pheromone communication in social insects: ants, wasps, bees and termites. Westview Press, Boulder, Colorado, p 293
11. Quicke DLJ (1997) Parasitic wasps. Chapman and Hall, London
12. Kainoh Y (1999) Parasitoids. In: Hardie J, Minks AK (eds) Pheromones of non-lepidopteran insects associated with agricultural plants. CABI Publishing, New York, p 383
13. Anderbrant O (1993) Pheromone biology of sawflies. In: Wagner MR, Raffa KF (eds) Sawfly life history adaptations to woody plants. Academic Press, San Diego, p 119
14. Anderbrant O (1999) Sawflies and seed wasps. In: Hardie J, Minks AK (eds) Pheromones of non-lepidopteran insects associated with agricultural plants. CABI Publishing, New York, p 199
15. Ayasse M, Paxton RJ, Tengo J (2001) Annu Rev Entomol 46:31
16. LaSalle J, Gauld ID (eds) (1993) Hymenoptera and biodiversity. C.A.B. International, London
17. Moreira JA, Corrêa AG (2000) J Brazil Chem Soc 11:614
18. Tai A, Syouno E, Tanaka K, Fujita M, Sugimura T, Higashiura Y, Kakizaki M, Hara H, Naito T (2002) Bull Chem Soc Jpn 75:111
19. Karlsson S, Högberg H-E (2000) Synthesis 2000:1863
20. Hedenström E, Edlund H, Lund S, Abersten M, Persson D (2002) J Chem Soc, Perkin Trans 1:1810
21. Hedenström E, Andersson F (2002) J Chem Ecol 28:1237

22. Karlsson S, Hedenström E (1999) Acta Chem Scand 53:620
23. Larsson M, Nguyen B-V, Högberg H-E, Hedenström E (2001) Eur J Org Chem 2001:353
24. Nakamura Y, Mori K (1999) Eur J Org Chem 1999:2175
25. Wassgren AB, Bergström G, Sierpinski A, Anderbrant O, Högberg H-E, Hedenström E (2000) Naturwissenschaften 87:24
26. Bergström G, Wassgren A-B, Anderbrant O, Ochieng SA, Östrand F, Hansson BS, Hedenström E, Högberg H-E (1998) Naturwissenschaften 85:244
27. Anderbrant O, Löfqvist J, Högberg H-E, Hedenström E, Baldassari N, Baronio P, Kolmakova G, Lyons B, Naito T, Odinokov V, Simandl J, Supatashvili A, Tai A, Tourianov R (2000) Entomol Exp Applic 95:229
28. Rieske LK, Townsend L, Anderbrant O, Hedenström E, Högberg H-E (2001) J Entomol Sci 36:67
29. Johansson BG, Anderbrant O, Simandl J, Avtzis ND, Salvadori C, Hedenström E, Edlund H, Högberg H-E (2001) J Chem Ecol 27:733
30. Martini A, Baldassari N, Baronio P, Anderbrant O, Hedenström E, Högberg H-E, Rocchetta G (2002) Agr For Entomol 4:195
31. Bartelt RJ, Cossé AA, Petroski RJ, Weaver DK (2002) J Chem Ecol 28:385
32. Cossé AA, Bartelt RJ, Weaver DK, Zilkowski BW (2002) J Chem Ecol 28:407
33. Cossé AA, Bartelt RJ, James DG, Petroski RJ (2001) J Chem Ecol 27:1841
34. Shibata C, Mori K (2004) Eur J Org Chem 2004:1083
35. Borg-Karlson A-K, Tengö J, Valterová I, Unelius CR, Taghizadeh T, Tolasch T, Francke W (2003) J Chem Ecol 29:1
36. Chen G-F, Zhang Q-H, Li Z-Y, Miao Z-W, Wang L-Z, Anderbrant O (1997) Chin J Biol Ctrl 13:61
37. Bergström G, Wassgren AB, Anderbrant O, Fägerhag J, Edlund H, Hedenström E, Högberg H-E, Geri C, Auger MA, Varama M, Hansson BS, Löfqvist J (1995) Experientia 51:370
38. Anderbrant O, Hansson BS, Hallberg E, Geri C, Varama M, Hedenström E, Högberg H-E, Fagerhag J, Edlund H, Wassgren AB, Bergström G, Löfqvist J (1995) J Insect Physiol 41:395
39. Anderbrant O, Zhang Q-H, Chu D (1997) J Appl Entomol 121:281
40. Anderbrant O, Löfqvist J, Högberg H-E, Hedenström E, Wassgren AB, Bergström G, Bengtsson M, Magnuson G (1992) Entomol Exp Applic 62:169
41. Anderbrant O, Bengtsson M, Löfqvist J, Baeckström P (1992) J Chem Ecol 18:1707
42. Powell W (1999) Parasitoid hosts. In: Hardie J, Minks AK (eds) Pheromones of non-lepidopteran insects associated with agricultural plants. CABI Publishing, Wallingford, p 405
43. Grasswitz TR (1998) J Insect Behav 11:539
44. Völkl W, Sullivan DJ (2000) Entomol Exp Applic 97:47
45. vanVeen FJF, Muller CB, Adriaanse ICT, Godfray HCJ (2002) J Animal Ecol 71:463
46. Petersen G, Matthiesen C, Francke W, Wyss U (2000) Eur J Entomol 97:545
47. Kainoh Y, Oishi Y (1993) J Chem Ecol 19:963
48. Kainoh Y, Nemoto T, Shimizu K, Tatsuki S, Kusano T, Kuwahara Y (1991) Appl Ent Zool 26:543
49. Hidoh O, Kawashima T, Fukami J-I, Kainoh Y (1992) Appl Entomol Zool 27:587
50. Swedenborg PD, Jones RL (1992) J Chem Ecol 18:1913
51. Swedenborg PD, Jones RL, Liu H-W, Krick TP (1993) J Chem Ecol 19:485
52. Shin I, Zhou H-W, Que NLS, Liu H-W, Swedenborg PD, Jones RL (1993) J Org Chem 58:2923
53. Swedenborg PD, Jones RL, Zhou H-Q, Shin I, Liu H-W (1994) J Chem Ecol 20:3373

54. Krokos FD, Konstantopoulou MA, Mazomenos BE (2001) J Chem Ecol 27:2169
55. Decker UM, Powell W, Clark SJ (1993) Entomol Exp Applic 69:33
56. Nazzi F, Powell W, Wadhams LJ, Woodcock CM (1996) J Chem Ecol 22:1169
57. Jewett DK, Carpenter JE (1998) Ann Entomol Soc Am 91:748
58. Shu S, Jones RL (1993) J Chem Ecol 19:2563
59. Finidori-Logli V, Bagnères A-G, Erdmann D, Francke W, Clément J-L (1996) J Chem Ecol 22:2063
60. Micha SG, Stammel J, Höller C (1993) Eur J Entomol 90:439
61. Turlings TCJ, Tumlinson JH, Heath RR, Proveaux AT, Doolittle RE (1991) J Chem Ecol 17:2235
62. Wickremasinghe MGV, Emden HFV (1992) Physiol Entomol 17:297
63. Park KC, Zhu J, Harris J, Ochieng SA, Baker TC (2001) Physiol Entomol 26:69
64. Schöller M, Prozell S (2002) J Stored Prod Res 38:177
65. Meiners T, Wäckers F, Lewis WJ (2003) Chem Senses 28:231
66. Syvertsen TC, Jackson LL, Blomquist GJ, Vinson SB (1995) J Chem Ecol 21:1971
67. Bin F, Wäckers F, Romani R, Isidoro N (1999) Intl J Insect Morphol Embryol 28:61
68. Isidoro N, Romani R, Pujade-Villar J, Ros-Farre P (1999) Zool Scripta 28:165
69. Guerrieri E, Pedata PA, Romani R, Isidoro N, Bin F (2001) J Natl Hist 35:41
70. Digilio MC, Isidoro N, Tremblay E, Pennacchio F (2000) J Insect Physiol 46:1041
71. Höller C, Williams HJ, Vinson SB (1991) J Chem Ecol 17:1021
72. Höller C, Bargen H, Vinson SB, Witt D (1994) J Insect Physiol 40:317
73. Hübner G, Völkl W, Francke W, Dettner K (2002) Biochem System Ecol 30:505
74. Arakaki N, Wakamura S, Yasuda T (1996) J Chem Ecol 22:1079
75. Arakaki N, Wakamura S, Yasuda T, Yamagishi K (1997) J Chem Ecol 23:153
76. Glinwood RT, Powell W, Tripahti CPM (1998) Biocontrol Sci Technol 8:607
77. Glinwood RT, Smiley DWM, Hardie J, Pickett JA, Powell W, Wadhams LJ, Woodcock CM (1999) Pest Sci 55:208
78. Glinwood RT, Du Y-J, Smiley DWM, Powell W (1999) J Chem Ecol 25:1481
79. Hilker M, Bläske V, Kobs C, Dippel C (2000) J Chem Ecol 26:2591
80. Reddy GVP, Holopainen JK, Guerrero A (2002) J Chem Ecol 28:131
81. Whitman DW, Eller FJ (1992) J Chem Ecol 18:1743
82. Li Y, Dickens JC, Steiner WWM (1992) J Chem Ecol 18:1761
83. Finidori-Logli V, Bagnères A-G, Clément J-L (1996) J Chem Ecol 22:541
84. Birkett MA, Chamberlain K, Guerrieri E, Pickett JA, Wadhams LJ, Yasuda T (2003) J Chem Ecol 29:1589
85. Mumm R, Schrank K, Wegener R, Schulz S, Hilker M (2003) J Chem Ecol 29:1235
86. Wegener R, Schulz S, Meiners T, Hadwich K, Hilker M (2001) J Chem Ecol 27:499
87. Gonzáles WL, Fuentes-Contreras E, Niemeyer HM (1999) J Appl Ent 123:413
88. Micha SG, Wyss U (1996) Chemoecology 7:132
89. Alborn HT, Turlings TCJ, Jones TH, Stenhagen G, Loughrin JH, Tumlinson JH (1997) Science 276:945
90. DeMoraes CM, Lewis WJ, Pare PW, Alborn HT, Tumlinson JH (1998) Nature 393:570
91. Glancey BM (1986) The queen recognition pheromone of *Solenopsis invicta*. In: Lofgren CS, Vander Meer RK (eds) Fire ants and leaf-cutting ants: biology and management. Westview Press, Boulder, p 223
92. Rocca JR, Tumlinson JH, Glancey BM, Lofgren CS (1983) Tetrahedron Lett 24:1889
93. Edwards JP, Chambers J (1984) J Chem Ecol 10:1731
94. Barbier J, Lederer E (1960) C R Seances Acad Sci Paris 251:1131
95. Callow RK, Johnston NC (1960) Bee World 41:152
96. Slessor KN, Kaminski L-A, King GGS, Borden JH, Winston ML (1988) Nature 332:354

97. Slessor KN, Foster LJ, Winston ML (1998) Royal flavours: honey bee queen pheromones. In: Vander Meer RK, Breed MD, Espelie KE, Winston ML (eds) Pheromone communication in social insects: ants, wasps, bees and termites. Westview Press, Boulder, Colorado, p 331
98. Pankiw T, Winston ML, Fondrk KM, Slessor KN (2000) Naturwissenschaften 87:487
99. Keeling CI, Slessor KN, Higo HA, Winston ML (2003) Proc Natl Acad Sci USA 100:4486
100. Hoover SER, Keeling CI, Winston ML, Slessor KN (2003) Naturwissenschaften 90:477
101. Blum MS (1996) Annu Rev Entomol 41:353
102. Cnaani J, Robinson GE, Bloch G, Borst D, Hefetz A (2000) Behav Ecol Sociobiol 47:346
103. Bloch G, Hefetz A (1999) J Chem Ecol 25:881
104. Dietemann V, Peeters C (2000) Ins Soc 47:223
105. Obin MS, Vander Meer RK (1994) J Entomol Sci 29:143
106. Vargo EL, Laurel M (1994) J Insect Physiol 40:601
107. Vargo EL (1999) Physiol Entomol 24:370
108. Vargo EL, Hulsey CD (2000) J Insect Physiol 46:1151
109. Klobuchar EA, Deslippe RJ (2002) Naturwissenschaften 89:302
110. Vander Meer RK, Alonso LE (2002) Behav Ecol Sociobiol 51:122
111. D'Ettorre P, Errard C, Ibarra F, Francke W, Hefetz A (2000) Chemoecology 10:135
112. Walter F, Fletcher DJC, Chautems D, Cherix D, Keller L, Francke W, Fortelius W, Rosengren R, Vargo EL (1993) Naturwissenschaften 80:30
113. Morgan ED (1990) Insect trail pheromones: a perspective in progress. In: McCaffery AR, Wilson ID (eds) Chromatography and isolation of insect hormones and pheromones. Plenum, New York, p 259
114. Traniello JFA, Robson SK (1995) Trail and territorial communication in social insects. In: Cardé RT, Bell WJ (eds) Chemical ecology of insects 2. Chapman and Hall, New York, p 241
115. Vander Meer RK, Morel L (1998) Nestmate recognition in ants. In: Vander Meer RK, Breed MD, Espelie KE, Winston ML (eds) Pheromone communication in social insects: ants, wasps, bees and termites. Westview Press, Boulder, Colorado, p 79
116. Dani FR, Jones GR, Destri S, Spencer SH, Turillazzi S (2001) Animal Behav 62:165
117. Nakata K, Tsuji K, Hölldobler B, Taki A (1998) J Insect Behav 11:869
118. Huang Z-Y, Plettner E, Robinson GE (1998) J Comp Physiol A 183:143
119. Stout JC, Goulson D, Allen JA (1998) Behav Ecol Sociobiol 43:317
120. Le Conte Y, Arnold G, Trouiller J, Masson C, Chappe B, Ourisson G (1989) Science 245:638
121. Arnold G, Le Conte Y, Trouiller J, Hervet H, Chappe B, Masson C (1994) C R Acad Sci Paris, Sciences de la vie/Life sciences 317:511
122. Mohammedi A, Paris A, Crauser D, Le Conte Y (1998) Naturwissenschaften 85:455
123. Pankiw T, Page RE Jr (2001) Behav Ecol Sociobiol 49:206
124. Le Conte Y, Mohammedi A, Robinson GE (2001) Proc R Soc Biol Sci B 268:163
125. Johnson CA (2000) J Insect Behav 13:711
126. Bergman P, Bergström G (1997) J Chem Ecol 23:1235
127. Bergström G (1981) Entomol Scand Supp 15:213
128. Hölldobler B, Wilson EO (1994) Journey to the ants: a story of scientific exploration. Harvard University Press, Cambridge, Massachusetts
129. Katzav-Gozansky T, Soroker V, Francke W, Hefetz A (2003) Ins Soc 50:20
130. Wossler TC (2002) Apidologie 33:139
131. Moritz RFA, Pflugfelder J, Crewe RM (2003) Naturwissenschaften 90:378
132. Akino T, Knapp JJ, Thomas JA, Elmes GW (1999) Proc R Soc Biol Sci B 266:1419
133. Elmes GW, Thomas JA, Wardlaw JC (1991) J Zool 223:447
134. Sasaki M, Ono M, Asada S, Yoshida T (1991) Experientia 47:1229

135. Ayasse M, Schiestl FP, Paulus HF, Lofstedt C, Hansson B, Ibarra F, Francke W (2000) Evolution Int J Org Evolution 54:1995
136. Schiestl FP, Ayasse M, Paulus HF, Lofstedt C, Hansson BS, Ibarra F, Francke W (2000) J Comp Physiol A 186:567
137. Naik DG, Banhatti P, Chadawa SS, Thomas D (2002) J Apic Res 41:59
138. Pankiw T, Huang Z-Y, Winston ML, Robinson GE (1998) J Insect Physiol 44:685
139. Kaatz HH, Hildebrandt H, Engels W (1992) J Comp Physiol B 162:588
140. Winston ML, Higo HA, Slessor KN (1990) Ann Entomol Soc Am 83:234
141. Engels W, Adler A, Rosenkranz P, Lübke G, Francke W (1993) J Comp Physiol B 163:363
142. Ledoux MN, Winston ML, Higo H, Keeling CI, Slessor KN, LeConte Y (2001) Ins Soc 48:14
143. Loper GM, Wolf WW, Taylor OR Jr (1993) J Chem Ecol 19:1929
144. Loper GM, Taylor OR Jr, Foster LJ, Kochansky J (1996) J Apic Res 35:122
145. Grozinger CM, Sharabash NM, Whitfield CW, Robinson GE (2003) Proc Natl Acad Sci USA 100:14519
146. Schmidt JO, Slessor KN, Winston ML (1993) Naturwissenschaften 80:573
147. Le Conte Y, Arnold G, Trouiller J, Masson C (1990) Naturwissenschaften 77:334
148. Mohammedi A, Crauser D, Paris A, Le CY (1996) C R Acad Sci Paris, Sciences de la vie/Life sciences 319:769
149. Le Conte Y, Sreng L, Poitout SH (1995) J Econ Entomol 88:798
150. Le Conte Y, Sreng L, Sacher N, Trouiller J, Dusticier G, Poitout SH (1994) Chemoecology 5/6:6
151. Wager BR, Breed MD (2000) Ann Entomol Soc Am 93:1329
152. Hunt GJ, Wood KV, Guzmán-Novoa E, Lee HD, Rothwell AP, Bonham CC (2003) J Chem Ecol 29:453
153. Becker MM, Brückner D, Crewe R (2000) J Apic Res 39:149
154. Engels W, Engels E, Lübke G, Schröder W, Francke W (1990) Entomol Gener 15:091
155. Engels W, Engels E, Francke W (1997) Invert Reprod Dev 31:251
156. Wittmann D, Redtke R, Zeil J, Lübke G, Francke W (1990) J Chem Ecol 16:631
157. Ayasse M, Engels W, Lübke G, Taghizadeh T, Francke W (1999) Behav Ecol Sociobiol 45:95
158. Do Nascimento RR, Morgan ED, Moreira DDO, Della Lucia TMC (1994) J Chem Ecol 20:1719
159. Oldham NJ, Morgan ED, Gobin B, Billen J (1994) Experientia 50:763
160. Hölldobler B, Oldham NJ, Morgan ED, König WA (1995) J Insect Physiol 41:739
161. Attygalle AB, Kern F, Huang Q, Meinwald J (1998) Naturwissenschaften 85:38
162. Hughes WOH, Howse PE, Vilela EF, Goulson D (2001) Physiol Entomol 26:165
163. Billen J, Beeckman W, Morgan ED (1992) Ethol Ecol Evol 4:197
164. Andryszak NA, Payne TL, Dickens JC, Moser JC, Fisher RW (1990) J Entomol Sci 25:593
165. Haak U, Hölldobler B, Bestmann HJ, Kern F (1996) Chemoecology 7:85
166. Bestmann HJ, Haak U, Kern F, Hölldobler B (1995) Naturwissenschaften 82:142
167. Übler E, Kern F, Bestmann HJ, Hölldobler B, Attygalle AB (1995) Naturwissenschaften 82:523
168. Kohl E, Hölldobler B, Bestmann HJ (2001) Chemoecology 11:67
169. Attygalle AB, Mutti A, Rohe W, Maschwitz U, Garbe W, Bestmann HJ (1998) Naturwissenschaften 85:275
170. Bestmann HJ, Janssen E, Kern F, Liepold B, Hölldobler B (1995) Naturwissenschaften 82:334
171. Tentschert J, Bestmann HJ, Hölldobler B, Heinze J (2000) Naturwissenschaften 87:377
172. Bestmann HJ, Kern F, Schaefer D, Witschel MC (1992) Angew Chem Int Ed Engl 31:795
173. Kern F, Bestmann HJ (1994) Z Naturforsch 49c:865

174. Blatrix R, Schulz C, Jaisson P, Francke W, Hefetz A (2002) J Chem Ecol 28:2557
175. Kern F, Klein RW, Janssen E, Bestmann H-J, Attygalle AB, Schäfer D, Maschwitz U (1997) J Chem Ecol 23:779
176. Steghaus-Kovâc S, Maschwitz U, Attygalle AB, Frighetto RTS, Frighetto N, Vostrowsky O, Bestmann HJ (1992) Experientia 48:690
177. Kern F, Bestmann HJ (1993) Naturwissenschaften 80:424
178. Janssen E, Übler E, Bauriegel L, Kern F, Bestmann HJ, Attygalle AB, Steghaus-Kovac S, Maschwitz U (1997) Naturwissenschaften 84:122
179. Greenberg L, Klotz JH (2000) J Econ Entomol 93:119
180. Kohl E, Hölldobler B, Bestmann HJ (2000) Naturwissenschaften 87:320
181. Janssen E, Bestmann HJ, Hölldobler B, Kern F (1995) J Chem Ecol 21:1947
182. Hölldobler B, Oldham NJ, Alpert GD, Liebig J (2002) Chemoecology 12:147
183. Hölldobler B, Janssen E, Bestmann HJ, Leal IR, Oliveira PS, Kern F, König WA (1996) J Comp Physiol A 178:47
184. Hölldobler B, Morgan ED, Oldham NJ, Liebig J (2001) J Insect Physiol 47:369
185. Janssen E, Hölldobler B, Kern F, Bestmann HJ, Tsuji K (1997) J Chem Ecol 23:1025
186. Morgan ED, Jackson BD, Ollett DG, Sales GW (1990) J Chem Ecol 16:3493
187. Cammaerts M-C, Cammaerts R (1994) Physiol Entomol 19:258
188. Dani FR, Jeanne RL, Clarke SR, Jones GR, Morgan ED, Francke W, Turillazzi S (2000) Physiol Entomol 25:363
189. Ono M, Terabe H, Hori H, Sasaki M (2003) Nature 424:637
190. Landolt PJ, Heath RR, Reed HC, Manning K (1995) Florida Entomol 78:101
191. Blomquist GJ, Vogt RG (eds) (2003) Insect pheromone biochemistry and molecular biology. Academic Press, New York
192. Plettner E, Slessor KN, Winston ML (1998) Insect Biochem Mol Biol 28:31
193. Plettner E, Slessor KN, Winston ML, Oliver JE (1996) Science 271:1851
194. Luxová A, Valterová I, Stránsky K, Hovorka O, Svatos A (2003) Chemoecology 13:81
195. Eisner T, Berenbaum M (2002) Science 295:1973
196. Whitfield CW, Band MR, Bonaldo MF, Kumar CG, Liu L, Pardinas JR, Robertson HM, Soares MB, Robinson GE (2002) Genome Res 12:555
197. Whitfield CW, Cziko AM, Robinson GE (2003) Science 302:296

Topics in Current Chemistry (2004) 239: 179–219
DOI 10.1007/b95453

Chemical Defense Strategies of Marine Organisms

Georg Pohnert

Max-Planck-Institut für chemische Ökologie, Hans-Knöll-Strasse 8, 07745 Jena, Germany
pohnert@ice.mpg.de

Abstract The diverse habitats of the marine environment and the adaptation required to live either in the open water or attached to a substrate resulted in various defense strategies. This review covers different ways how organisms from the plankton can maintain a chemical defense as well as the dynamic chemical defense strategies of benthic organisms. It opens with a detailed discussion of recent studies of chemical defenses among organisms floating in the open water. These defenses include the production of toxins by harmful algal blooms as well as the rapid production of defensive metabolites from non-toxic precursors upon cell disruption. A comprehensive account of recent advances in the field of dynamic chemical defense strategies of benthic organisms is then presented. It includes the known examples of activated defense based on the enzymatic transformation of storage metabolites as well as induced chemical defense strategies. These strategies depend on the recognition of signals from an herbivore or pathogen, followed by the up-regulation of the biosynthesis of metabolites involved in the chemical defense.

Keywords Marine chemical defense · Activated defense · Induced defense · Plankton · Chemical ecology

List of Abbreviations

DMS	Dimethylsulfide
DMSP	Dimethylsulfoniopropionate
DSP	Diarrhetic shellfish poisoning
PSP	Paralytic shellfish poisoning
PUFA	Polyunsaturated fatty acid

1 Introduction

The ocean is a rich source of natural products, and their structural variety has inspired numerous chemists and chemical ecologists. Comprehensive reviews exist on the structure and occurrence, biosynthesis, and biological activity of marine natural products (see, e.g., [1–5]). An excellent book published in 2001 that deals exclusively with marine chemical ecology highlights different aspects of chemical communication and chemical defense [6]. It would be impossible to cover all aspects of chemical defense in this environment, and thus this contribution will focus on a few selected aspects that have been under intense research and discussion in recent years.

One major objective of this review is to summarize findings from the growing field of chemical defense of pelagic organisms (living in or frequenting the open ocean). This topic is stimulating controversial but fruitful discussion about methods for elucidating chemical defensive principles and modeling their mode of action.

From the large body of literature about chemical defense in the benthic environment (benthos: the bottom of the sea and the littoral zones), only a few aspects can be highlighted here. The selection of examples from the benthos will focus on dynamic defense reactions including fast wound-activated and induced defense strategies. This field is well established in the terrestrial environment, while the investigation of these processes in marine organisms is just beginning.

Other work in this field like the well investigated effects of whole organism or tissue extracts against potential herbivores or pathogens will not be covered here but are reviewed elsewhere (see, e.g., [6]). Several studies have addressed the biosynthesis of defensive compounds by questioning precursors or mechanistic aspects of biosynthesis but rarely has the regulation of product formation been a focus (for reviews on the biosynthesis of marine natural products see, e.g., [3, 7]). Our knowledge of the production, storage, and transformation of microbial metabolites is also advanced but will not be reviewed here. Questions include, for example, the role of symbiotic associations in the defense of sponges or the effects of secondary metabolites on the structure of microbial communities (e.g., [8–10]).

2
Chemical Defense of Plankton

Plankton comprises the free-floating organisms that live in the open water. Global CO_2 fixation resulting from primary production is shared equally between terrestrial contributions and organisms from the sea [11, 12]. Of the marine organisms that are able to perform photosynthesis, those belonging to the plankton (phytoplankton) are by far the most important. Diatoms and small flagellates dominate this group. Other algal taxa, such as coccoliths and larger dinoflagellates, can contribute significantly to primary production as well. All these phototrophic unicellular organisms are at the bottom of the marine food chain and serve as a major food source for herbivores such as crustaceans, ciliates, and protozoan feeders. These heterotrophic organisms represent the base for higher trophic levels, including planktivorous and piscivorous fish (Fig. 1) [13].

Although the environment in the open ocean or the coastal waters appears homogenous, thousands of different species can co-exist in a cubic meter of the water column. The apparent contradiction between species-richness and limited factors for competitive exclusions resulted early on in the formulation of the "paradox of the plankton" [14]. The diversity can be explained at least partially by oscillating resource limitations [15]. In addition, pelagic evolution is seen to be ruled by the development of different defense strategies. This "watery arms race" [16] includes the protection of cells by remarkably strong silicified cell walls as found in diatoms [17], by the formation of large colonies that cannot be absorbed by the comparably small herbivores, and by the generation of voluminous gelatinous cell walls [18]. Besides these mechanical strategies, the defense of phytoplankton can also be accomplished by noxious chemicals. In contrast to the benthic environment (where species interactions can often be followed by observation, and laboratory bioassays are well established), little is

Fig. 1 Generalized pelagic food web [13]. The old view of a simple food chain (*vertical line in this figure*) with phytoplankton (mainly diatoms) at the base, herbivorous mesozooplankton (mainly copepods) at the second trophic level, and planktivorous fish has been extended to a pelagic food web including nanoplankton (<20 µm), picoplankton (<2 µm), and their protozoan feeders (*lower left*). Herbivorous tunicates and jellyfish as primary carnivores also play a role, as do mixotrophic flagellates. The main pathway of energy flow depends on the nutrient scenarios [13]. DOC=dissolved organic carbon, HNF=heterotrophic nanoflagellates. From [13] with permission of Kluwer Academic Press

known about the chemical ecology of plankton. This is partly due to the fact that interactions are very complex and occur among numerous microscopically small species. Accordingly, appropriate bioassays reflecting the situation of pelagic life forms in their environment are difficult to design. Moreover, the methodological limitations that are encountered during elucidation of the nature of dilute signal molecules in the free water hamper the understanding of chemically mediated dynamics in the plankton. Even after structure elucidation, isolation, or synthesis of the active metabolites, it is difficult to conduct ecologically realistic bioassays on their effects in the ecosystem. The compounds are often held in the cells and not released into the water. Bioassays reflecting this situation would make use of artificial diets coated or loaded with specific toxins, yet they are not a routine procedure in plankton ecology. Lacking appropriate methods, bioassays based on the direct administration of compounds to the water are the main source of direct information about the toxicity, feeding deterrent properties, and allelopathic potential of chemicals. Most current studies give thus only indirect evidence for the involvement of signal molecules in the plankton obtained after monitoring population dynamics in the field or in the laboratory.

The following sections summarize our sparse knowledge about chemical defense in plankton and outlines the role that defensive metabolites can play in this ecosystem.

2.1 Constitutive Chemical Defense of Phytoplankton

2.1.1 Toxic Marine Phytoplankton

Harmful algal blooms, namely mass occurrences of phytoplankton species that produce compounds toxic to fish and other top-consumers such as human beings, are increasing in frequency worldwide [19,20]. These blooms can have dramatic impacts on ecosystems by causing massive fish kills. After accumulating in, for example, shellfish, these compounds are also frequently poisoning marine mammals or humans [21,22]. Harmful algal blooms can thus have substantial economic impact by forcing fisheries to shut down or by discouraging tourism. Due to these sometimes dramatic effects of algal toxins, numerous studies have been performed on the structure and toxicity of bioactive metabolites from cyanobacteria [19], dinoflagellates, and other phytoplankton species [21]. In addition, several pharmacologically interesting properties of harmful algal bloom toxins have encouraged an intense search for new lead-structures and led to the discovery of numerous interesting metabolites. Although these toxins have drastic effects on higher consumers, there is little evidence to suggest they poison herbivorous zooplankton and thus contribute directly to reduced grazing pressure [20, 23]. Ingested toxins might interfere with digestive processes or cause an enhanced energy expenditure of algal consumers due to requirement for detoxification [24], but often these organisms are unaffected by the toxins. Since some zooplankton species can accumulate toxins from phytoplankton primary consumers, they might even serve as vector for these metabolites to higher trophic levels [25–27]. Thus, in many cases the ecological reasons for phytoplankton toxicity are not clear.

Discussion about the action of toxins is ongoing. Post-ingestion toxic effects can reduce the population or fitness of herbivores, but only the whole algal community is protected, not the single individuals or species. The single individual would only benefit from a defensive mechanism that is based on the release or exposure of feeding deterrent metabolites. Once these are detected by the herbivores, behavioral changes can be induced, including cessation of feeding or a selection for non-toxin producers. Only this strategy provides direct advantages without requiring food uptake by the consumers.

Due to the lack of bioassays that mimic the situation of toxin-containing food particles, most of the work has been performed on herbivores that feed on diets of phytoplankton species with different degrees of toxicity. This has led to multiple variant parameters, not allowing the direct comparison of effects due to the varying toxin contents. In a few cases, these studies have the advantage of being able to directly compare different clones of one species with variable toxicities but otherwise comparable biochemical composition. There, the effect of the toxins can be determined without overlaying effects due to different nutritional quality of the food.

2.1.1.1
Paralytic Shellfish Poisoning Producers

Some dinoflagellates of the genus *Alexandrium* produce neurotoxic compounds known as paralytic shellfish poisoning (PSP) toxins. Because these toxins can contaminate filter-feeding shellfish they may threaten public health and create economic problems for fisheries. PSP-toxins include at least a dozen saxitoxins, neosaxitoxins, and gonyautoxins (Scheme 1).

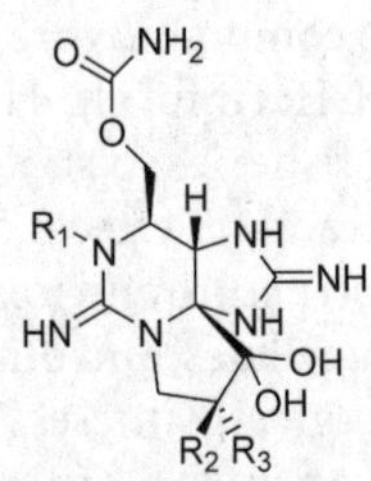

saxitoxin: R_1, R_2, R_3 = H (**1**)
neosaxitoxin: R_1 = OH; R_2, R_3 = H (**2**)
gonyautoxins: R_1 = H or OH;
R_2, R_3 = H or OSO_3^- (**3, 4**)

Scheme 1 Paralytic shellfish poisoning toxins

These heat-stable Na^+ channel blockers inhibit nerve conduction and cause relaxation of smooth muscles and paralysis in mammals. In contrast, their effect on herbivores feeding directly on these dinoflagellates appears to be highly species dependant. Certain copepods have been observed to ingest high rates of toxic *Alexandrium* spp. with no apparent physiological effects [24, 28]. Although toxic effects are not fatal for the herbivores the feeding process can be influenced by PSP-toxins. If offered a choice between toxic and non-toxic *Alexandrium* spp. that are virtually identical in respects other than the toxin levels, copepods can perform discriminatory feeding on the non-toxic strain based on chemosensoric recognition [29]. Rejection or avoidance of certain toxic dinoflagellates including *Alexandrium* spp has also been observed in other studies [30, 31] PSP toxins can disable coordinated feeding by copepods as well, thereby reducing food uptake. That these effects can strongly depend on the species and history of the copepod was demonstrated by Colin and Dam. They compared the response of copepods from different geographic locations, one of which had most likely never been exposed to toxic blooms, while the other had regularly encountered toxin containing *Alexandrium* [32, 33]. Toxic *Alexandrium* spp. reduced the feeding rates of grazers from populations native to *Alexandrium* blooms but not those of grazers from populations that are regularly exposed to the blooms. The authors conclude that a resistance to the toxins might have evolved in the previously exposed copepods, while direct toxicity causes the reduced food uptake in the native species.

2.1.1.2
Polyether-Toxin Producers

Polyether polyketides are an unusual class of complex metabolites from dinoflagellates (Scheme 2). These *all-trans*-fused linear ring systems exhibit interesting structures with specific pharmacological activity. Brevetoxins like **5** or **6** were the first members of this structure class to be discovered [34, 35]. Brevetoxins are lipid-soluble neurotoxins that exert their biological effects by, e.g., binding to the sodium channels of neurons, keeping them open and thereby causing depolarization of the cell membrane.

The toxic or feeding deterrent effects of polyether toxins towards herbivores have not been directly investigated with pure compounds. Nevertheless, indirect evidence exists that these polyketides can directly influence interactions between dinoflagellates and their consumers. In a comparative study Huntley et al. found that copepods rejected four out of fourteen dinoflagellate species tested. Among them, the brevetoxin (**5**, **6**)-producing *Gymnodinium brevis* and the yessotoxin (**7**)-producing *Protoceratium reticulatum* [30, 36]. If taken up, *G. brevis* caused elevated heart rates and loss of motor control in the herbivores. Ingestion of *P. reticulatum* initiated regurgitation of the gut content. This rejection is obviously based on the recognition of chemical cues, since it was also induced by aqueous extracts of the rejected dinoflagellate species [30, 36].

Additionally, *G. brevis* can have deleterious effects on subsequent generations of herbivores. Copepod development was severely impaired when *G. brevis*-containing diets were administered. Feeding on these diets by adults resulted in offspring with impaired development past the first feeding stage [37]. Whether or not this effect is due to the Na^+ channel activator brevetoxins (**5**) and (**6**) has not been addressed.

Food uptake of the copepod *Eurytemora affinis* was not reduced in the presence of bloom-forming *Prymnesium patelliferum*, known to produce the remarkably complex prymnesins 1 (**8**) and 2 (**9**). However, mortality of copepods was high when elevated concentrations of *P. patelliferum* were present. Since the effect was related not to ingestion rates but to the number of cells of the alga in the assay volume, toxic exudates were concluded to be the cause [38]. This toxicity is again highly specific for different isolates, since toxin-level in *P. patelliferum* varies in relation to geographic origin and growth conditions [39].

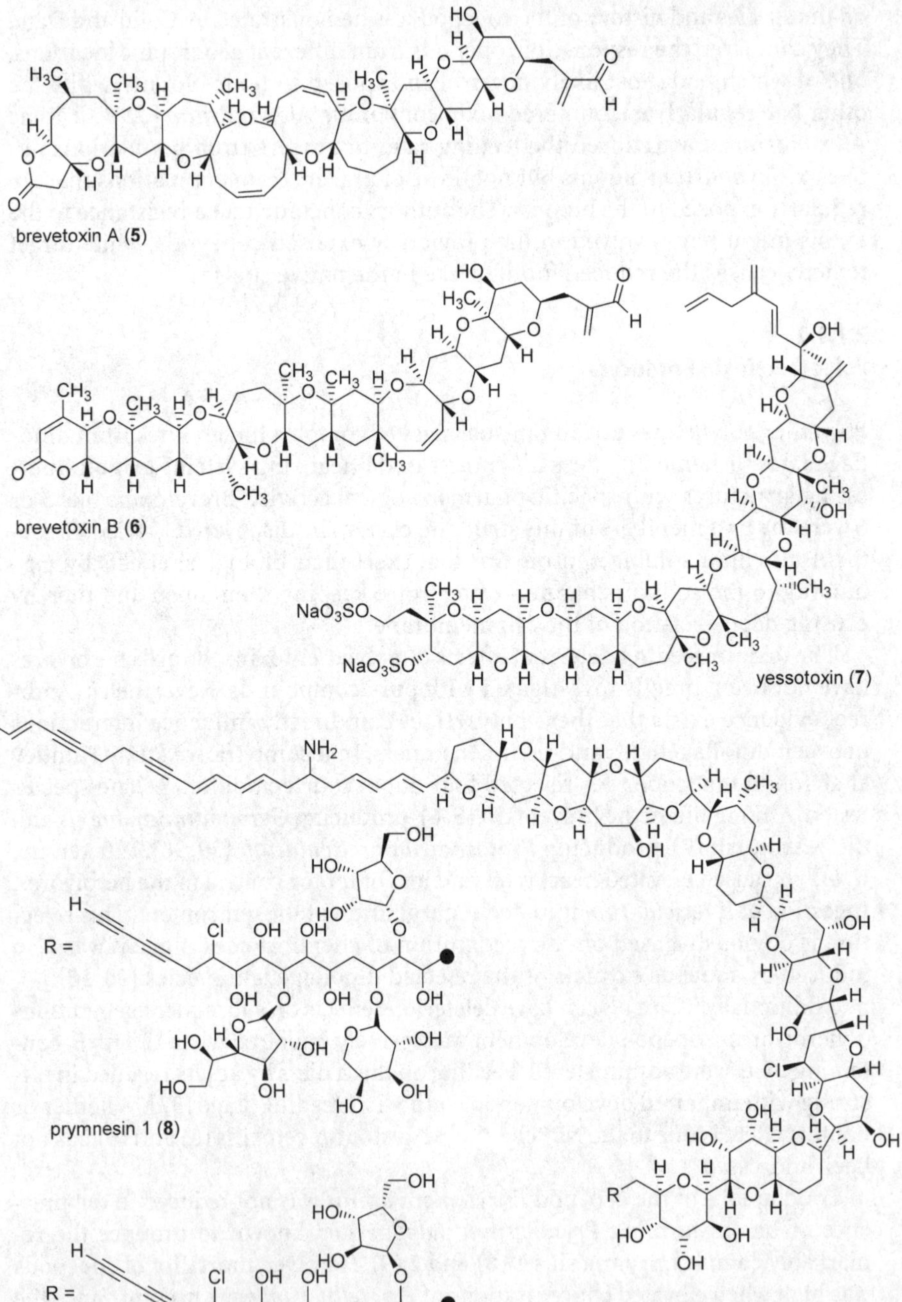

Scheme 2 Polyether toxins form red tide dinoflagellates

2.1.1.3
Studies with Purified Metabolites

Surprisingly few studies have been performed with purified toxins. When added externally to the water, toxins of various origins were tested on the copepod *Tigriopus californicus*. The protein phosphatase inhibitor okadaic acid (**17**) from red tide dinoflagellates [22] and the neuronal depolarizing agent domoic acid (**10**) from diatoms [40, 41] had different effects on the herbivores (Scheme 3). Micromolar concentrations of okadaic acid (**17**) acted both as toxin

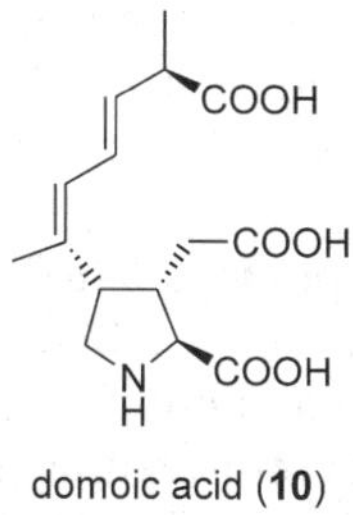

Scheme 3 The diatom-toxin domoic acid

and feeding deterrent. Comparable concentrations of domoic acid (**10**) were not effective as feeding deterrent yet were toxic to the copepods [42]. In the same setup, the apo-fucoxanthinoids **11–14**, isolated from the diatom *Phaeodactylum tricornutum* [43, 44], acted solely as feeding deterrents but exhibited no direct toxicity (Scheme 4). Since structurally unrelated metabolites caused

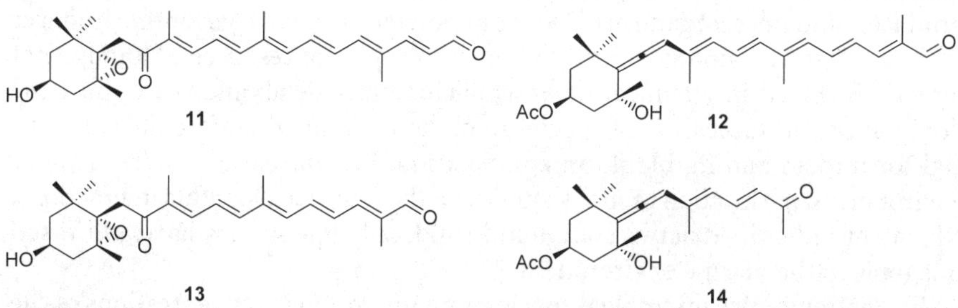

Scheme 4 Structures of apo-fucoxanthinoids isolated from the diatom *Phaeodactylum tricornutum*. **11** apo-10′-fucoxanthinal, **12** apo-12′-fucoxanthinal, **13** apo-12-fucoxanthinal, **14** apo-13′-fucoxanthinone

comparable behavioral changes, this comparison illustrates the ability of copepods to perceive different chemical signals from their environment.

The observed toxicity seems to depend strongly on the form of toxin exposure. In contrast to the findings obtained with dissolved toxins, no feeding responses or toxicity in copepods were observed when the diatom-toxin domoic acid (**10**) was administered with particulate matter [45]. More indirect information about the influence of **10** was obtained by comparison of different species of the diatom *Pseudo-nitzschia* which were administered to krill. Different feeding responses were observed in the presence of the nontoxic *P. pungens* compared to the domoic acid-producing *Pseudo-nitzschia multiseries*. However, no direct toxic effects to the herbivores were observed. The krill species *Euphausia pacifica* responded to the presence of non-toxin-producing *P. pungens* with constant feeding rates, while domoic-acid-producing *P. multiseries* resulted in an alternating feeding/non-feeding pattern. For the first 6 h interval, *E. pacifica* ingested as many toxic *P. multiseries* as non-toxic *P. pungens*, but for the next 6 h interval they stopped feeding. This alternating food uptake was continued for 24 h. The influence of the toxin itself is not clear, since no assays with administered toxins in neutral particulate matter were performed [46].

In summary, evidence of how phytoplankton toxins function as direct chemical defense against grazers is still contradictory and the ecological functions of these toxins are still poorly understood. The highly variable findings among studies with different herbivores indicate that adaptations against toxic compounds might have evolved in specific cases and that thus no general effects of toxins can be deducted from single studies.

2.2
Allelopathic Effects of Algal Toxins

Other roles for noxious metabolites produced by certain phytoplankton species include mediation of allelopathic interactions [47]. Allelopathy covers biochemical interactions among different primary producers or between primary producers and microorganisms. These can provide an advantage for the producer in the competition among different photoautotrophs for resources. Although not directly involved in chemical defense, allelopathic metabolites can affect the dominance and succession of species in phytoplankton; therefore they are crucial for understanding plankton composition. In contrast to the fresh water environment, the location of many studies on the role of allelopathic interactions, which have identified active compounds [47], only few studies have addressed this topic in the marine environment.

Incubation of five microalgal species with micromolar concentrations of the diarrhetic shellfish poisoning (DSP) toxins okadaic acid (**17**) and dinophysistoxin 1 (DTX-1 (**17a**)) inhibited the growth of four non-toxin-producing species at micromolar concentrations. In contrast, the producer of **17** and **17a**, the dinoflagellate *Exuviaella lima* (or *Prorocentrum lima*), was not affected even at elevated concentrations [48]. Despite the fact that the concentrations used were very high compared to the nanomolar toxin level in stationary *E. lima* cultures [49], the different effects on producers and non-producers suggests that toxins may play an allelopathic role and raises questions regarding strategies for avoiding autotoxicity [48].

In *E. lima* the storage form of the toxins is the hydrophilic trisulfated DTX-4 (**15**) [50, 51], which can be transformed within minutes into the okadaic acid diol ester (**16**) by culture filtrates of *E. lima* or intact *E. lima* cells (Scheme 5). Both compounds are less active than okadaic acid (**17**) itself, which is slowly released by esterases acting on okadaic acid diol ester (**16**) after prolonged incubation [52]. In independent experiments, culture medium of *E. lima* added to a dinoflagellate assemblage proved to be an active growth inhibitor for all species. This effect could not be attributed exclusively to okadaic acid (**17**), because active fractions obtained by HPLC did not contain any of the okadaic precursors **15**, **16** or okadaic acid (**17**) itself [49]. The nature of the signal will have to be verified by bioassay-guided structure elucidation of the unknown active principles.

Interestingly, okadaic acid diol ester (**16**) in the culture medium is oxidatively transformed into the more hydrophilic metabolites **18**–**20** by intact cells of the diatom *Thalassiosira weissflogii* [52, 53]. This transformation is speculated to change the toxic metabolite with allelopathic activity into a less toxic transport form that can be better excreted by the diatom.

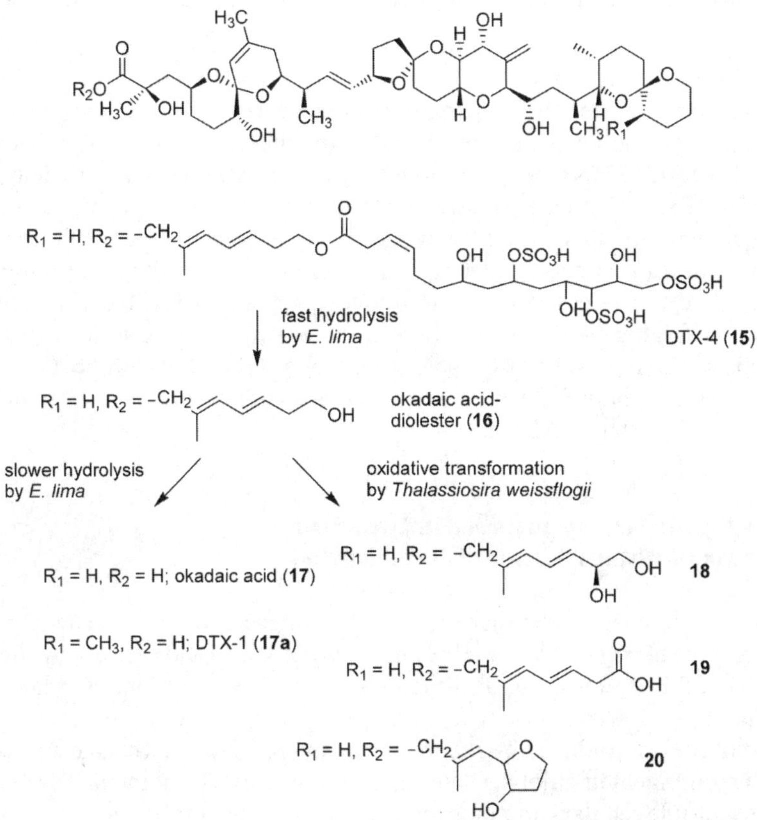

Scheme 5 Release of okadaic acid (**17**) from less toxic precursors by *E. lima* (*left*). Oxidative transformation of okadaic acid diol ester (**16**) by the diatom *Thalassiosira weissflogii* (*right*)

2.3
Activated Chemical Defense in Phytoplankton

In contrast to the above-mentioned toxins that have adverse effects on human health, several more subtle ways of maintaining a defense against grazers are known from phytoplankton [18, 54]. These defense strategies rely not on heavily toxic metabolites but rather on a dynamic realization of the chemical defense, according to which defensive compounds are produced from storage forms upon cell disruption. This strategy was initially identified in higher plants and can be seen as a wound-activated defense. Pre-formed metabolites are rapidly transformed into more toxic or highly reactive defensive principles after tissue disruption [55]. Activated defense has to fulfill three criteria defined by Paul and Van Alstyne [56]:

1. A less potent stored metabolite is transformed into a more potent one
2. The process takes place within seconds or few minutes
3. One or a few enzymes usually mediate the conversion; slight structural changes in the metabolites can markedly enhance potency of chemical defenses

In higher plants mechanisms of this type of defensive reactions are investigated in detail. Most plants rely on a compartmentalization of the storage forms of aggressive metabolites and enzymes, releasing the defensive metabolites after tissue disruption. This strategy diminishes the potential risk of self-toxicity and allows the use of labile, aggressive metabolites that cannot be stored in the tissue. In plankton this strategy may also be a way to overcome the limitations through dilution effects of defensive metabolites, since the compounds in question are only released locally around feeding herbivores. In contrast to the reliance of higher plants on pre-toxins, the examples hitherto known from phytoplankton represent cases in which metabolites that play important roles as primary metabolites during normal cell growth are transformed to defensive principles after cell disruption.

2.3.1
β-Dimethylsufoniopropionate and its Lyase Products as a Grazing-Activated Defense of Phytoplankton

A dynamic defense model that seems to be widespread among marine algae involves the cleavage of β-dimethylsufoniopropionate (DMSP) (**21**) by the enzyme DMSP-lyase resulting in dimethylsulfide (DMS) (**22**) and acrylate (**23**) (Scheme 6).

DMSP (**21**) is produced by many marine micro- and macroalgae and is especially prominent in dinoflagellates and haptophytes. The nontoxic DMSP (**21**) fulfills multiple cellular functions including cryoprotection, the involvement as osmolyte in osmoregulation and as a methyl donor in transmethylation reactions [18].

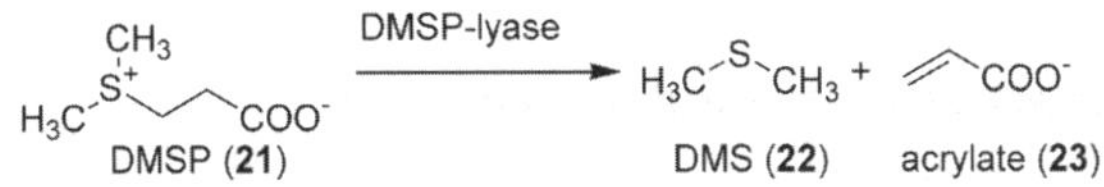

Scheme 6 The DMSP-lyase reaction

Lysis of ingested prey cells initiates mixing of algal DMSP (**21**) and the enzyme DMSP-lyase. The release of the volatile DMS (**22**) into the medium signals the occurrence of this activity. If the unicellular haptophyte alga *Emiliania huxleyi* was fed to the protozoan predator *Oxyrrhis marina*, DMS-release resulting from the lyase reaction could be directly correlated with food uptake. During the experiment, DMS (**22**) was not detected until *E. huxleyi* cells were ingested, indicating that potentially harmful acrylate apparently forms inside predator food vacuoles after uptake of the unicellular algae.

Using this correlation between DMS (**22**)-production and feeding on DMSP-lyase rich algae Wolfe et al. demonstrated in an elegant series of experiments the involvement of the lyase reaction in a wound-activated chemical defense [57]. Selection of five *E. huxleyi* strains all containing comparable amounts of DMSP (**21**), but exhibiting different DMSP-lyase activity allowed one to compare the response of the grazers to this attribute. Since it was not possible to distinguish visually between the *E. huxleyi* cells of varying DMSP-lyase content DMS, production was used to detect grazing on high-activity strains. Feeding correlated directly with DMS-production when DMSP-lyase rich strains were administered to *O. marina*. When mixed diets were administered, food uptake remained constant during the assay period, but the onset of DMS-release was observed after the low-producing algae had been selectively consumed, offering direct proof that predators prefer to feed on DMSP-lyase poor strains. Even grazers that tolerate high activity strains avoided them when offered a choice of prey. In addition to direct protection the effect on grazer specificity can shift grazing pressure to other prey species, thereby reducing the competition for nutrients. *E. huxleyi* cells need not be fully ingested to deter predators, since their recognition after contact can already lead to avoidance behavior. Nano- to micromolar concentrations of DMSP (**21**) can reduce feeding in different protists, acting as a non-toxic chemical defense signal that could indicate the potential toxicity of DMSP-rich strains [58]. The DMSP-lyase reaction can thus function directly as a grazing-activated chemical defense of the individual cells without the need of ingestion since its presence is perceived by chemical cues [59].

In addition to their role in chemical defense, DMSP-lyase products may also function as chemical cue in more complex trophic cascades. In the natural environment DMS-production is related to zooplankton herbivory [60] and can thus act as an indicator for the availability of food for planktivorous birds. Indeed, some Antarctic *Procellariiform* seabirds can detect DMS (**22**) and are highly attracted to the cue, as was shown with DMS-scented oil slicks on the ocean surface [61]. The odors released during zooplankton grazing (DMS) as well as those of zooplankton itself (e.g., trimethylamine and pyrazines) are attractive to birds [62], thus assisting vertebrate search behavior.

2.3.2 Oxylipins in the Chemical Defense of Diatoms

Diatoms are the major bloom-forming algae in temperate waters and belong to the most important primary producers [12]. They provide the bulk of food for herbivores and only few toxic species of diatoms are known. Among the secondary metabolites of diatoms only domoic acid (**10**) is toxic to consumers located higher in the food chain [40]. In consequence, these algae have been traditionally regarded as beneficial to the growth and survival of marine organisms. In recent years evidence that certain diatom diets have an insidious effect on copepod reproduction has accumulated, calling this view into question. Because copepods usually dominate the zooplankton in diatom-rich regions, the interaction between these herbivores and diatoms is of special interest. In the early 1990s key experiments showed that some diatom species are harmful to copepods not through direct toxicity or by reducing feeding rates but rather through indirect effects on herbivores. After administering the diatom *Thalassiosira rotula* to the copepod *Calanus helgolandicus* Poulet et al. observed that this diatom has evolved an unusual antipredation strategy: it reduces copepod population levels by inhibiting their reproductive success [63]. When adult females were fed the diatom, both total egg production and hatching success were significantly lower than when fed a dinoflagellate control diet. This effect must be due to a chemical defense, since embryonic development was also arrested when eggs of copepods fed with a neutral diet were exposed to diatom but not dinoflagellate extracts. Subsequently several other diatom species were identified which negatively affected copepod egg-hatching success and/or fecundity; others were discovered which did not exhibit any adverse effect ([64, 65], reviewed in [66]). In field studies Miralto et al. showed that the hatching success of copepods feeding on a diatom-dominated bloom was very low. They identified 2,4-decadienal (**24**) and 2,4,7-decatrienal (**27**) as metabolites that could be responsible for this observed effect [67]. Administered in high (micromolar) concentrations to the water, these compounds arrested cell division of freshly spawned copepod eggs but had no negative effect on the feeding copepods themselves. Both, **24** and **27** from diatoms thus elicited effects comparable to those observed in the field. Since this study relied on the addition of micromolar amounts of purified compounds to the water, an ecological implication cannot be directly deduced. The identification of two *T. rotula* strains that are genetically closely related and morphologically nearly identical but different in their ability to produce the unsaturated aldehydes allowed a comparative laboratory study to be performed. While the aldehyde-producing *T. rotula* strain inhibited egg hatching, the related non-producing strain lacked this ability further demonstrating the involvement of these compounds in the indirect chemical defense [68].

Detailed investigation of *T. rotula* and other diatom species revealed that 2,4-decadienal (**24**) and 2,4,7-decatrienal (**27**) represent only members of a structurally diverse class of α,β,γ,δ-unsaturated aldehydes found in several diatom species (Scheme 7) [68–73]. In a structure activity test with synthetic compounds, Adolph et al. showed that a general requirement for biological activity is the existence of a α,β-unsaturated aldehyde group, and that besides **30**, the entire structurally diverse class of unsaturated aldehydes from diatoms exhibited activity in sea urchin egg cleavage assays. Saturated volatile aldehydes, also detected in diatoms, were essentially inactive [74].

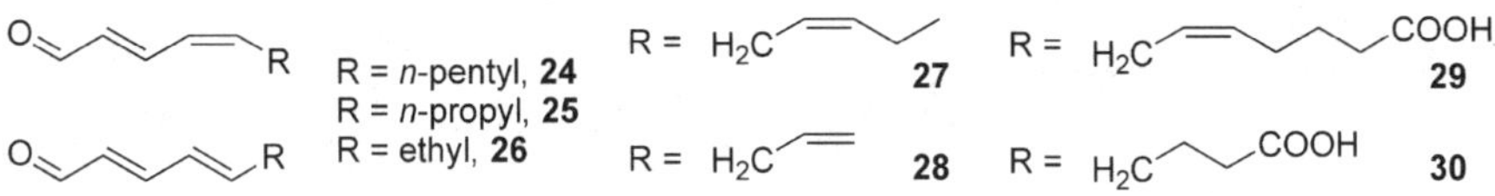

Scheme 7 α,β,γ,δ-unsaturated aldehydes from diatoms

Interestingly, quantification experiments focused on the unsaturated aldehydes from *T. rotula* and other diatoms showed that intact cells did not contain any of the metabolites **24–30**. These compounds are only released within seconds after mechanical damage which could happen during the grazing process. Elevated amounts of these compounds can thus be locally present during grazing and targeted efficiently against the herbivores [69]. Investigation of the biosynthesis of these aldehydes revealed how this grazing-activated defense is regulated by the unicellular algae. The biosynthesis of these aldehydes in diatoms (reviewed in [54]) depends on a lipoxygenase-mediated oxidation of fatty acids followed by a subsequent cleavage of intermediate hydroperoxides by hydroperoxide lyase activity (Scheme 8) [69, 70]. This transformation has been investigated in detail with the diatom *Gomphonema parvulum*. This alga transforms eicosapentaenoic acid (**32**) via the intermediate 9-hydroperoxyeicosapentaenoic acid (**35**) into the 9-oxo acid **30** and the hydrocarbon hormosirene (**36**) as a second fragment [70]. Interestingly, **36** was previously reported as a brown algal pheromone [75]. The availability of free precursor fatty acids, such as eicosapentaenoic and arachidonic acid for decatrienal (**27**) and decadienal (**24**) and hexadecatrienoic acid for octadienal (**25**) [76, 77], controls the production of these aldehydes. Intact diatom cells do not contain detectable amounts of free polyunsaturated fatty acids (PUFAs), these lipoxygenase substrates are released after cell damage by a phospholipase A_2 [77]. Cellular resources are therefore invested in the production of phospholipids that can be fast and efficiently transformed into a multi-enzyme-dependent reaction whenever required.

Interestingly, the release of PUFAs without subsequent action of a lipoxygenase can act as a wound-activated defense diatom in rich fresh water biofilms. This reaction could be directly associated with a chemical defense against the grazer *Thamnocephalus platyurus* [78].

The role of antiproliferative effects of an activated defense reducing the next generation of herbivores compared with the straightforward action of toxins or feeding deterrent metabolites in the evolutionary arms race is still under discussion. The proposed mechanism suggests natural selection at the group level,

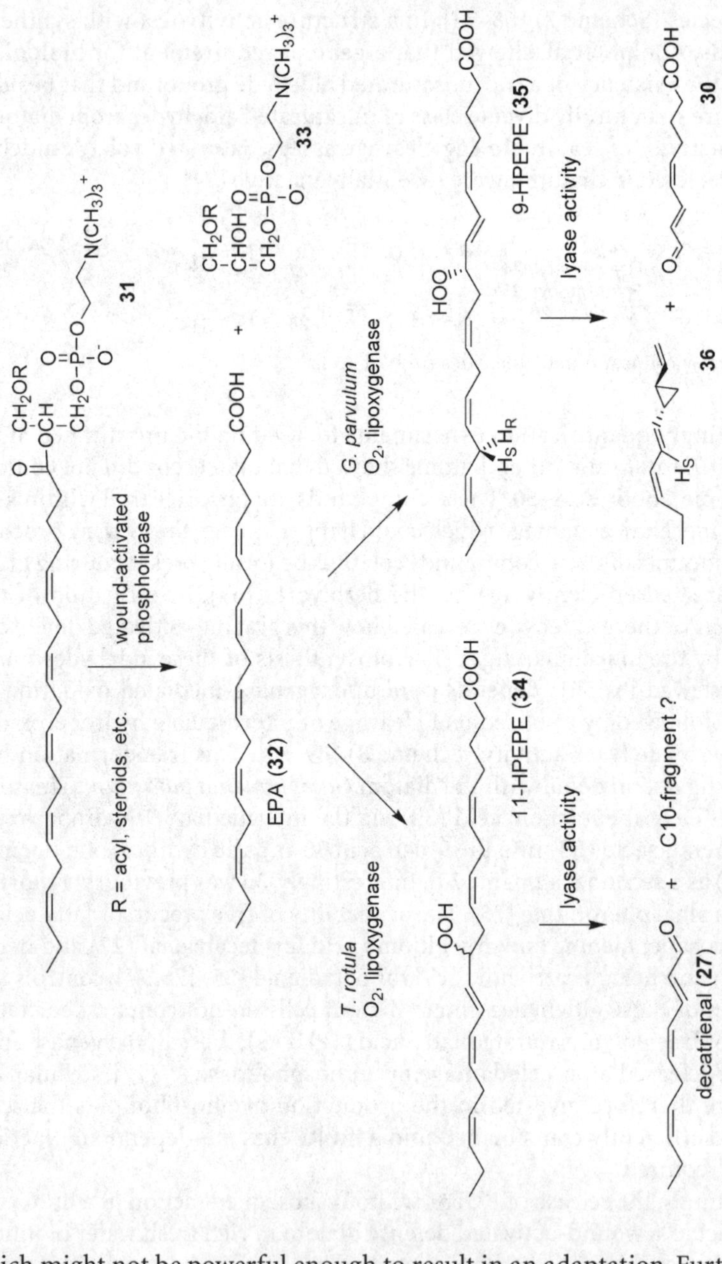

Scheme 8 Enzymatic cascade leading to the release of α,β,γ,δ-unsaturated aldehydes after tissue disruption of *Thalassiosira rotula* (*left*) and *Gomphonema parvulum* (*right*)

which might not be powerful enough to result in an adaptation. Further studies in the field will be required to proof the concept.

The above findings would challenge the classical view of marine food web energy flow from diatoms to fish by means of copepods if this defense would be an universal feature of diatoms. In field experiments monitoring diatom abundance (but not food uptake) and hatching success of copepods, only a few diatom-rich situations resulted in hatching failure [79]. These observations correlate well with the fact that only a small fraction of investigated diatom species is capable of producing the antiproliferative aldehydes (Wichard and Pohnert, unpublished results). One way to verify the existence diatoms' activated defense strategy might be to directly correlate results from chemical analysis and field observations. A detailed re-investigation in field experiments should follow the food uptake of copepods and the potential toxicity of the consumed diatoms. With data on the reproductive success of copepods, diatom effects on herbivores could be evaluated.

The effects of decadienal (**24**) as well as of diatom extracts are not restricted to the reduction of copepod success. Low, micromolar concentrations of this compound also inhibits fertilization, embryogenesis, and hatching success in polychaetes and echinoderms. Crude diatom extracts as well as purified aldehydes inhibited these processes in a dose-dependent manner [80].

2.4
Chemical Defense of Other Plankton Species

Only a few studies on defined chemical defense metabolites from non-phytoplankton organisms that spend their entire life history in the water column have been reported to date [81].

One example for a chemically defended zooplankton species is the Antarctic pteropod *Clione antarctica*. This shell-less pelagic mollusk offers a potentially rich source of nutrients to planktivorous predators. Nonetheless fish do not prey on this organism, due to its efficient chemical defense. In a bioassay-guided structure elucidation, pteroenone **37** could be isolated and characterized as the main defensive principle of *C. antarctica* [82, 83]. If embedded in alginate, this compound is a feeding-deterrent in nanomolar concentrations. This unusual metabolite is likely to be produced by *C. antarctica* itself and not accumulated from its food, since its major food sources did not contain any detectable quantities of **37**.

In addition to protecting the pteropod, pteroenone **37** can also serve invertebrates as an indirect chemical defense (Fig. 2). The amphipod *Hyperiella dilatata* gains protection by carrying the pteropod *Clione antarctica* on its back. That this carrying behavior provides efficient protection is clear, since in all cases

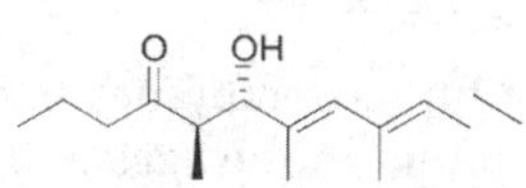

pteroenone (**37**)

Fig. 2 *Left*: scanning electron micrograph showing the amphipod *H. dilatata* carrying the chemically defended pteropod *C. limacina*. Magnification is about 50% [84]. With permission of the Nature publishing group. *Right*: pteroenone (**37**), the defensive principle of *C. limacina*

investigated in the laboratory, amphipods that were not carrying pteropods were readily eaten by fish, whereas almost all of those carrying pteropods were rejected. Although fish violently ejected the carrying individuals from their mouths, the amphipod-pteropod pairs were usually not separated and both organisms survived [84].

2.4.1 Chemical Defense of Invertebrate Meroplankton

Meroplankton comprise organisms that spend only part of their life in the plankton. Primarily eggs and larvae of benthic or actively swimming adults fulfill this criterion, and the study of their chemical defense is often supported by the knowledge of defensive metabolites of their adult life stages.

A comparative study of the saponin-content [85] of the starfish *Acanthaster planci* in different developmental stages showed that comparable amounts of the same chemical type of saponins like **38** are contained in eggs, ovaries, larvae and adults (Scheme 9). As the chemical analysis predicts, the eggs and larvae of *A. planci* were not eaten by fish. The predators were able to select against the larvae prior to ingestion. Saponin-rich extracts that were supplied to planktivorous fish with gelatinous food particles had a similarly inhibiting effect on food uptake since the predators discriminated against food particles containing saponins [86].

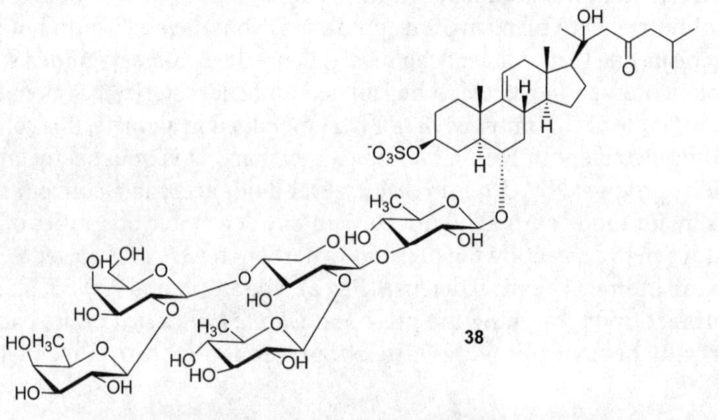

38

Scheme 9 The major saponin from the sea star *Acanthaster planci*

The defensive furanocembranes from the adult soft coral *Sinularia polydactyla* are also found in the eggs and larvae of their offspring. HPLC-comparison of extracts from offspring with those from adults showed that pukalide (**39**) and 11β-acetoxypukalide (**40**) (Scheme 10) [87, 88] are contained in both the blastula and planula of the soft coral [89]. Feeding assays with the larvae and their extracts showed that fish rejected the larvae due to chemical factors – most likely the pukalides. The pukalide-concentrations in the extracts of

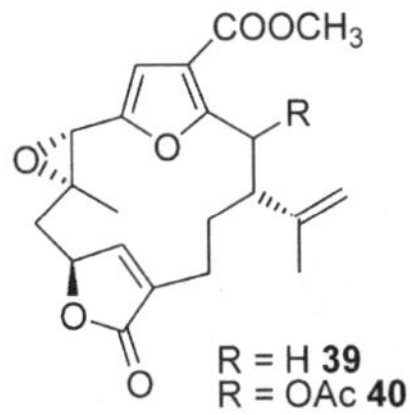

Scheme 10 Pukalides from the soft coral *Sinularia polydactyla*

rejected larvae were similar to those required to induce rejection of artificial food [90].

Cowart et al. have shown that the marine polychaete larvae *Streblospio benedicti* and *Capitella* sp. contain halogenated metabolites like **41–43** that are also found in adults (Scheme 11) [91]. The concentrations of halocarbons in *S. benedicti* increased with the developmental stage, which was interpreted as a result of the continuous biosynthesis of the metabolites during development. This might be seen as a strategy to reduce potential autotoxicity during sensitive developmental transitions. Allocation of potential defense compounds to offspring varied as a function of species: surprisingly pre-release larvae of *Capitella* sp. contained higher concentration of total halogenated metabolites compared post-release larvae, although the pre-release larvae are still protected by the adults [91].

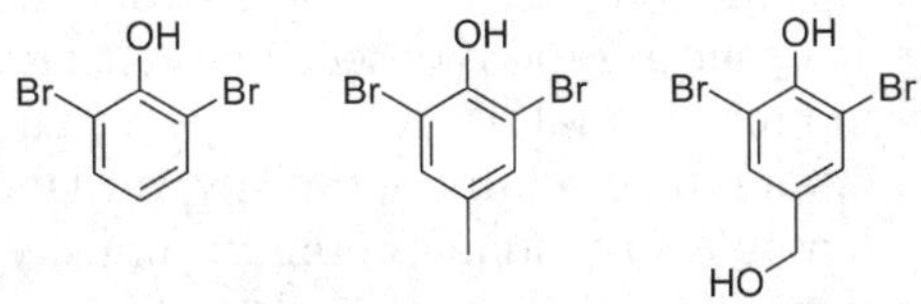

41 **42** **43**

Scheme 11 Halogenated aromatic metabolites involved in the chemical defense of the polychaete *Capitella sp.*

Larvae of the tunicate *Trididemnum solidum* are most likely protected by didemnins like **44–46**, cyclic depsipeptides that were initially identified from adults of this species [92]. The larvae are highly unpalatable to the wrasse *Thalassoma bifasciatum*, which rejected the larvae while readily consuming krill eyes that served as larval mimics. Coating one krill eye with the lipid soluble compounds from a single *T. solidum* larva rendered it as unpalatable as the larvae themselves. Reduced feeding was also observed when didemnin B (**45**) and nordidemnin B (**46**) were administered to reef fishes in the field (Scheme 12) [92].

A remarkable indirect strategy for chemical defense is found in embryos of the shrimp *Palaemon macrodactylus*, which are resistant to infection by the pathogenic fungus *Lagenidium callinectes*, a known pathogen of many crustaceans. This resistance is caused by bacteria on the surface of the embryos

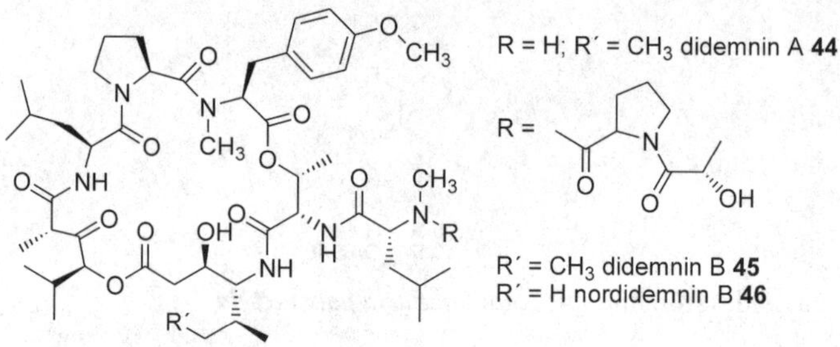

Scheme 12 Didemnins from the tunicate *Trididemnum solidum*

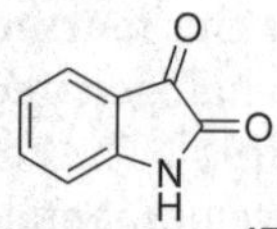

47

Scheme 13 Istatin produced by symbiotic bacteria defends crustacean embryos from a pathogenic fungus

which produce and release the antifungal compound 2,3-indolinedione, also known as isatin (**47**) (Scheme 13) [93].

Besides these few reports on toxic, feeding-repellent, and allelopathic properties of metabolites from pelagic organisms, we know little about chemical factors that can influence the complex species interaction in plankton. This is partly due to the fact that concentrations of secondary metabolites available in the water column are in most natural situations very low, hampering the identification of signal compounds. In addition, transferring laboratory results to field situations is often problematic. Major points of controversy are the use of appropriate concentrations and the selection of adequate predators in bioassays. Mainly the complex situation with numerous coexisting species in a small volume hampers the appropriate selection of target organisms and does not facilitate definition of a specific ecological role of the secondary metabolites. Plankton-released compounds face considerable dilution, so production and excretion needs to be sufficient to overcome this limitation. Also, surface-associated metabolites and wound-activated production of secondary metabolites seem to play an important role in the species interaction in this environment.

3
Chemical Defense in the Benthos

In contrast to our sparse knowledge about the action of chemicals as defense in plankton, the identified metabolites with defined defensive activities from benthic organisms are numerous. Nearly every habitat from the Antarctic [94] to coral reefs [95] has been extensively investigated – often motivated by the interesting structural and pharmacological properties of marine natural products. In fact, most of the 14,500 marine natural products listed in the database marinLit [96] are derived from benthic organisms. Given this enormous amount of information any review on marine chemical defense in this habitat has to select a few topics. This chapter will primarily focus on dynamic chemical defense strategies in the benthos with special attention paid to algal chemical defense.

3.1 Chemical Defense Strategies of Macroalgae

The chemical defense of macroalgae has been extensively studied. Defense principles range from simple acidification of the tissue ($pH<1$) maintained by stored sulfuric acid in vacuoles of *Desmarestia munda* [97] to complex polyketides [98]. Several review articles on different aspects of algal/herbivore and algal/pathogen interactions provide a comprehensive treatment of many processes and compounds involved (e.g., [54, 95, 99–105]). During recent years a field has emerged dealing with dynamic chemical defenses of algae, signal recognition, and hormonal regulation of defensive reactions that will be reviewed here.

It is common knowledge in chemical ecology and plant physiology that higher plants do not rely exclusively on static chemical defensive mechanisms, namely, a chemical defense based on pre-formed or continuously released metabolites. Strategies that avoid the need for such constitutive defense compounds are known from plants (see, e.g., [55, 106, 107]), but our understanding of comparable processes in the aquatic is still at the beginning. Since algae have emerged as a variety of independent lineages early in the evolution of eukaryotes, it is of particular interest if their basic mechanisms for pathogen recognition and defense resemble those of terrestrial plants and animals or if these essential cell functions have arisen independently in the different lineages.

3.1.1 Activated Chemical Defense in Macroalgae

3.1.1.1 The Wound-Activated Transformation of Halimedatetraacetate

In 1992, Paul and Van Alstyne reported on the processes that occur after tissue disruption in different species of the calcified green seaweed *Halimeda* [56]. After wounding, these algae transform their major secondary metabolite, the *bis*-enoylacetate diterpene halimedatetraacetate (**48**), into halimedatrial (**50**) and epihalimedatrial (**51**). The structural relationship between the educt and the reaction products suggests that the transformation occurs by a combination of solvolysis and hydrolysis reactions as indicated in Scheme 14 [108].

The wound response occurs within minutes after the actual damage suggesting the enzymatic transformation of the tetraacetate **48** as a likely mechanism of activation. A similar reaction does not occur with purified halimedate-

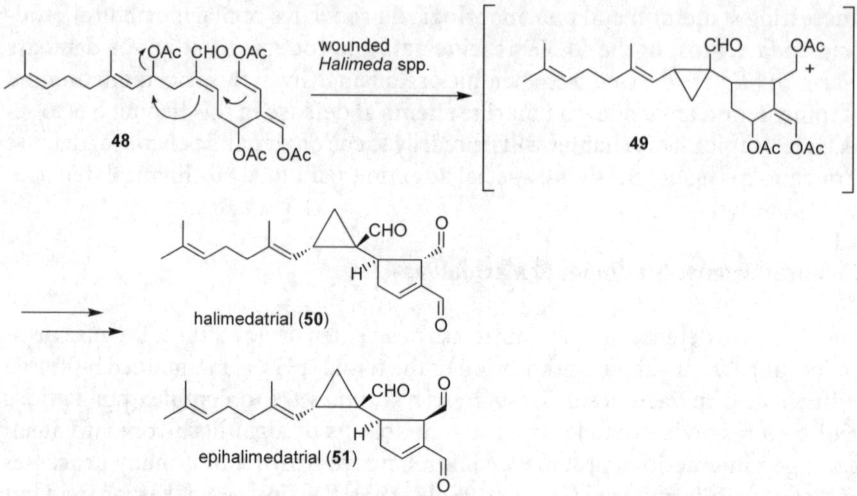

Scheme 14 The wound-activated transformation of halimedatetraacetate (**48**) to halimedatrial (**50**) and the unstable epihalimedatrial (**51**) increases the defensive potential of *Halimeda* spp.

traacetate (**48**) or during chemical transformations of **48**. Extracts from injured algae contained higher amounts of halimedatrial (**50**) and were more deterrent toward natural populations of herbivorous fishes than extracts from algae in which the formation of halimedatrial (**50**) was inhibited.

The ability to respond so rapidly should be advantageous when herbivore-grazing is intense but extremely variable over short periods of time. The "defense on demand" could thus result in high levels of the defensive metabolites halimedatrial (**50**) and the unstable epihalimedatrial (**51**) only in the presence of actively feeding herbivores. Since grazing was also reduced significantly more by purified halimedatrial (**50**) than by halimedatetraacetate (**48**) this reaction fulfills all three criteria defined for a wound-activated defense (see above).

The transformation of halimedatetraacetate (**48**) remained the only reported example of an activated defense in the field of macroalgal ecology for nearly a decade.

In addition to this transformation, other structural relationships between metabolites from green algae of the family of Udoteaceae are likely to result from biotransformations of stored precursors. It has been concluded that udoteatrial (**52**) can arise out of the transformation of udoteal (**53**) (Scheme 15), and other

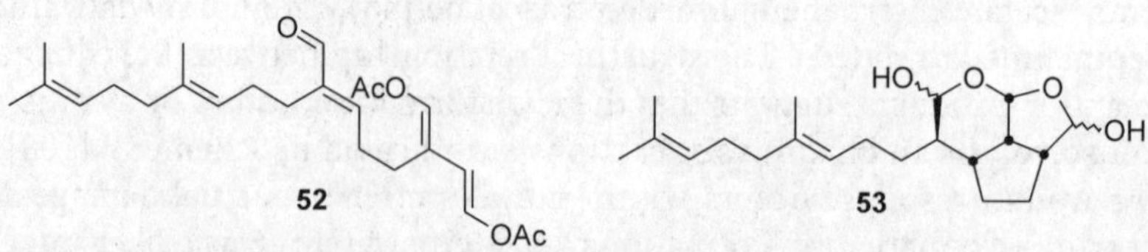

Scheme 15 Udoteatrial (**53**) is assumed to be formed from udoteal (**52**)

deacetylated aldehydic metabolites exhibit a direct relationship to the corresponding *bis*-enoylacetates [109].

3.1.1.2
The Wound-Reaction of Caulerpa spp

A fast dynamic reaction after wounding which is based on the deacetylation of functionalized terpenes is also observed in *Caulerpa taxifolia*. This invasive alga originates in the tropics and has spread rapidly after its accidental introduction into the Mediterranean. It is now also found on the Pacific coast of North and Central America [110–112]. The success of the alga, which outcompetes the natural seagrass assemblages, is attributed to, among other factors, its efficient chemical defense. *C. taxifolia* is not accepted as a food source by most Mediterranean herbivores. This was considered to be mainly caused by the activity of its major secondary metabolite caulerpenyne (**54**). This sesquiterpene is found in concentrations up to 1.3% of the wet weight in the tissue of this alga [113]. Besides caulerpenyne (**54**), several related metabolites bearing a 1,4-*bis*-enoylester unit, or structural elements formally derived from this motive after deacetylation, rearrangement, or transesterification, have been identified from Caulerpales (Scheme 16) [114–120].

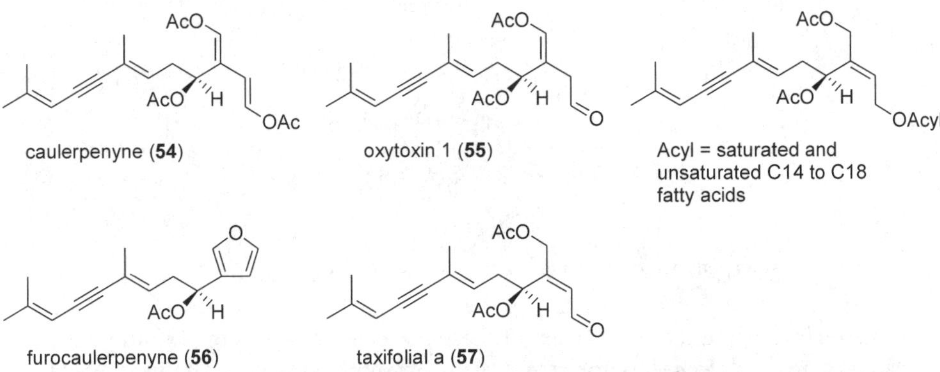

Scheme 16 Caulerpenyne (54) and caulerpenyne-related products from *Caulerpa* spp.

An investigation of the wound reaction of *Caulerpa* spp. showed that the aldehyde **55** found as minor component in algal extracts is indeed an intermediate in a wound-activated transformation of **54**. This sesquiterpene is degraded within seconds after tissue disruption to form the reactive aldehydes **55** and **59-64** that were characterized after trapping reactions (Scheme 17) [121].

By use of model substrates and inhibitor studies, an esterase that is reactive in unbuffered sea water as well as in the disrupted algal tissue from *C. taxifolia* was identified which mediates cleavage of the acetyl residues of caulerpenyne (**54**) [121]. After complete deacetylation, oxytoxin 2 (**64**) appears as an unstable end-product. Due to the lack of an appropriate assay procedure for labile metabo-

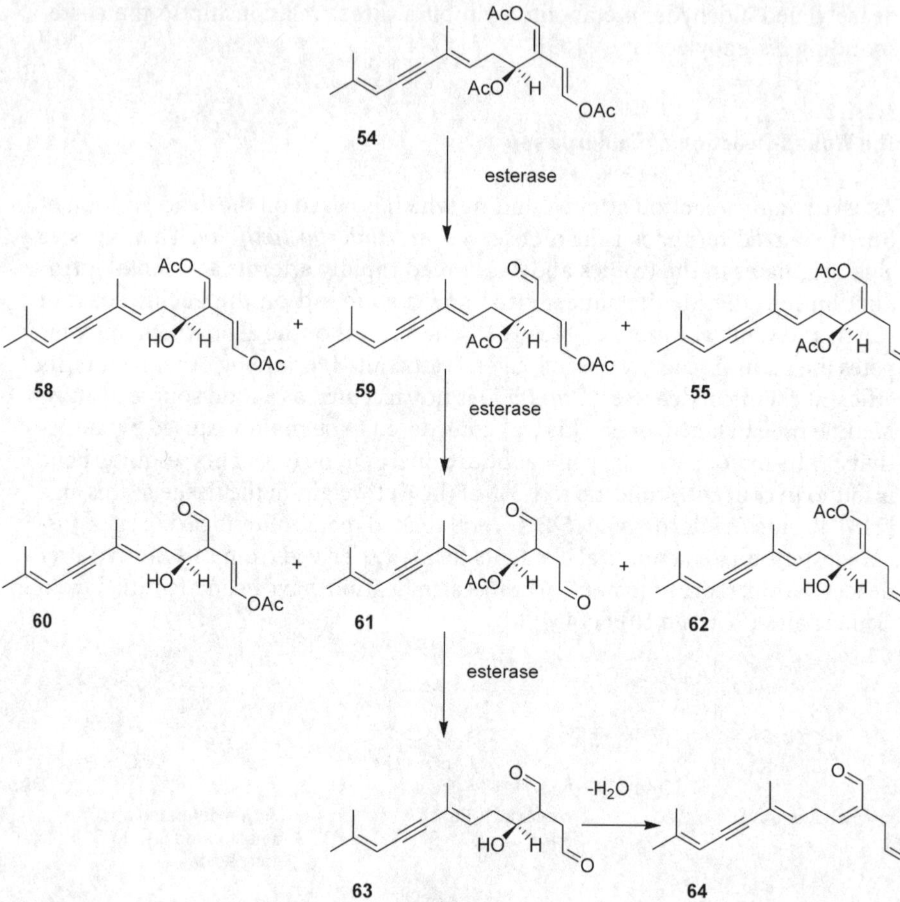

Scheme 17 Caulerpenyne-transformation by wounded *Caulerpa* spp.

lites, the involvement of the release of these reactive aldehydes in a wound-activated chemical defense has not been directly proven. Comparison of invasive and non-invasive *Caulerpa* spp. in the Mediterranean revealed that all investigated species contain comparable amounts of caulerpenyne (**54**) and can perform the fast esterase-mediated transformation to reactive down-stream products [122]. This wound-activated reaction is thus not the key for the success of the invasive species but a common feature of the chemically well-protected *Caulerpa* spp.

The 1,4-*bis*-aldehyde **64** has previously been identified together with oxytoxin 1 (**55**) from the herbivorous sacoglossan *Oxynoe olivacea*, which is specialized on *Caulerpa* spp. [123].

Production of these aldehydes from algal-derived caulerpenyne (**54**) was thought to be carried out by the mollusk [123, 124], but since the alga has been found to perform this transformation, re-investigation of the biosynthesis of this metabolite in mollusks is advised. The impressive ability of mollusks to exploit and transform chemicals from their diet for defense purposes is reviewed elsewhere (see, e.g., [7, 125, 126]).

3.1.1.3
DMSP-Lyase in the Activated Defense of Macroalgae

Remarkably, macroalgae use an activated defense strategy initially identified from phytoplankton [127]. Dimethylsulfide (DMS) **22**-release was observed during sea urchin grazing on different macro algae containing dimethylsulfoniopropionate (DMSP) **21**, suggesting a comparable reaction to that observed after phytoplankton cell damage (Scheme 6). A DMSP-lyase dependent reaction results in the release of DMS (**22**) and acrylate (**23**) around the feeding sites. In choice assays it was observed that DMS-containing diets were avoided in favor of the untreated control. Algae with high internal DMSP-content were avoided if offered together with others of low DMSP-content. Results from monitoring the DMS release during the feeding process demonstrated that the products of activation are actually generated during consumption of the algae and also that post-ingestive processes can cause the release of the volatile sulfide [127].

3.1.1.4
A Survey of Activated Defenses in Seaweeds

That these few examples of activated defenses do not reflect unique strategies of selected algal species was demonstrated by Cetrulo and Hay [128]. In a broad survey they addressed the hypothesis that activated defense strategies are more widely distributed among algae than had been expected, judging from the few reports in the literature. To test the frequency of potentially activated defenses in seaweeds, these authors incorporated pairs of algal extracts from 42 different species into artificial food. Direct solvent extraction of intact tissue was used to avoid enzymatic reactions that could occur during wounding. The other set of samples was prepared from ground tissue of the same algae, whose cellular compartmentalization had been disrupted and wound-activated reactions were allowed for 30 s before extraction. These diets were offered fish and sea urchins in feeding assays. Extracts from seven out of 42 species resulted in feeding patterns consistent with an activated defense. In four cases the artificial food containing the extract of wounded algae was preferred. The reduced defensive potential of the extract-loaded artificial food might also be explained with an activated defense: Highly reactive products that cannot be transferred without degradation into the artificial diet could be released after tissue disruption.

Extracts of the 42 species before and after wounding were also compared using thin-layer chromatography. The highly dynamic character of algal wound-reactions was confirmed, since 70% of the extracts differed before and after wounding of the tissue.

The existence of only a few well-understood examples of activated defenses of seaweeds seems thus to be due more to a restricted methodological approach by marine chemical ecologists focusing on constitutive chemical defense than to a limited distribution of this strategy among algae.

3.1.2
Induced Chemical Defense of Macroalgae

Predator-induced morphological defenses have long been recognized in several marine taxa, including seaweeds [129–131]. In contrast, descriptions of induced chemical defenses are comparably rare. The induced host-herbivore interaction relies on the recognition of the attacker by the host as a first step. This is followed by the increased production of defensive metabolites which is often regulated by hormones. Recognition may involve the perception of signals from either the attacker (so-called exogenous elicitors) or of endogenous signals, such as degradation products from the host cell wall itself. A combination of endogenous and exogenous factors can also be active in selected signaling processes. While we have a detailed understanding of these events in higher plants, little is known about induced defense in algae. Some examples of pathogen recognition have been identified in algae, but only few signals have been isolated. The field of algal hormones is still completely open.

3.1.2.1
Brown Algae

3.1.2.1.1
Polyphenol-Based Defense in *Fucus* jspp.

The first report on a grazing-induced chemical defense of algae was reported for *Fucus distichus* [132]. Van Alstyne damaged the seaweed manually by clipping tissue from the branches and found in response an up-regulation of phenolics within two weeks after damage. Herbivorous snails initially had a preference for clipped *F. distichus*; however, over a two-week period snails shifted their preference to uninjured algae [132].

The plant hormone methyl jasmonate, which can act as a chemical cue to induce defensive reactions in terrestrial plants [133], was suggested as a putative signal involved in the regulation of phlorotannin production in *Fucus spp.* After brief exposure of *Fucus vesiculosus* to exogenous methyl jasmonate (**71**) (Scheme 19), an increase in the mean polyphenolic content of up to 158% of the control was recorded [134]. This result raises several questions: Are there specific oxylipin-signaling pathways in marine brown algae? Is methyl jasmonate (**71**) a true signal that plays a role in the natural environment? No published studies are known which demonstrate the presence of methyl jasmonate (**71**) or jasmonic acid in this division. Can other oxylipin-mediated signaling cascades be active as well? The activity of methyl jasmonate (**71**) used for the induction of *F. vesiculosus* might be a result of structural similarities to the true signal.

3.1.2.1.2
Induced Chemical Defense of *Ascophyllum nodosum*

All above examples demonstrate that algae are able to respond to mechanical damage of their tissue or the feeding of a herbivore. The question arose if also recognition of a chemical signal from herbivores can elicit defensive responses, as it is known from higher plants. The first proof that seaweeds have indeed evolved a mechanism to sense and respond to chemical signals indicating the presence of actively feeding herbivores was offered by Toth and Pavia [135]. While the phlorotannin-level of the brown seaweed *Ascophyllum nodosum* was unaffected by mechanical damage, a significant increase in compounds of this class was recorded when signals from feeding flat periwinkles were present. In bioassays flat periwinkles were offered a choice between artificial diets containing seaweed material from the induction experiments and untreated material. A preference was observed for the food containing seaweed material from control experiments compared to the diet that previously received the cues from the feeding herbivores, indicating the involvement of the newly formed metabolites in the chemical defense [135]. Comparison of the phlorotannin-content after feeding by different herbivores revealed that this up-regulation is a specific process. In contrast to the feeding of flat periwinkles, herbivory by the isopod *Idotea granulosa* does not result in the up-regulation of phlorotannins in *A. nodosum* and does not lead to increased resistance [136].

3.1.2.1.3
Grazer Induced Up-Regulation of the Terpenoid Level in Brown Algae

That terpenoid metabolites can also be up-regulated in response to herbivory was demonstrated with the brown alga *Dictyota menstrualis*. The generalist amphipod *Amphithoe longimana* induced increased concentrations of the defensive diterpenes **65–67** in this alga, making it less palatable (Scheme 18).

Compared to undamaged controls, amphipod-damaged seaweeds had up to 34% higher contents of the diterpenes dictyol E (**65**), dictyodial (**66**), and pachydictyol A (**67**) of which **65** deters feeding by the amphipods [137]. Remarkably this study not only relies on laboratory experiments but provides field data. In their natural environment previously heavily attacked algae were less susceptible to herbivory compared to those that had a record of low grazing.

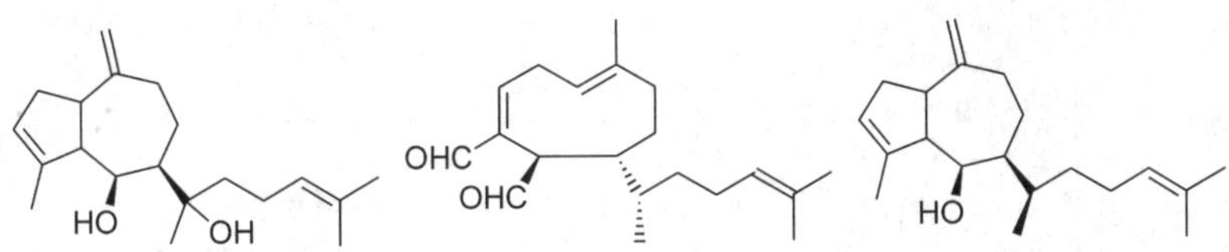

dictyol E (**65**) dictyodial (**66**) pachydictyol (**67**)

Scheme 18 Grazing-induced diterpenes from *Dictyota menstrualis*

In the brown seaweed *Sargassum muticum*, induced defenses are found to be tissue specific. Young tissue, which is responsible for future growth, is dynamically defended against the amphipod *A. longimana*. Comparison of the antifeedant properties of the algae with dried, ground and subsequently reconstituted tissue in an agar matrix showed that this is due to undefined chemicals. In contrast, older tissue at the base of the alga, which link the other parts of the alga to the substrate, is defended constitutively by virtue of its toughness rather than via deterrent chemistry [138].

3.1.2.1.4
The Induced Oxidative Burst in Brown Algae

All the above examples demonstrate that complex secondary metabolites can be produced de novo upon recognition of attackers. That also direct release of very simple but aggressive metabolites can defend algae was recently observed during the investigation of certain red and brown algae. Sporophytes of the brown alga *Laminaria digitata* react upon cell wall degradation by increased production of O_2^- and H_2O_2 [139]. This release of reactive oxygen species has first been found in higher plants and is termed oxidative burst [105, 139]. The production of these aggressive defensive principles in *L. digitata* is initiated upon perception of cell wall oligosaccharides as endogenous elicitors and not through simple mechanical damage [139, 140].

The search for the inducing signals led to oligosaccharides that were identified based on principle biochemical considerations. Alginate, the major brown algal cell wall polysaccharide consists of linear anionic copolymers of mannuronic acid and of its C_5 epimer guluronic acid. Upon hydrolysis oligoalginates result of which only a fraction is recognized by *L. digitata* (Scheme 19). The homo-oligomeric guluronates (**68**) elicited this defense reaction, while homo-oligomeres from mannuronic acid or heterooligomer saccharides exhibited

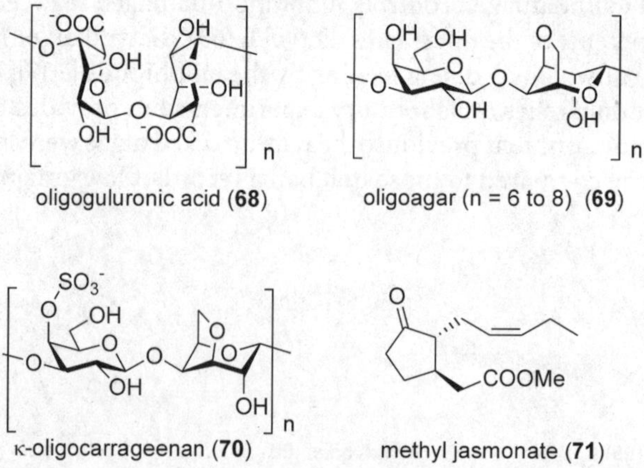

Scheme 19 Elicitors of brown algal and red algal defense

reduced activity. These endogenous elicitors result from the depolymerization of the alginate from *L. digitata* by alginate-degrading enzymes that are found in different attackers, such as marine mollusks as well as bacteria associated with the algae. Inhibitor studies suggest that protein kinases, a phospholipase A_2 as well as potassium, calcium, and anion channels, are involved in the response, which features some conserved steps with the oxidative burst that is found in higher plants [140].

In a broad survey 45 species of brown algae were tested for their ability to respond with an oxidative burst to oligoalginates [141]. A total of 15 investigated species reacted with a rapid strong oxidative burst, all of them belonging to an alginate-rich group with complex thallus morphology. This reaction is thus widely distributed among brown algae of different taxa.

That the oxidative burst is directly involved in the chemical defense of these algae is clear. This reaction can be inhibited by diphenyleneiodonium, a suicide inhibitor of NADPH-oxidase which suppresses both the production of reactive oxygen species and the natural resistance to epiphytic bacteria. In addition a role in the defense against endophytes was indicated, since pre-treatment with oligomeric guluronates resulted in decreased infection of *L. digitata* with the pathogen *Laminariocolax tomentosoides* [141].

3.1.2.2
Red Algae

3.1.2.2.1
The Oxidative Burst in *Gracilaria conferta*

Gracilaria conferta, a red alga of commercial interest due to its agar content, suffers a tip bleaching disease after bacterial infection. The symptom rather indicates a hypersensitive response of the alga than successful bacterial infection.

In hypersensitive responses, the defense of the host is so intense that it not only suppresses pathogens but causes damages to host tissue as well. Host cells that are close to an intruding pathogen die and their decompatimentation is followed by accumulation of toxic defense agents that contribute to the death of the pathogen. Unraveling the signal events involved in the hypersensitive reaction of the red alga gave the first evidence that algae are able to recognize oligosaccharides as endogenous elicitors. In a series of experiments on *G. conferta*, Weinberger et al. found that the alga responds with an oxidative burst and with rapid increase of respiration and halogenating activity when agar, agarose, or agarose-degradation products are added to the medium [142]. In structure-activity tests, oligoagar elicitors had the highest activity. The molecular size of the elicitors is of major importance for the recognition. Those fragments with 6 to 8 disaccharide repeating units were the most active (**69**) (Scheme 19) [143]. Infective bacteria contain agar-degrading enzymes, accordingly the products released during the infection process serve as elicitors. This reaction is also triggered by undefined bacterially excreted peptides as exogenous elicitors, indicating that the alga has developed a way to recognize various molecular signals. These allow it to respond to the maceration of its cell wall as well as to the presence of certain epiphytic organisms [144]. The H_2O_2 released by *G. conferta* not only caused the elimination of epiphytic agar-decaying bacteria [145] but also induced the hypersensitive response damaging healthy algae. This reaction was identified as the cause for the tip-bleaching disease often observed after bacterial infections.

3.1.2.2.2
The Interaction of *Chondrus crispus* with the Endophyte *Acrochaete operculata*

One of the best understood examples of an induced defense in a host-pathogen interaction is the system of the red alga *Chondrus crispus* and the endophytic green alga *Acrochaete operculata*. *C. crispus* is of commercial interest due to its cell wall saccharide carrageenan which serves the food industry as raw material. During a certain, sporophytic developmental phase of the host only the pathogen green alga is able to invade the tissue of *C. crispus*. After settling and germination the spores of the pathogen form vegetative filaments that completely invade the host, resulting in severe tissue damage. In contrast, the internal tissue of gametophytes (another developmental phase of *C. crispus*) is not infected, and the alga can co-exist with the pathogen over years [146, 147]. The early defense reaction of resistant gametophytes of *C. crispus* is to release micromolar concentrations of H_2O_2 into the surrounding medium, amounts sufficient to kill penetrating pathogens. In strong contrast to the reactions described above, this H_2O_2-production is not sensitive to diphenyliodonium inhibition and does not feature other key characteristics of an oxidative burst [148]. Also the signals involved differ from those identified in other algal/pathogen interactions. H_2O_2-production of the gametophytes is not triggered by alginates or carrageenanes but initiated by L-asparagine. This amino acid is released by the green algal pathogen and is rapidly transformed by an amino acid oxidase from *C. crispus* to give the corresponding keto-acid, ammonium, and the defensive principle H_2O_2 (Fig. 3). L-Asparagine represents a specific signal in the host pathogen interaction, since neither D-asparagine, L-glutamine, nor other proteinogenous amino acids were accepted as substrates of the amino acid oxidase.

The content of extracellular asparagine in cultures of *A. operculata* has been found to strongly depend on signals from the host. In accordance with the observation that only gametophytes produced elevated amounts of H_2O_2 in the presence of the pathogen, increased asparagine release was found only when cell wall fragments of *C. crispus* were added to the green alga. These cell wall fragments can result from the carrageenolytic activity of *A. operculata*, which is able to break down the host cell wall polysaccharides. The gametophyte cell wall κ-carrageenans are cleaved to κ-oligocarrageenans, which are responsible for the induction of L-asparagine-release by the pathogen (**70**) (Scheme 19) [149]. The corresponding λ-oligocarrageenans from the sporophytes did not trigger comparable L-asparagine-release from *A. operculata*.

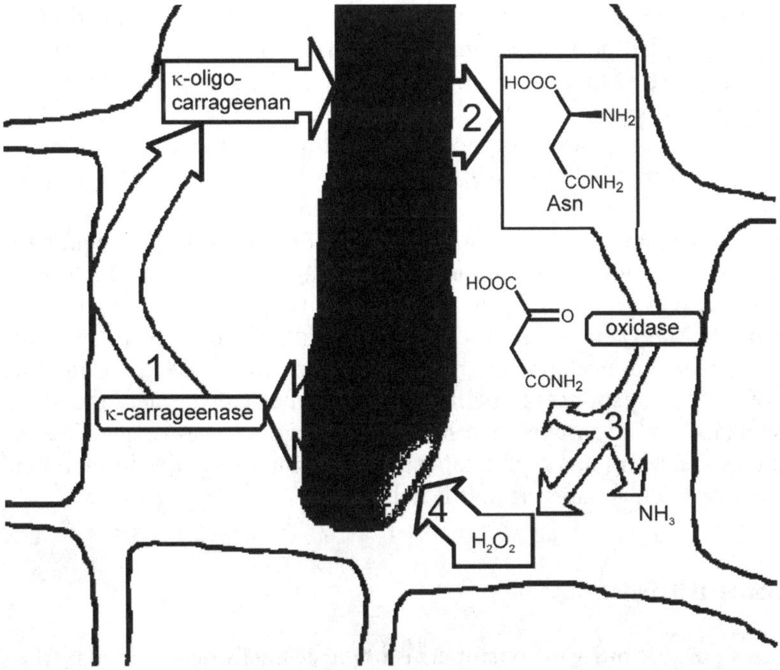

Fig. 3 Model for the involvement of κ-carrageenan and asparagine in the interaction of *C. crispus* and *A. operculata*. After penetration of the green algal endophyte, the carrageenans of the red algal tissue are broken down (1). The resulting oligocarrageenans trigger the release of asparagine from the pathogen (2). This amino acid is then converted to H_2O_2 and α-ketosuccinamid by an aminoacid oxidase from the red algal host (3). This oxidant causes cell death of the pathogen (4). From [148] with permission of Wiley VCH

The interaction of the host and the pathogen thus involves a complex communication (Fig. 3). After successful penetration of the outer cell layers of resistant, gametophytic *C. crispus*, filaments of the pathogen will reach the inner tissue of the host and macerate its κ-carrageenan cell wall, releasing κ-oligocarrageenans. These trigger the release of the signal L-asparagine from the pathogen which finally serves as a substrate for the production of the defensive principle H_2O_2 [148]. The induced signal of the attacker thus serves directly as a substrate for the production of a chemical defense metabolite.

This initial defensive reaction is not the only aspect of the interplay of *C. crispus* and *A. operculata*. Besides the release of asparagine, the pathogen responds with the de novo synthesis of proteins and altered virulence after oligocarrageenan-treatment [149]. The gametophytes do not rely on the aminoacid oxidase as their only line of defense. A second diphenyleneiodonium-sensitive release of H_2O_2 featuring the properties of an oxidative burst occurs in the gametophytes after attack of *A. operculata*. This is followed by stimulation of protein biosynthesis and increased resistance to infection [149]. Interestingly, algal polysaccharide carrageenans are also used commercially to elicit defense responses in higher plants. Defensive gene expression of sesquiterpene cyclases, chitinases, and proteinase inhibitors as well as up-regulation of signaling pathways mediated by ethylene, jasmonic acid, and salicylic acids are observed after treatment of tobacco with sulfated λ-carrageenan [150, 151].

Like higher plants, macroalgae rely heavily on the use of activated and induced defenses. Some of the immune traits of algae are conserved in other eukaryotic lineages. For example, initial reactions to pathogens with an oxidative burst are strikingly similar to those found in higher plants, suggesting that the underlying biochemical machinery arose early in evolution. On the other hand most of the later responses to pathogens or herbivory make clear that distinct mechanisms lacking parallels in other lineages originate from algae. The search for signals and hormonal regulation will reveal more about the development of communication mechanisms in this field.

3.1.3 Associational Defense

A species' susceptibility to consumers is not necessarily determined by the deterrent capabilities of the organism itself, but can be influenced by metabolites produced from associated organisms. This concept is well understood for host-epibiont associations that alter the fate of the host with respect to herbivore susceptibility or resistance to pathogens [152, 153]. Often the chemical reasons for these interactions have not been addressed. The association of host seaweeds with symbiotic bacteria might also influence their resistance. That this could be due to the production of secondary metabolites by a symbiont was suggested by Kubanek et al. [98]. Pathogenic fungi can be fended off by a 22-membered cyclic lactone, lobophorolide (**72**) of the brown alga *Lobophora variegata* (Scheme 20).

Lobophorolide (**72**) is an exceptional metabolite of brown algae in many ways. Most seaweed secondary metabolites that possess antimicrobial or antifeedant activities have been isolated at concentrations of 0.1–10% of algal dry weight. In contrast, the amount of (**72**) reached only around 1.2×10^{-4}% of the dry mass. Nevertheless it exhibited potent activity at and below this concentration. The isolation of a polycyclic polyketide from a brown alga is surprising, since this class of secondary metabolites has not been previously known to occur in marine macrophytes [98]. Lobophorolide (**72**) has striking structural similarities with tolytoxin (**73**), since both products share the cyclic lactone structure. The carbon skeleton is also the basis for the 44-membered lactone swinholide (**75**). While tolytoxin (**73**) was isolated from terrestrial cyanobacteria [154], swinholide (**75**) can be extracted from marine sponges of the genus *Theonella* [155]. Because of the presence of large populations of cyanobacteria in *Theonella* spp., swinholide (**75**) has been assumed to be a cyanobacterial product. However, sepa-

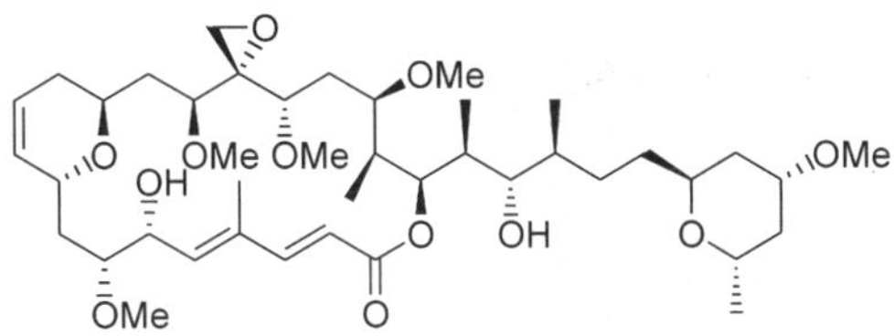

lobophorolide (**72**)

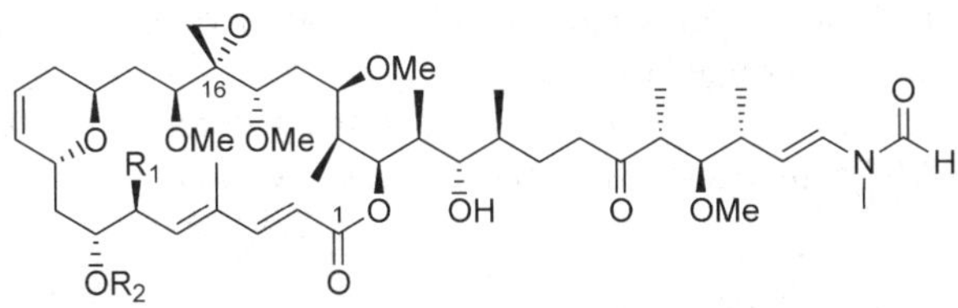

tolytoxin R_1 = OH, R_2 = Me (**73**)

scytophycin C R_1=R_2=H and 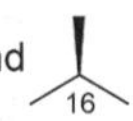 at C16 (**74**)

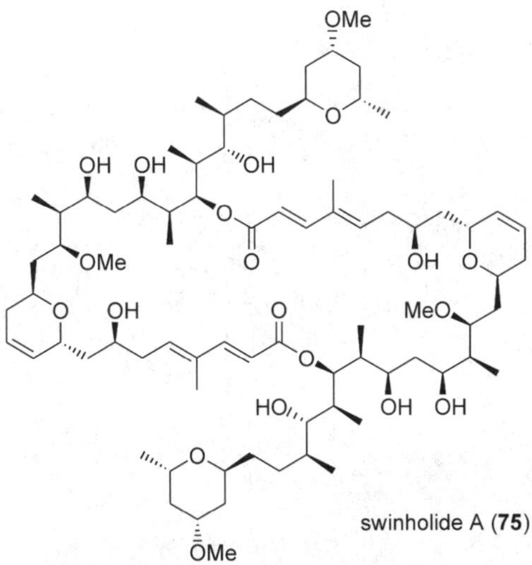

swinholide A (**75**)

Scheme 20 Lobophorolide (72) from the brown alga *Lobophora variegata*, and structurally similar natural products from cyanobacteria and the sponge *Theonella swinhoei*

ration of sponge, cyanobacterial and eubacterial cells of *Theonella swinhoei* provided (**75**) only in the eubacterial fraction [156]. The structural similarities of lobophorolide (**72**) to cyanobacterial, and eubacterial metabolites suggest that this unusual polyketide could be the product of a microbial symbiont rather than the alga itself. This is supported by the fact that ephiphytic bacteria were found on *L. variegate* surfaces [98].

3.2
Dynamic Defense Reactions in Other Phyla

3.2.1
Activated Defense

The only known example of an activated defense in sponges arises from investigations of the defensive chemistry of *Aplysina aerophoba*. A wound-activated transformation of stored pro-toxins was found after disruption of compartmentalization of this sponge [157, 158].

Upon tissue disruption, the isoxazoline alkaloids **76** and **78** are converted to aeroplysinin-1 (**79**), which subsequently provides the dienone (**80**) (Scheme 21).

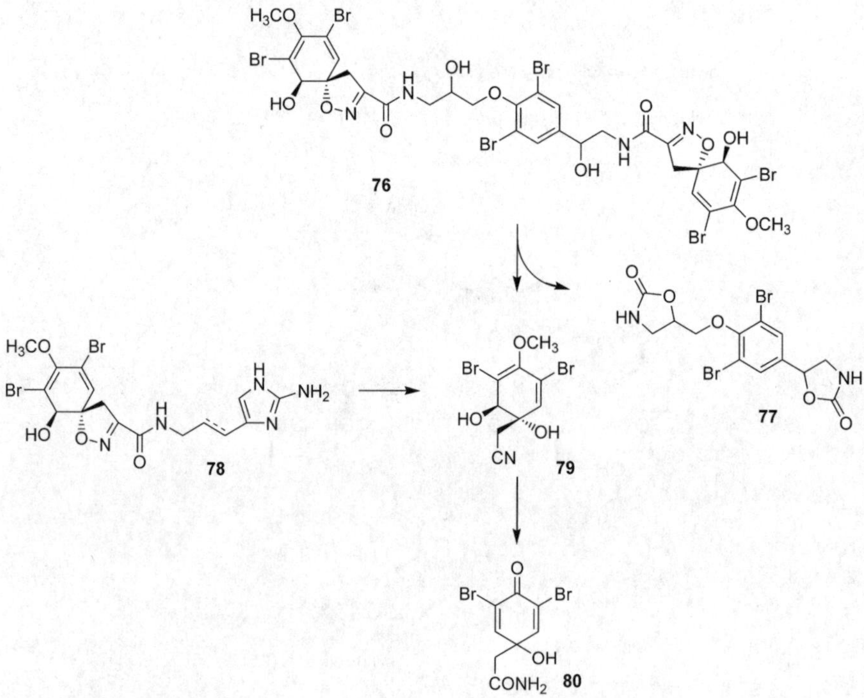

Scheme 21 Wound-activated bioconversion of brominated isoxazoline alkaloids by the sponge *Aplysina aerophoba*

Neither **79** nor **80** are detected in the intact tissue of the sponge implying that substrates and enzymes for the conversion are stored in different cellular compartments or different cell types. When isofistularin-3 (**76**) is the substrate, **77** is detected as a second cleavage product of this reaction. This wound-activated bioconversion is paralleled by a marked increase of biological activity of the products **79** and **80** compared to the isoxazolines. In feeding assays the aqueous extracts containing **79** and **80** as major constituents were significantly more deterrent towards the wrasse *Thalassoma bifasciatum* than the methanolic extracts high in isoxazolines. As in other wound-activated defense reactions, this transformation occurs soon after wounding: an almost complete turnover of **78** was observed within 50 s while about 50% of the starting material was recovered even after 2 min in the case of **76**. The enzymatic activity responsible for the conversion has also been detected in Caribbean *Aplysina* sponges [158], but no indication for the existence of a wound-activated defense was found in a re-investigation with living specimens of two species from the Caribbean. In both short (150 s) and long (120 min) time-course experiments, no conversion to aeroplysinin-1 (**79**) or dibromocyclohexadienone (**80**) was found indicating that the wound-activated defense is not a universal feature of *Aplysina* sponges [159].

Marine hydroids are commonly defended from predation by nematocysts that are capable of penetrating the tissue of predators, and injecting proteinaceous venom. However, lipophilic secondary metabolites also protect many hydroid species. The hydroid *Tridentata marginata* is chemically defended by tridentatols A-D (**85–88**), of which tridentatol A (**85**) is a potent deterrent to fish predation (Scheme 22) [160].

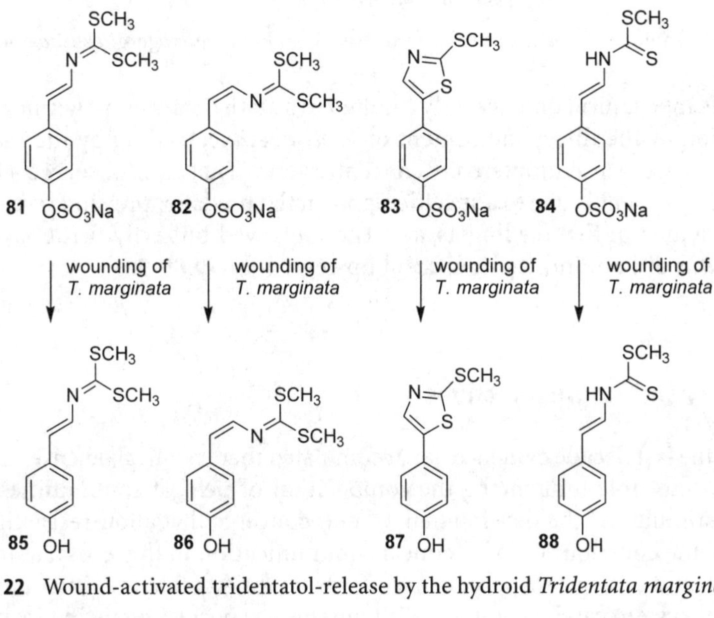

Scheme 22 Wound-activated tridentatol-release by the hydroid *Tridentata marginata*

A re-examination of the defensive chemistry of *T. marginata* revealed that the tridentatols (**85–88**) represent the transformation products of less active sulfate ester storage forms (**81–84**) [161]. Crushing the tissue as an attacking predator would do results in the rapid conversion of the sulfate esters within seconds after wounding. This suggests that this reaction plays a role in a wound-activated defense.

3.2.2
Induced Defense in Corals

A recent example demonstrates that corals rely on induced biosynthesis of terpenes as a dynamic defense strategy as well. The induction of terpenoid secondary metabolites was observed in the sea whip *Pseudopterogorgia elisabethae* [162]. Levels of pseudopterosins **89–92**, a group of diterpene glycosides with anti-inflammatory and analgesic properties (Scheme 23) [163–165], are increased in response predation by the mollusk *Cyphoma gibbosum*. First bioassays indicate that these natural products are involved in the chemical defense.

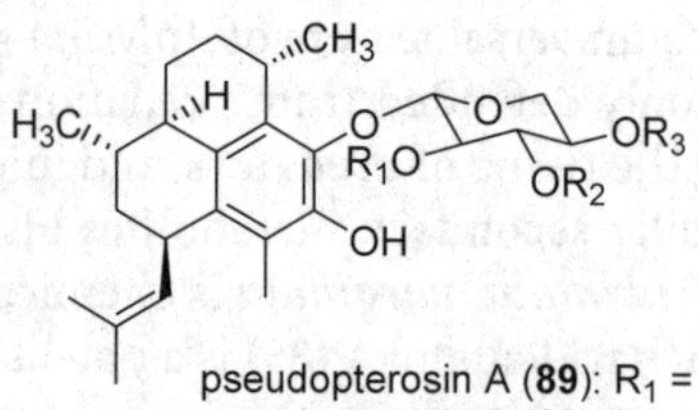

pseudopterosin A (**89**): $R_1 = R_2 = R_3 = H$
pseudopterosin B (**90**): $R_1 = Ac$; $R_2 = R_3 = H$
pseudopterosin C (**91**): $R_2 = Ac$; $R_1 = R_3 = H$
pseudopterosin D (**92**): $R_1 = R_2 = H$; $R_3 = Ac$

Scheme 23 Inducible pseudopterosins in the sea whip *Pseudopterogorgia elisabethae*

While mechanical damage only resulted in a statistically non-significant up-regulation of the diterpene content of *P. elisabethae*, feeding by the mollusk greatly elevated pseudopterosin concentrations. That this observed change diterpene-content is due to a specific signal in the predator-prey interaction became obvious in fish feeding assays. The four-eyed butterfly fish *Chaetodon capistratus* did not induce significant up-regulation of **89-92.**

4
Summary and Future Directions

During the last decade evidence has accumulated that chemical defense can play an important role influencing the composition of pelagic communities. This view is stimulating the development of new concepts that allow researchers to evaluate the contribution of chemical communication in this ecosystem. Both chemical and ecological approaches suffer from the fact that the situation in the intact ecosystem can be complex, with numerous species coexisting in a small volume of water. Approaches to unraveling certain aspects of this complex interaction will require method development of chemists and ecologists working in close association. These investigations should not only focus on constitutive defenses, since also dynamic chemical defense strategies are important for pelagic organisms. They allow to overcome both the problem of dilution of secondary metabolites and the risk of autotoxicity.

These dynamic strategies also play important roles in the chemical defense of benthic organisms. Only in recent years have marine scientists started to apply methods that allow these defensive strategies to be monitored. This opens the field for the identification of signals that play a role in species-species interactions in the marine environment and for the elucidation of hormones responsible for regulating the production of secondary metabolites.

Acknowledgements I thank Prof. Dr. Stefan Schulz for the invitation to write this review. Prof. Wilhelm Boland is gratefully acknowledged for support during the preparation of this manuscript. I thank Emily Wheeler, Dr. Martin Heil, Janine Rattke, Dr. Anne Busch, Dr. Uli Lion, Thomas Wichard, Sven Adolph, Theresa Wiesemeier, Christoph Beckmann, and Johann Pohnert for helpful discussion during the preparation of this manuscript. The Max-Planck-Society is acknowledged for funding.

References

1. Blunt JW, Copp BR, Munro MHG, Northcote PT, Prinsep MR (2003) Nat Prod Rep 20:1
2. Faulkner DJ (2002) Nat Prod Rep 19:1 and previous articles of this series
3. Moore BS (1999) Nat Prod Rep 16:653
4. Proksch P, Edrada RA, Ebel R (2002) Appl Microbiol Biotechnol 59:125
5. Faulkner DJ (2000) Nat Prod Rep 17:1
6. McClintock JB, Baker BJ (2001) (eds) Marine chemical ecology. CRC Press, Boca Raton
7. Garson MJ (2001) In: McClintock JB, Baker BJ (eds) Marine chemical ecology. CRC Press, Boca Raton
8. Slattery M, Rajbhandari I, Wesson K (2001) Microb Ecol 41:90
9. Piel J (2002) Proc Natl Acad Sci USA 99:14002
10. Lee YK, Lee JH, Lee HK (2001) J Microbiol 39:254
11. Field CB, Behrenfeld MJ, Randerson JT, Falkowski P (1998) Science 281:237
12. Smetacek V (1999) Protist 150:25
13. Sommer U, Stibor H, Katechakis A, Sommer F, Hansen T (2002) Hydrobiologia 484:11
14. Hutchinson GE (1961) Am Nat 95:137
15. Huisman J, Weissing FJ (1999) Nature 402:407
16. Smetacek V (2001) Nature 411:745
17. Hamm CE, Merkel R, Springer O, Jurkojc P, Maier C, Prechtel K, Smetacek V (2003) Nature 421:841
18. Wolfe GV (2000) Biol Bull 198:225
19. Patocka J (2001) Acta Mecica 44:69
20. Turner JT, Tester PA (1997) Limnol Oceanogr 42:1203
21. Landsberg JH (2002) Rev Fish Sci 10:113
22. Van Dolah FM (2000) Environ Health Perspect 108:133
23. Hay ME, Kubanek J (2002) J Chem Ecol 28:2001
24. Dutz J (1998) Mar Ecol-Prog Ser 175:97
25. Maneiro I, Frangopulos M, Guisande C, Fernandez M, Reguera B, Riveiro I (2000) Mar Ecol-Prog Ser 201:155
26. Tester PA, Turner JT, Shea D (2000) J Plankton Res 22:47
27. Teegarden GJ, Cembella AD, Capuano CL, Barron SH, Durbin EG (2003) J Plankton Res 25:429
28. Teegarden GJ, Cembella AD (1996) J Exp Mar Biol Ecol 196:145

29. Teegarden GJ (1999) Mar Ecol-Prog Ser 181:163
30. Huntley M, Sykes P, Rohan S, Marin V (1986) Mar Ecol-Prog Ser 28:105
31. Ives JD (1987) J Exp Mar Biol Ecol 112:131
32. Colin SP, Dam HG (2003) Mar Ecol-Prog Ser 248:55
33. Colin SP, Dam HG (2002) Limnol Oceanogr 47:1430
34. Shimizu Y, Chou HN, Bando H, Vanduyne G, Clardy JC (1986) J Am Chem Soc 108:514
35. Lin YY, Risk M, Ray SM, Vanengen D, Clardy J, Golik J, James JC, Nakanishi K (1981) J Am Chem Soc 103:6773
36. Sykes PF, Huntley ME (1987) Mar Biol 94:19
37. Huntley ME, Ciminiello P, Lopez MDG (1987) Mar Biol 95:103
38. Koski M, Rosenberg M, Viitasalo M, Tanskanen S, Sjolund U (1999) ICES J Mar Sci 56:131
39. Larsen A, Bryant S (1998) Sarsia 83:409
40. Bates SS (2000) J. Phycology 36:978
41. Mos L (2001) Environ Toxicol Pharmacol 9:79
42. Shaw BA, Andersen RJ, Harrison PJ (1997) Mar Biol 128:273
43. Shaw BA, Andersen RJ, Harrison PJ (1995) Mar Biol 124:467
44. Shaw BA, Harrison PJ, Andersen RJ (1995) Mar Biol 124:473
45. Lincoln JA, Turner JT, Bates SS, Leger C, Gauthier DA (2001) Hydrobiologia 453:107
46. Bargu S, Marinovic B, Mansergh S, Silver MW (2003) J Exp Mar Biol Ecol 284:87
47. Gross EM (2003) Crit Rev Plant Sci 22:313
48. Windust AJ, Wright JLC, McLachlan JL (1996) Mar Biol 126:19
49. Sugg LM, VanDolah FM (1999) J Phycology 35:93
50. Hu TM, Curtis JM, Walter JA, Wright JLC (1995) J Chem Soc-Chem Commun 597
51. Hu TM, Curtis JM, Walter JA, McLachlan JL, Wright JLC (1995) Tetrahedron Lett 36:9273
52. Windust AJ, Hu TM, Wright JLC, Quilliam MA, McLachlan JL (2000) J Phycology 36:342
53. Hu TM, Burton I, Curtis JM, Quilliam MA, Walter JA, Windust AJ, Wright JLC (1999) Tetrahedron Lett 40:3981
54. Pohnert G, Boland W (2002) Nat Prod Rep 19:108
55. Wittstock U, Gershenzon J (2002) Curr Op Plant Biol 5:300
56. Paul VJ, Van Alstyne KL (1992) J Exp Mar Biol Ecol 160:191
57. Wolfe GV, Steinke M, Kirst GO (1997) Nature 387:894
58. Strom S, Wolfe G, Slajer A, Lambert S, Clough J (2003) Limnol Oceanogr 48:230
59. Strom S, Wolfe G, Holmes J, Stecher H, Shimeneck C, Lambert S, Moreno E (2003) Limnol Oceanogr 48:217
60. Archer SD, Stelfox-Widdicombe CE, Malin G, Burkill PH (2003) J Plankton Res 25:235
61. Nevitt GA, Veit RR, Kareiva P (1995) Nature 376:680
62. Nevitt GA (2000) Biol Bull 198:245
63. Poulet SA, Ianora A, Miralto A, Meijer L (1994) Mar Ecol-Prog Ser 111:79
64. Ianora A, Poulet SA, Miralto A (1995) Mar Biol 121:533
65. Ban SH, Burns C, Castel J, Chaudron Y, Christou E, Escribano R, Umani SF, Gasparini S, Ruiz FG, Hoffmeyer M, Ianora A, Kang HK, Laabir M, Lacoste A, Miralto A, Ning XR, Poulet S, Rodriguez V, Runge J, Shi JX, Starr M, Uye S, Wang YJ (1997) Mar Ecol-Prog Ser 157:287
66. Paffenhofer GA (2002) Mar Ecol-Prog Ser 227:305
67. Miralto A, Barone G, Romano G, Poulet SA, Ianora A, Russo GL, Buttino I, Mazzarella G, Laabir M, Cabrini M, Giacobbe MG (1999) Nature 402:173
68. Pohnert G, Lumineau O, Cueff A, Adolph S, Cordevant C, Lange M, Poulet S (2002) Mar Ecol-Prog Ser 245:33
69. Pohnert G (2000) Angew Chem-Int Edit 39:4352

70. Pohnert G, Boland W (1996) Tetrahedron 52:10073
71. Wendel T, Jüttner F (1996) Phytochemistry 41:1445
72. d'Ippolito G, Romano G, Iadicicco O, Miralto A, Ianora A, Cimino G, Fontana A (2002) Tetrahedron Lett 43:6133
73. d'Ippolito G, Iadicicco I, Romano G, Fontana A (2002) Tetrahedron Lett 43:6137
74. Adolph S, Poulet SA, Pohnert G (2003) Tetrahedron 59:3003
75. Müller DG, Clayton MN, Gassmann G, Boland W, Marner FJ, Schotten T, Jaenicke L (1985) Naturwissenschaften 72:97
76. d'Ippolito G, Romano G, Caruso T, Spinella A, Cimino G, Fontana A (2003) Org Lett 5:885
77. Pohnert G (2002) Plant Physiol 129:103
78. Jüttner F (2001) J Phycol 37:744
79. Irigoien X, Harris RP, Verheye HM, Joly P, Runge J, Starr M, Pond D, Campbell R, Shreeve R, Ward P, Smith AN, Dam HG, Peterson W, Tirelli V, Koski M, Smith T, Harbour D, Davidson R (2002) Nature 419:387
80. Caldwell GS, Olive PJW, Bentley MG (2002) Aquat Toxicol 60:123
81. McClintock JB, Baker BJ, Steinberg DK (2001) In: McClintock JB, Baker BJ (eds) Marine chemical ecology. CRC Press, Boca Raton, p 195
82. Bryan PJ, Yoshida WY, McClintock JB, Baker BJ (1995) Mar Biol 122:271
83. Yoshida WY, Bryan PJ, Baker BJ, McClintock JB (1995) J Org Chem 60:780
84. McClintock JB, Janssen J (1990) Nature 346:462
85. Kitagawa I, Kobayashi M (1977) Tetrahedron Lett:859
86. Lucas JS, Hart RJ, Howden ME, Salathe R (1979) J Exp Mar Biol Ecol 40:155
87. Missakian MG, Burreson BJ, Scheuer PJ (1975) Tetrahedron 31:2513
88. Slattery M, Paul VJ, Van Alstyne KL, Wylie CR (1995) J Cell Biochem 19B:34
89. Slattery M, Hines GA, Starmer J, Paul VJ (1999) Coral Reefs 18:75
90. Slattery M, Starmer J, Paul VJ (2001) Mar Biol 138:1183
91. Cowart JD, Fielman KT, Woodin SA, Lincoln DE (2000) Mar Biol 136:993
92. Lindquist N, Hay ME, Fenical W (1992) Ecol Monogr 62:547
93. Gilturnes MS, Hay ME, Fenical W (1989) Science 246:116
94. Amsler CD, Iken KB, McClintock JB, Baker BJ (2001) Marine chemical ecology. CRC Press, Boca Raton
95. Hay ME (1997) Coral Reefs 16:S67
96. http://www.chem.canterbury.ac.nz/research/marinlit.htm
97. Pelletreau KN, Muller-Parker G (2002) Mar Biol 141:1
98. Kubanek J, Jensen PR, Keifer PA, Sullards MC, Collins DO, Fenical W (2003) Proc Natl Acad Sci USA 100:6916
99. Paul VJ, Cruz-Rivera E, Thacker RW (2001) In: McClintock JB, Baker BJ (eds) Marine chemical ecology. CRC Press, Boca Raton, p 227
100. Hay ME (1996) J Exp Mar Biol Ecol 200:103
101. Van Alstyne KL, Dethier MN, Duggins DO (2001) In: McClintock JB, Baker BJ (eds) Marine chemical ecology. CRC Press, Boca Raton, p 227
102. Cronin G (2001) In: McClintock JB, Baker BJ (eds) Marine chemical ecology. CRC Press, Boca Raton, p 227
103. de Carvalho LR, Roque NF (2000) Quim Nova 23:757
104. Williams SI, Walker DI (1999) Oceanography and marine biology, vol 37, p 87
105. Potin P, Bouarab K, Salaun J-P, Pohnert G, Kloareg B (2002) Curr Op Plant Biol 5:308
106. Karban RK, Baldwin IT (1997) Induced responses to herbivory. The University of Chicago Press, Chicago
107. Heil M, Baldwin IT (2002) Trends Plant Sci 7:61

108. Paul VJ, Fenical W (1984) Tetrahedron 40:3053
109. Paul VJ, Fenical W (1987) Bioorg Mar Chem 1:1
110. BellanSantini D, Arnaud PM, Bellan G, Verlaque M (1996) J Mar Biol Assoc UK 76:235
111. Meinesz A, Hesse B (1991) Oceanol Acta 14:415
112. Kaiser J (2000) Science 289:222
113. Amade P, Lemee R (1998) Aquat Toxicol 43:287
114. Paul VJ, Fenical W (1986) Mar Ecol-Prog Ser 34:157
115. Denapoli L, Fattorusso E, Magno S, Mayol L (1981) Experientia 37:1132
116. Denapoli L, Magno S, Mayol L, Novellino E (1983) Experientia 39:141
117. Smyrniotopoulos V, Abatis D, Tziveleka LA, Tsitsimpikou C, Roussis V, Loukis A, Vagias C (2003) J Nat Prod 66:21
118. Guerriero A, D'Ambrosio M (1999) Eur J Org Chem:1985
119. Guerriero A, Meinesz A, D'Ambrosio M, Pietra F (1992) Helv Chim Acta 75:689
120. Guerriero A, Marchetti F, D'Ambrosio M, Senesi S, Dini F, Pietra F (1993) Helv Chim Acta 76:855
121. Jung V, Pohnert G (2001) Tetrahedron 57:7169
122. Jung V, Thibaut T, Meinesz A, Pohnert G (2002) J Chem Ecol 28:2091
123. Cimino G, Crispino A, Dimarzo V, Gavagnin M, Ros JD (1990) Experientia 46:767
124. Gavagnin M, Mollo E, Montanaro D, Ortea J, Cimino G (2000) J Chem Ecol 26:1563
125. Gavagnin M, Fontana A (2000) Curr Org Chem 4:1201
126. Cimino G, Sodano G (1993) Topi Curr Chem 167:77
127. Van Alstyne KL, Houser LT (2003) Mar Ecol-Prog Ser 250:175
128. Cetrulo GL, Hay ME (2000) Mar Ecol-Prog Ser 207:243
129. Steneck RS, Adey WH (1976) Bot Marina 19:197
130. Lewis SM, Norris JN, Searles RB (1987) Ecology 68:636
131. Van Alstyne KL (1989) Mar Ecol-Prog Ser 56:169
132. Van Alstyne KL (1988) Ecology 69:655
133. Farmer EE, Ryan CA (1990) Proc Natl Acad Sci USA 87:7713
134. Arnold TM, Targett NM, Tanner CE, Hatch WI, Ferrari KE (2001) J Phycol 37:1026
135. Toth GB, Pavia H (2000) Proc Natl Acad Sci USA 97:14418
136. Pavia H, Toth GB (2000) Ecology 81:3212
137. Cronin G, Hay ME (1996) Ecology 77:2287
138. Taylor RB, Sotka E, Hay ME (2002) Oecologia 132:68
139. Potin P, Bouarab K, Kupper F, Kloareg B (1999) Curr Opin Microbiol 2:276
140. Kupper FC, Kloareg B, Guern J, Potin P (2001) Plant Physiol 125:278
141. Kupper FC, Muller DG, Peters AF, Kloareg B, Potin P (2002) J Chem Ecol 28:2057
142. Weinberger F, Friedlander M, Hoppe HG (1999) J Phycol 35:747
143. Weinberger F, Richard C, Kloareg B, Kashman Y, Hoppe HG, Friedlander M (2001) J Phycol 37:418
144. Weinberger F, Friedlander M (2000) J Appl Phycol 12:139
145. Weinberger F, Friedlander M (2000) J Phycol 36:1079
146. Correa JA, McLachlan JL (1992) Mar Ecol-Prog Ser 81:73
147. Correa JA, McLachlan JL (1991) J Phycol 27:448
148. Weinberger F, Pohnert G, Kloareg B, Potin P (2002) Chembiochem 3:1260
149. Bouarab K, Potin P, Correa J, Kloareg B (1999) Plant Cell 11:1635
150. Mercier L, Lafitte C, Borderies G, Briand X, Esquerre-Tugaye MT, Fournier J (2001) New Phytologist 149:43
151. Klarzynski O, Descamps V, Plesse B, Yvin JC, Kloareg B, Fritig B (2003) Mol Plant-Microbe Interact 16:115
152. Wahl M, Hay ME (1995) Oecologia 102:329

153. Wahl M, Mark O (1999) Mar Ecol-Prog Ser 187:59
154. Carmeli S, Moore RE, Patterson GML (1990) J Nat Prod 53:1533
155. Carmely S, Kashman Y (1985) Tetrahedron Let 26:511
156. Bewley CA, Holland ND, Faulkner DJ (1996) Experientia 52:716
157. Teeyapant R, Proksch P (1993) Naturwissenschaften 80:369
158. Ebel R, Brenzinger M, Kunze A, Gross HJ, Proksch P (1997) J Chem Ecol 23:1451
159. Puyana M, Fenical W, Pawlik JR (2003) Mar Ecol-Prog Ser 246:127
160. Stachowicz JJ, Lindquist N (1997) Mar Ecol-Prog Ser 155:115
161. Lindquist N (2002) J Nat Prod 65:681
162. Thornton RS, Kerr RG (2002) J Chem Ecol 28:2083
163. Look SA, Burch MT, Fenical W, Zheng QT, Clardy J (1985) J Org Chem 50:5741
164. Look SA, Fenical W, Matsumoto GK, Clardy J (1986) J Org Chem 51:5140
165. Look SA, Fenical W, Jacobs RS, Clardy J (1986) Proc Natl Acad Sci USA 83:6238

Author Index Volumes 201–239

Author Index Vols. 101–150 see Vol. 150
Author Index Vols. 151–200 see Vol. 200
Author Index Vols. 26–50 see Vol. 50
Author Index Vols. 51–100 see Vol. 100

The volume numbers are printed in italics

Subject Index

The New Springer Global Website
Be the first to know
▸ Benefit from new practice-driven features.
▸ Search all books and journals – now faster and easier than ever before.
▸ Enjoy big savings through online sales.
springeronline.com – the innovative website with you in focus.
springeronline.com
The interactive website for all Springer books and journals
Springer
010984x

Printing: Saladruck, Berlin
Binding: Stein+Lehmann, Berlin